STRUCTURE REPORTS

for 1975

Volume 41A

Structure Reports is prepared under the guidance of a Commission of the International Union of Crystallography. The members of the Commission sometime concerned with the preparation of this volume are listed below.

STRUCTURE REPORTS

for 1975

Volume 41 A

METALS AND INORGANIC SECTIONS

General editor

J. Trotter

Section editors

L. D. Calvert *and* J. Trotter

Published for the

Springer Science+Business Media, B.V.

by

BOHN, SCHELTEMA & HOLKEMA, UTRECHT

First published in 1977

ISBN 978-94-017-3129-4 ISBN 978-94-017-3127-0 (eBook)
DOI 10.1007/978-94-017-3127-0

Koninklijke Drukkerij Van de Garde BV., Zaltbommel

TABLE OF CONTENTS

SYMBOLS

The letters a, b, c, α, β, γ are used consistently for the edges and angles of the unit cell. Other letters used consistently are:

D_m	Measured density in g/cm^3
Z	Number of formula units per unit cell
x, y, z	Atomic coordinates as fractions of cell edge
A, B, C	Types of layer in layer structures
M, A, B	Variable metal atom(s) in a sequence of related structures
X	Variable non-metals, usually halogen, in a sequence of related structures
Ln	Lanthanon, rare-earth
R	Discrepancy factor (for diffractometer data unless otherwise indicated); also variable organic radical

LIMITS OF ERROR

Errors are quoted as standard deviations in units of the last place. Thus 1.542(3) Å means 1.542 Å, standard deviation 0.003 Å. Usually only an indication of bond length standard deviation is given.

TRANSLITERATION OF RUSSIAN

Transliteration is in accordance with draft recommendation no. 6 of the International Organization for Standardization, and the scheme is reproduced in previous volumes of Structure Reports (see e.g. Vol. 29, page VI). There are some apparent inconsistencies, since the names of Russian authors which appear in English language Journals are given as in the original; only one form is listed in the Author Index.

INTRODUCTION

In the past the aim of Structure Reports has been to present critical reports on all work of crystallographic structural interest, whether it is derived directly from X-ray, electron, or neutron diffraction, or even indirectly from other experiments. The reports were intended to be critical and not mere abstracts, except in some cases when a brief indication of the content of a paper of related interest was included in the form of an abstract. In selecting topics for reporting, the criterion 'of structural interest' was freely interpreted in terms of what was topically interesting. However, the amount of literature covering matters of structural interest became so large that this policy could no longer be followed, and from Volume 28 onwards, critical reports are given only on actual structure determinations. Only in this way was it possible to keep yearly volumes to a fairly uniform and usable size.

Starting with Volume 30, Structure Reports is produced in a new format by photo-offset printing from typed manuscript with unjustified lines. At the time when the decision for this change was taken, the cost of setting the manuscript in type was becoming so high as to render the cost of individual subscription prohibitive. At that time automatic typing methods giving justified lines, etc. for photo-offset reproduction did not offer any saving over type-setting, but hand typing of the manuscript could give a considerable saving in production costs. In the belief that a publication that is too expensive to buy is of little value, the format was changed, sacrificing elegance to availability. The new format did not lead to increased length of the volumes since the information content of the typed and typeset pages is practically identical. However, the amount of work to be reported demanded the eventual separation of Structure Reports into two volumes, A. Metals and Inorganic, and B. Organic, and it was convenient to introduce this change also at Volume 30.

Ideally, the reports have been prepared in such a way that no further structural information would be gained by consulting the original paper, although from Volume 21 onwards, atomic positional parameters are not generally reproduced for structures containing more than about 30 independent atoms. The two main reasons for this are that such tables occupy a great deal of space, making the volumes very bulky, and that the chance of including typographical errors in reproducing extensive tables of data is such, that anybody wishing to make detailed use of them would in any case consult the primary reference.

The phenomenal increase in the number of structural papers in the late 1960's made it necessary to pursue these policies to their final conclusion, in order to keep the volumes to a usable size. The beginning of a new decade afforded an opportunity to consolidate the various economies in space and format. Starting with Volumes 37, the main economy is omission of tables of atomic positional parameters (usually meaningless numbers in themselves) except for simple structures (where atoms are often in special positions). This means that atomic parameters are usually given in full in the Metals section, infrequently in the Inorganic section, and very seldom in the Organic section. If the actual values of the atomic parameters are required, it is now therefore often necessary to consult the original paper (or in the future computer data banks). To compensate for the omission of this tabular material, efforts have been made to describe the structures more fully, with illustrations if appropriate. In addition, interatomic distances (especially in the Organic section) are usually not given in full, but the significant ones are discussed in the descriptions of the structures. Details of the structure analyses are now given only briefly, since methods have become fairly standardized; unless otherwise stated it may be assumed that the analyses involve single-crystal diffractometer data, Patterson or direct methods of structure solution, and least-squares refinement.

Although the data must be presented as briefly as possible, every effort is made to avoid jargon, so that the information is readily understandable by the non-crystallographer as well as the crystallographer (a copy of the International Tables for X-ray Crystallography, Volume I, Kynoch Press, Birmingham, would, nevertheless, be useful). The arrangement in individual reports is generally: name, formula, paper(s) reported, unit cell and space group data, brief details of analysis, atomic positions (if given in full), interatomic distances and angles (if given in full), description and discussion of the structure (with diagrams if appropriate), and additional references. Editorial comments are enclosed in square brackets, and it may be assumed that material not distinguished in this manner is based directly on the papers reported. The Volumes are divided into three main sections: Metals, Inorganic Compounds (in the A volumes), and Organic Compounds (including organometallic compounds, in the B volumes). The arrangement in the Metals section is roughly alphabetical, and that of the Inorganic and Organic sections is roughly in order of increasing complexity of composition, related substances and related structures being kept together as far as possible. The Subject Indexes are arranged alphabetically by the names printed as the headings of the reports, and also include other common names and information. The Formula Index in the A volumes is arranged in alphabetical order of the chemical symbols; in the B volumes the classification is by the number of carbon atoms and secondary classification by the number of hydrogen atoms; other constituents then follow alphabetically.

University of British Columbia
Vancouver, Canada

J. TROTTER

16 November 1976

STRUCTURE REPORTS

SECTION I

METALS

Edited by

L. D. Calvert

(National Research Council of Canada)

with the assistance of

J. K. Byron

ARRANGEMENT

As in previous volumes the arrangement in the Metals section is approximately, but not strictly, alphabetical, and to find particular substances the subject index or formula index should be used.

ALUMINUM BARIUM

$Al_{13}Ba_7$

I. M.L. FORNASINI and G. BRUZZONE, 1975. J. Less-Common Metals, 40, 335-340.

Trigonal, P$\bar{3}$m1, a = 6.099, c = 17.269 Å, c/a = 2.83, D_m = 3.87, Z = 1. Cu and Mo radiations, R = 0.10 for 418 reflexions, photographic data with allowance for anomalous dispersion. See also 1, 2.

Atomic positions

	x	y	z
Ba(1) in 1(a)	0	0	0
Ba(2) in 2(d)	1/3	2/3	0.8942
Ba(3) in 2(d)	1/3	2/3	0.6792
Ba(4) in 2(c)	0	0	0.3920
Al(1) in 2(d)	1/3	2/3	0.0962
Al(2) in 6(i)	0.1565	-0.1565	0.2127
Al(3) in 2(d)	1/3	2/3	0.3362
Al(4) in 3(f)	1/2	0	1/2

Interatomic distances (Å)

Ba(1)	- 6 Ba	3.97	Al(1)	- 3 Al	2.75
	- 12 Al	3.89, 4.03		- 7 Ba	3.49 - 3.89
Ba(2)	- 7 Ba	3.71 - 5.08	Al(2)	- 6 Al	2.75 - 3.24
	- 10 Al	3.49 - 3.57		- 6 Ba	3.51 - 4.03
Ba(3)	- 4 Ba	3.71, 3.73	Al(3)	- 6 Al	2.84, 3.33
	- 12 Al	3.53 - 3.58		- 6 Ba	3.53, 3.65
Ba(4)	- 4 Ba	3.73	Al(4)	- 6 Al	3.05, 3.33
	- 12 Al	3.51 - 3.65		- 6 Ba	3.56, 3.58

The structure is characterized by zigzag chains of Ba atoms, similar to those found in the Laves phase $MgNi_2$ (Fig. 1); average C.N. for Ba is 16.6 and for Al is 11.7.

Al_5Ba_4

II. M.L. FORNASINI, 1975. Acta Cryst., B31, 2551-2532.

Hexagonal, $P6_3/mmc$, a = 6.092, c = 17.782 Å, c/a = 2.92, D_m = 3.88, Z = 2. Fe and Mo radiations, R = 0.085 for 279 observed reflexions, photographic data corrected for absorption and anomalous dispersion; see also 3.

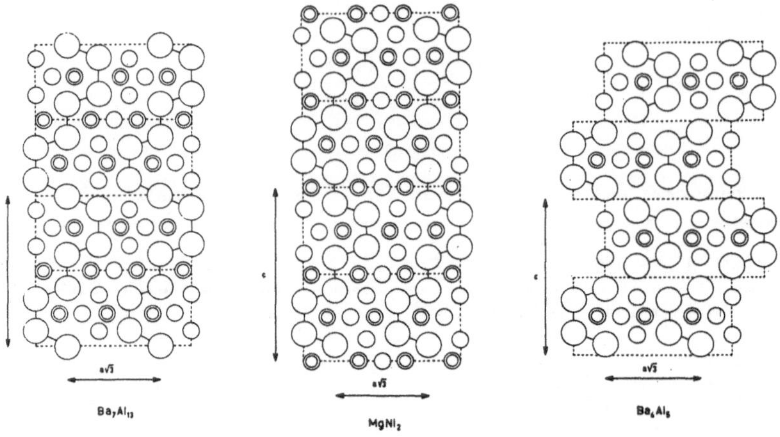

Fig. 1. A comparison of the (110) sections of hexagonal $Al_{13}Ba_7$, $MgNi_2$, and Al_5Ba_4, showing the zigzag chains common to these structures.

Atomic positions

	x	y	z
Ba(1) in 4(f)	1/3	2/3	0.0700
Ba(2) in 4(e)	0	0	0.1429
Al(1) in 4(f)	1/3	2/3	0.6347
Al(2) in 6(h)	0.4909	-0.4909	1/4

Interatomic distances (Å)

Ba(1)	- 6 Ba	3.75, 4.31	Al(1)	- 7 Ba	3.52 - 3.70
	- 7 Al	3.61 - 3.70		- 3 Al	2.77
Ba(2)	- 4 Ba	3.75, 3.81	Al(2)	- 6 Ba	3.59, 3.61
	- 9 Al	3.52, 3.59		- 6 Al	2.77 - 3.21

The structure of Ba_4Al_5 is closely related to those of the Ba_7Al_{13} and of the $MgNi_2$ phases (Fig. 1). The three structures are characterized by the common presence of three-dimensional blocks, with the largest atoms (Ba or Mg) arranged in a distorted tetrahedral close-packing. [There are also similarities with Ti_4S_5 (4).]

1. Structure Reports, 29, 97.
2. This volume, p. 119 [Al_2Sr_3].
3. E.M. FLANIGAN, 1952. Thesis, Syracuse Univ., Syracuse, New York.
4. Structure Reports, 35A, 142.

ALUMINUM BARIUM IRON

ALUMINUM BARIUM NICKEL

ALUMINUM COBALT STRONTIUM

Al_9BaFe_2
Al_9BaNi_2
Al_9Co_2Sr

K. TURBAN and H. SCHÄFER, 1975. J. Less-Common Metals, 40, 91-96.

Hexagonal, P6/mmm, Z = 1. Mo radiation, diffractometer data. Ba or Sr in 1(a): 0,0,0; Fe, Co, or Ni in 2(c): 1/3, 2/3, 0; Al(1) in 3(f): 1/2, 0, 0; Al(2) in 6(m): x, 2x, 1/2.

	a(Å)	c(Å)	c/a	D_m	R	x(Al(2))
Al_9BaFe_2	8.04	3.89	0.48	3.64	0.177	0.212
Al_9BaNi_2	7.93	3.96	0.50	3.80	0.125	0.213
Al_9Co_2Sr	7.91	3.96	0.50	3.49	0.118	0.212

Interatomic distances (Å)

Al_9BaFe_2			Al_9BaNi_2			Al_9Co_2Sr		
Ba	- 2 Ba	3.89	Ba	- 2 Ba	3.96	Sr	- 2 Sr	3.96
	- 18 Al	3.54, 4.02		- 18 Al	3.53, 3.97		- 18 Al	3.52, 3.96
Fe	- 2 Fe	3.89	Ni	- 2 Ni	3.96	Co	- 2 Co	3.96
	- 9 Al	2.32, 2.58		- 9 Al	2.29, 2.58		- 9 Al	2.28, 2.59
Al(1)	- 2 Ba	4.02	Al(1)	- 2 Ba	3.97	Al(1)	- 2 Sr	3.96
	- 2 Fe	2.32		- 2 Ni	2.29		- 2 Co	2.28
	- 14 Al	2.85 - 4.02		- 14 Al	2.85 - 3.97		- 14 Al	2.85, 3.96
Al(2)	- 2 Ba	3.54	Al(2)	- 2 Ba	3.53	Al(2)	- 2 Sr	3.52
	- 2 Fe	2.58		- 2 Ni	2.58		- 2 Co	2.59
	- 10 Al	2.85 - 3.89		- 10 Al	2.85 - 3.96		- 7 Al	2.85 - 3.96

ALUMINUM BERYLLIUM BORON

$Al_{0.06}B_{3.05}Be$

R. MATTES, K.-F. TEBBE, H. NEIDHARD and H. RETHFELD, 1975. Z. anorg. Chem., 413, 1-9.

Hexagonal, P6/mmm, a = 9.800, c = 9.532 Å, c/a = 0.973, D_m = 2.43, A = 109-110. Mo radiation, diffractometer data, R = 0.081 for 685 reflexions; there are 13 site-sets. This is the 'BeB_2' phase of 1, 2; see 3 for earlier work.

This complex structure contains B_{12} icosahedra and Be_3B_{12} polyhedra with D_{3h} symmetry. The polyhedra are linked in a way similar to that of β-rhombohedral boron. The Al atoms occupy interstitial sites. The mean value for B-B bonds in icosahedra is 1.786 Å; for B-B and Be-B in the Be_3B_{12} polyhedra 1.810 and 2.008 Å, respectively.

1. L.Ja. MARKOVSKIJ, Ju.A. KONDRAŠEV and G.V. KAPUTOVSKAJA, 1955. Ž. Obšč. Khim., 25, 1045.
2. Structure Reports, 26, 57.
3. Ibid., 39A, 27.

ALUMINUM GADOLINIUM GERMANIUM

AlGdGe

T.I. RJABOKON, 1974. Vest. L'vov. Univ. Ser. Chem., 15, 26-28.

Tetragonal, α-$ThSi_2$ type (1), $I4_1/amd$, a = 4.152, c = 14.421 Å, c/a = 3.47, D_m = 6.91, Z = 4. R = 0.103, photographic data.

Atomic positions

	x	y	z
Gd in 4(a)	0	3/4	1/8 †
4Al + 4Ge* in 8(e)	0	1/4	0.290†

* [Misprinted as Si.]
† [Given as 0, 0, 0 and 0, 0, 0.290, but the values above, referred to a centre as origin, are required to derive the α-$ThSi_2$ structure.]

[Interatomic distances (Å)]

Gd - 8 Gd	4.15, 4.16		Al,Ge - 3 Al,Ge	2.38, 2.45
- 12 Al,Ge	3.16, 3.18		- 6 Gd	3.16, 3.18

[The structure is conveniently described as a 3-connected 3-dimensional network of small atoms (Al,Ge) with the larger atoms (Gd) in the interstices. The short distances, 2.38, 2.45 Å, occur in this network.]

1. Structure Reports, 9, 121.

ALUMINUM MOLYBDENUM SULPHUR

GALLIUM MOLYBDENUM SULPHUR

$A_xMo_2S_4$ (A = Al, Ga)

J.M. VANDENBERG and D. BRASEN, 1975. J. Solid State Chem., 14, 203-208.

Cubic, $Ga_{0.5}Mo_2S_4$ type, $F\bar{4}3m$, Z = 8. Cu radiation, powder diffractometer data with allowance for anomalous dispersion; see also 1.

$A_xMo_2S_4$	a(Å)	R(I)	reflexions
$Al_{0.55}Mo_2S_4$	9.726	0.134	41
$Ga_{0.52}Mo_2S_4$	9.739	0.103	41

Atomic positions

				x	y	z
4A	in	4(a)		0	0	0
*	in	4(c)		1/4	1/4	1/4
16Mo	in	16(e)	Al	0.605	0.605	0.605
			Ga	0.603	0.603	0.603
16S(1)	in	16(e)	Al	0.363	0.363	0.363
			Ga	0.360	0.360	0.360
16S(2)	in	16(e)	Al	0.868	0.868	0.868
			Ga	0.865	0.865	0.865

* Occupancies are 0.40Al and 0.16Ga respectively.

Interatomic distances (Å)

	$Al_{0.55}Mo_2S_4$	$Ga_{0.52}Mo_2S_4$
A - 4 S	2.22	2.28
Mo - 6 S	2.37, 2.58	2.42, 2.59
- 3 Mo	2.89	2.84
S(1) - 3 S	3.11	3.03
S(2) - 3 S	3.25	3.17

The ferro-magnetic compounds are defect versions of the 'normal' spinel structure with the 8-fold tetrahedral position of Fd3m split into two 4-fold positions and the vacancies ordered in 4(c). The Mo atoms are displaced towards these vacancies from the ideal positions thus forming tetrahedral clusters (Mo-Mo ∿ 2.85 Å). Similar clusters, with Mo-Mo bonding, occur in Mo halides, chalcogenides, and pnictides, but d-electrons are delocalized giving rise to metallic or semi-conducting character. In $Ga_xMo_2S_4$ individual clusters are isolated, being more than 4 Å apart, giving the observed localized moment with a weak ferro-magnetic interaction.

1. H. BARZ, 1973. Mater. Res. Bull., 8, 983.

ALUMINUM PLATINUM

GALLIUM PLATINUM

T. CHATTOPADHYAY and K. SCHUBERT, 1975. J. Less-Common Metals, 41, 19-32.

$AlPt_2$ (h.t.)
Orthorhombic, Ni_2Si [Co_2Si] type (1), Pnma, a = 5.4007, b = 4.0547, c = 7.8985 Å, Z = 4. Cu radiation, R = 0.16, powder photographic data.

$GaPt_3$ (l.t.)
Tetragonal, $GaPt_3$ type, P4/mbm, a = 5.4723, c = 7.886 Å, c/a = 1.44, Z = 4. Cu radiation, R = 0.19, powder photographic data. See also 2.

Atomic positions

$AlPt_2$ [transformed to Pnma from Pbnm]	x	y	z
Pt(1) in 4(c)	0.033	1/4	0.213
Pt(2) in 4(c)	0.169	1/4	0.570
Al in 4(c)	0.720	1/4	0.59
$GaPt_3$			
Pt(1) in 4(g)	0.231	0.731	0
Pt(2) in 4(h)	0.290	0.790	1/2
Pt(3) in 4(e)	0	0	0.251
Ga in 4(f)	0	1/2	0.258

[Interatomic distances (Å)]

$AlPt_2$			$GaPt_3$		
Pt(1)	- 8 Pt	2.76 - 2.98	Pt(1)	- 8 Pt	2.74 - 2.77
	- 5 Al	2.60 - 2.88		- 4 Ga	2.71, 2.91
Pt(2)	- 8 Pt	2.82 - 2.98	Pt(2)	- 8 Pt	2.77
	- 5 Al	2.43 - 2.98		- 4 Ga	2.51, 2.95
Al	- 10 Pt	2.46 - 2.88	Pt(3)	- 8 Pt	2.77
				- 4 Ga	2.74
			Ga	- 12 Pt	2.51 - 2.95

Pt_3Ga is a tetragonal distortion of the Cu_3Au structure, with all atoms 12-coordinated.

1. Structure Reports, 16, 123; 19, 124.
2. Ibid., 24, 115.

ALUMINUM RARE-EARTHS

Al_3Ln (Ln = La, Gd, Th, Ce, Y, Ho)

E.E. HAVINGA, 1975. J. Less-Common Metals, 41, 241-254.

Data for hexagonal Ni_3Sn type (1) compounds, $P6_3/mmc$, Z = 2, are given in the following table, with atoms placed: Ln in 2(c): 1/3, 2/3, 1/4; Al in 6(h): x, 2x, 1/4; stacking sequence h.

Compound	$LaAl_3$	$GdAl_3$	$ThAl_3$	$CeAl_3$
a (Å)	6.667	6.332	6.499	7.043
c (Å)	4.616	4.600	4.626	5.451
c/a	0.6923	0.7246	0.712	0.774
x	0.8627	0.8552	0.8646	0.8238
Total reflexions	63	57	61	31
Observed reflxns.	53	49	54	22
R(I)	0.085	0.094	0.139	0.064

Al_3Y, trigonal, $BaPb_3$ type (2), $R\bar{3}m$, a = 6.195, c = 21.137 Å, c/a = 3.412, Z = 9, R(I) = 0.113; stacking sequence hhc.

Al_3Ho, trigonal, Al_3Ho type (3), $R\bar{3}m$, a = 6.059, c = 35.86 Å, c/a = 5.918, Z = 15, R(I) = 0.066; stacking sequence hchcc.

For all compounds, Cu radiation, powder diffraction data with allowance for anomalous dispersion.

Atomic positions

Al_3Y	x	y	z
Y(1) in 3(a)	0	0	0
Y(2) in 6(c)	0	0	0.2180
Al(1) in 9(e)	1/2	1/2	0
Al(2) in 18(h)	0.1864	-0.1864	0.1108
Al_3Ho*			
Ho(1) in 3(a)	0	0	0
Ho(2) in 6(c)	0	0	0.1316
Ho(3) in 6(c)	0	0	0.3998
Al(1) in 9(e)	1/2	1/2	0
Al(2) in 18(h)	0.504	-0.504	0.1332
Al(3) in 18(h)	0.4880	-0.4880	0.3990

* Results differ from those of 3.

[Interatomic distances (Å)]

	Al_3La	Al_3Gd	Al_3Th	Al_3Ce
Al - 4 Ln	3.23, 3.35	3.09, 3.18	3.21, 3.27	3.33, 3.52
- 6 Al	2.75, 2.80	2.75, 2.79	2.64, 2.77	3.32, 3.47
Ln - 8 Ln	4.49, 4.62	4.32, 4.60	4.41, 4.63	

	Al_3Y	Al_3Ho
Al(1) - 4 Ln	3.02, 3.10	2.99, 3.03
- 8 Al	2.89, 3.10	2.94 - 3.10
Al(2) - 4 Ln	3.02 - 3.10	3.03 - 3.02
- 8 Al	2.73 - 3.46	2.94 - 3.10

	Al_3Y	Al_3Ho
Al(3) - 4 Ln		3.01 - 3.03
- 8 Al		2.81 - 3.25
Ln(1) - 6 Ln	4.33	4.23
- 12 Al	3.08, 3.10	3.01, 3.03
Ln(2) - 6 Ln	4.18, 4.33	4.21, 4.31
- 12 Al	3.02, 3.11	2.99 - 3.03
Ln(3) - 6 Ln		4.21, 4.23
- 12 Al		2.96 - 3.03

The c/a values and the atomic coordinates were calculated on a model based on the short-range repulsive energy between nearest neighbours, with reasonable agreement resulting from this model.

1. Structure Reports, 20, 29.
2. Ibid., 29, 30.
3. Ibid., 31A, 12.

ANTIMONY BISMUTH NICKEL SULPHUR

(HAUCHECORNITE)

$(Bi_{1.3}Sb_{0.7})Ni_9S_8$

V. KOCMAN and E.W. NUFFIELD, 1974. Canad. Miner., 12, 269-274.

Tetragonal, P4/mmm, a = 7.300, c = 5.402 Å, c/a = 0.74, D_m = 6.47, Z = 1. Mo radiation, R = 0.067 for 271 reflexions, diffractometer data corrected for absorption; see also 1, 2, 3. From the Friedrich Mine, Westphalia.

Atomic positions

			x	y	z
Bi	in	1(a)	0	0	0
*M	in	1(d)	1/2	1/2	1/2
Ni(1)	in	1(b)	0	0	1/2
Ni(2)	in	8(t)	0.18076	1/2	0.25249
S(1)	in	4(m)	0.31274	0	1/2
S(2)	in	4(j)	0.26961	0.26961	0

* M = $Bi_{0.3}Sb_{0.7}$

Interatomic distances (Å)

Bi - 4 S	2.78	Ni(1) - 4 S	2.28
- 2 Ni	2.70	- 2 Bi	2.70
M - 8 Ni	2.69	Ni(2) - 4 S	2.26, 2.32
		- 1 M	2.69

S(1) - 8 Ni	2.28, 2.32	S(2) - 4 Bi	2.78
		- 4 Ni	2.26

The structure consists of double Ni(2)-S ribbons (Fig. 1) linked by Bi-Ni(1) chains (Fig. 2), all parallel to [001]. M (0.7Sb + 0.3Bi) sites occur along the line of intersection of four double ribbons. Bi and Ni(1) are octahedrally coordinated. Ni(2) is in distorted square-planar coordination with 4 S. The M site is coordinated by 8 Ni(2). The closest Ni-Ni distances in the double ribbons are 2.639, 2.674, and 2.728 Å, suggesting orbital interaction.

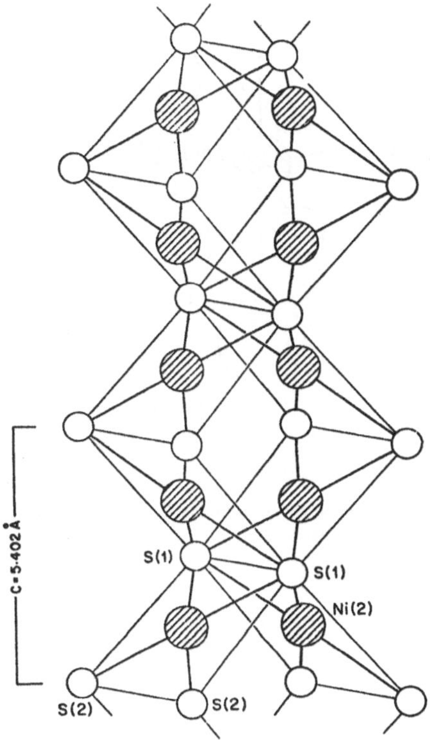

Fig. 1. The double Ni-S ribbons which occur in the tetragonal hauchecornite, $Ni_9(Bi_{1.3}Sb_{0.7})S_8$, structure.

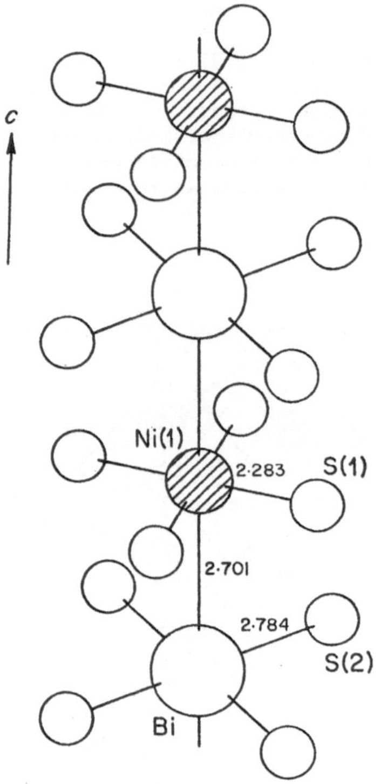

Fig. 2. The Bi-Ni chains in hauchecornite, in which both Bi and Ni are octahedrally coordinated.

1. Structure Reports, 13, 284.
2. R. SCHIEBE, 1893. Jb. Konigl. Preuss. geol. Landesanst. Bergakad., 12, 91 (for 1891).
3. R.I. GAIT and D.C. HARRIS, 1972. Canad. Miner., 11, 819.

ANTIMONY CALCIUM

Ca_2Sb

C. HAMON, R. MARCHAND, P. L'HARIDON and Y. LAURENT, 1975. Acta Cryst., B31, 427-430.

Tetragonal, I4mm, a = 4.69, c = 16.39 Å, c/a = 3.49, D_m = 3.65, Z = 4. Mo radiation, R = 0.069 for 244 reflexions. Diffractometer data with allowance for anomalous dispersion. Contrary to 1, which gives a similar structure in I4/mmm [La_2Sb type (2)] for Ca_2Sb; see also 2, 3 for earlier related work.

Atomic positions

Ca(3) in 4(b), all others in 2(a)

	x	y	z
Sb(1)	0	0	0.1322
Sb(2)	0	0	0.8549
Ca(1)	0	0	0.3238
Ca(2)	0	0	0.6580
Ca(3)	0	1/2	0

Interatomic distances (Å)

Sb(1)	- 9 Ca	3.14 - 3.34
Sb(2)	- 9 Ca	3.23 - 3.36
Ca(1)	- 5 Sb	3.14, 3.36
	- 4 Ca	3.72
Ca(2)	- 5 Sb	3.28, 3.34
	- 4 Ca	3.49
Ca(3)	- 4 Sb	3.19, 3.34
	- 8[6] Ca	3.32, 3.49

The structure can be described as an arrangement of cubes with antimony atoms at each corner; the faces are centred by calcium atoms. These cubes build infinite layers along the a and b axes; the corners of the upper cubes project at the centre of the faces of the lower cubes.

1. Structure Reports, 40A, 8.
2. Ibid., 35A, 12.
3. G. BRAUER and O. MÜLLER, 1961. Angew. Chem., 73, 169.

ANTIMONY CHROMIUM IRON

$(Cr_{1-x}Fe_x)_{1+\delta}Sb$

E. HELLNER, G. HEGER, D. MULLEN and W. TREUTMANN, 1975. Mater. Res. Bull., 10, 91-94.

Hexagonal, $P6_3/mmc$, partially filled Ni_2In type (1), X-ray and neutron diffractometer data; see also 2.

Composition		a (Å)	c (Å)	c/a	Refl.	R
$(Cr_{0.30}Fe_{0.70})_{1.22}Sb$	(X)	4.147	5.283	1.274	130	0.07
	(N)	4.156	5.272	1.269	80	
$(Cr_{0.53}Fe_{0.47})_{1.20}Sb$	(X)	4.149	5.372	1.295	130	0.07
	(N)	4.140	5.378	1.299	80	
$(Cr_{0.52}Fe_{0.48})_{1.23}Sb$	(X)	4.159	5.373	1.292	130	0.07
	(N)	4.158	5.365	1.290	80	
$(Cr_{0.48}Fe_{0.52})_{1.26}Sb$	(X)	4.166	5.374	1.290	130	0.07
	(N)	4.135	5.334	1.290	80	

(X) = X-ray results (N) = Neutron results

Atomic positions

2(a): 0, 0, 0; 0, 0, 1/2; 2(d)*: 2/3, 1/3, 1/4; 1/3, 2/3, 3/4; [Sb in 2(c): 1/3, 2/3, 1/4; 2/3, 1/3, 3/4]

Composition	Occupancy		Cr:Fe ratios	
	2(a) site	2(d) site	2(a) site	2(d) site
$(Cr_{0.30}Fe_{0.70})_{1.22}Sb$	(1.000)	0.22±0.030	30:70	30:70
$(Cr_{0.53}Fe_{0.47})_{1.20}Sb$	(1.000)	0.20±0.027	56:44	37:63
$(Cr_{0.52}Fe_{0.48})_{1.23}Sb$	(1.000)	0.23±0.044	60:40	17:83
$(Cr_{0.48}Fe_{0.52})_{1.26}Sb$	(1.000)	0.26±0.022	50:50	44:56

* [given as 2(c)]

1. Structure Reports, 9, 91.
2. K. YAMAGUCHI, H. YAMAMOTO, Y. YAMAGUCHI and H. WATANABE, 1972. J. Phys. Soc. Japan, 33, 1292.

ANTIMONY LEAD SULPHUR

$Pb_3S_{15}Sb_8$ (fülöppite)

I. E.W. NUFFIELD, 1975. Acta Cryst., B31, 151-157.

Monoclinic, C2/c, a = 13.441, b = 11.726, c = 16.930 Å, β = 94.71°, D_m = 5.22, Z = 4. Mo radiation, R = 0.06 for 1568 reflexions, diffractometer data corrected for absorption, with allowance for anomalous dispersion, 14 site-sets; see also 1, 2.

Interatomic distances (Å)

Sb(1) - 3 S	2.48 - 2.56		Pb(1) - 8 S	2.84 - 3.67
Sb(2) - 3 S	2.43 - 2.51		Pb(2) - 7 S	2.76 - 3.32
Sb(3) - 4 S	2.48 - 2.86			
- 1 S	3.14			
Sb(4) - 3 S	2.48, 2.49			

The structure can be resolved into two kinds of interleaving and interlocked Pb-Sb-S complexes, both of which extend parallel to [110]. The first kind, of composition $Pb_2Sb_4S_6$, is similar to the groups that form the chains in stibnite (Sb_2S_3) but have Pb bonded to both ends. The second kind has the composition $PbSb_4S_9$ and consists of a string of four SbS_3 polyhedra symmetrically arranged about a central twofold rotation axis and a Pb atom. Three of the four Sb atoms have three close (2.43-2.56 Å) S neighbours; the other has five (2.48-3.14 Å). The two Pb atoms are irregularly coordinated by six and seven S atoms at distances of 2.76 to 3.32 Å. The structure (Fig. 1) is compared to those of stibnite, plagionite, and semseyite; a somewhat different description is given in 2.

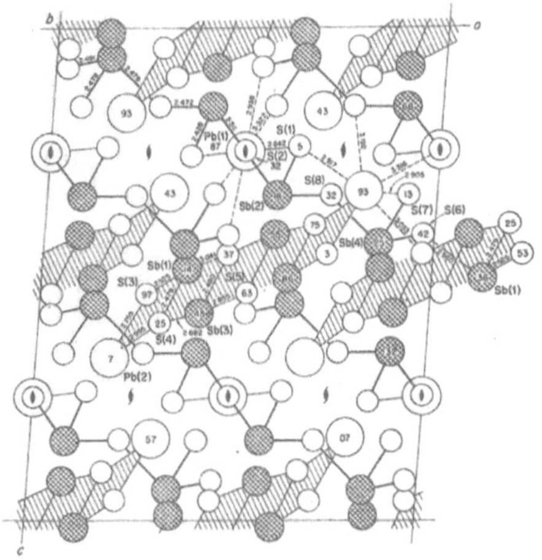

Fig. 1. The fülöppite structure projected onto (010). The $Pb_2Sb_4S_6$ complexes (hatched) are elongated parallel to (102); the $PbSb_4S_9$ complexes are parallel to ($\bar{1}$02).

$Pb_6S_{27}Sb_{14}$ (zinckenite)

II. J.C. PORTHEINE and W. NOWACKI, 1975. Z. Kristallogr., 141, 79-96.

Hexagonal, $P6_3$, a = 22.148, c = 4.333 Å, c/a = 0.196, D_m = 5.36, Z = 1.5. Cu radiation, R = 0.13 for 956 observed reflexions, diffractometer data corrected for absorption; 14 site-sets are given; $P6_3/m$ was rejected; diffuse layer lines confirm that the true cell has c' = 2c ∿ 8.6 Å (3, 4), see also 5, 6.

Interatomic distances (Å)

Pb(1) - 8 S	2.67 - 3.28	Sb(1) - 6 S	2.46 - 3.45
Pb(2) - 9 S	2.73 - 3.76	Sb(2) - 6 S	2.33 - 3.17
Pb(3) - 9 S	2.75 - 3.85	Sb(3) - 6 S	2.42 - 3.68
Pb/Sb - 8 S	2.66 - 3.56		

The structure has three pure Pb positions and one mixed (Pb,Sb) position. Pb(1) has an eight coordination (comprising a bicapped trigonal prism), while Pb(2) and Pb(3) are surrounded by 9 S nearest neighbours. The Pb(2) and Pb(3) positions are 20% occupied. The average Pb-S distances for Pb(1), Pb(2), and Pb(3) are 3.14, 3.23, and 3.24 Å, respectively. The coordination of (Pb,Sb) is similar to that of Pb(1); the average (Pb,Sb)-S distance is 3.09 Å. All three Sb ions are coordinated by 6 S ions in a distorted octahedron. Considering only Sb-S distances smaller than 2.7 Å, Sb(1) has a trigonally pyramidal coordination, while Sb(2) and Sb(3) have planar threefold coordinations. The SbS_3 groups are joined in endless chains which form spirals around the 2_1 axes. These spirals are held together by the Pb ions, and are characteristic of an S/Sb ratio (ϕ) of the type V ($1 < \phi < 2$), in this case $\phi = 1.93$.

1. Structure Reports, 11, 340.
2. Ibid., 40A, 11.
3. Strukturbericht, 7, 156.
4. H. TAKEDA and H. HORIUCHI, 1971. J. Miner. Soc. Japan, 10, 283.
5. Structure Reports, 11, 342.
6. D.C. HARRIS, 1965. Canad. Miner., 8, 381.

ANTIMONY LITHIUM ZINC

LiSbZn

G. SCHROEDER and H.-U. SCHUSTER, 1975. Z. Naturforsch., 30B, 978-979.

Hexagonal, GaGeLi type (1), $P6_3mc$, a = 4.431, c = 7.157 Å, c/a = 1.61, D_m = 5.222, Z = 2. R = 0.093, diffractometer data; see also 1.

Atomic positions

	x	y	z
Sb in 2(b)	1/3	2/3	0.385
Zn in 2(b)	1/3	2/3	0
Li in 2(a)	0	0	0.196

Interatomic distances (Å)

Zn - 4 Sb	2.69, 2.75	Li - 6 Sb	2.89, 3.39
		- 6 Zn	2.92, 3.36

The Zn and Sb atoms occupy the positions of the Zn and S atoms in wurtzite with the Li atoms occupying octahedral holes.

1. Structure Reports, 35A, 61; 38A, 168; this volume, p. 65.

ANTIMONY MERCURY SULPHUR

(LIVINGSTONITE)

HgS_8Sb_4

T. SRIKRISHNAN and W. NOWACKI, 1975. Z. Kristallogr., 141, 174-192.

Monoclinic, A2/a, a = 30.567, b = 4.015, c = 21.465 Å, β = 103.39°, D_x= 4.89, Z = 8. Cu radiation, R = 0.073 for 596 observed reflexions, diffractometer data corrected for absorption, with allowance for anomalous dispersion; 14 site-sets; in agreement with 1, 3, contrary to 2. The composition was formerly (1, 2) given as $HgSb_4S_7$.

Interatomic distances (Å)

Hg(1) - 6 S	2.52 - 3.34	Sb(3) - 3 S	2.30 - 2.85
Hg(2) - 6 S	1.90 - 3.41	Sb(4) - 4 S	2.40 - 2.98
Sb(1) - 4 S	2.38 - 2.93	S(4) - S(5)	2.06
Sb(2) - 5 S	2.58 - 2.94	S-S distances range from 3.15-3.99.	

There is an S_2 group with an S-S distance of 2.06 Å. There are two kinds of layers running parallel to the c axis in the structure. The S-S bond joins two Sb_2S_4 double chains. The other Sb_2S_4 double chains are joined together by the Hg atoms. The coordination of the Hg atoms is octahedral; two of the S atoms are strongly and linearly bonded, as in cinnabar. There are four independent Sb atoms in the structure. Two of them have a coordination of four sulphur atoms which could be described as distorted trigonal pyramid plus one additional sulphur atom. The other two Sb atoms have a square pyramidal coordination and the familiar trigonal pyramidal coordination, respectively. There is good agreement with 3. Livingstonite can be classified as type IV.a_1.

1. Strukturbericht, 4, 150.
2. Structure Reports, 15, 244.
3. Ibid., 21, 347.

ANTIMONY POTASSIUM SULPHUR

KS_2Sb

H.A. GRAF and H. SCHÄFER, 1975. Z. anorg. Chem., 414, 211-219.

Monoclinic, C2/c, a = 8.75, b = 8.98, c = 6.84 Å, β = 121.6°, D_m = 3.19, Z = 4. Cu radiation, R = 0.121 for 1700 reflexions. Diffractometer data.

Atomic positions

	x	y	z
Sb in 4(e)	0	0.5926	1/4
K in 4(e)	0	0.1397	1/4
S in 8(f)	0.2026	0.4243	0.2221

Interatomic distances (Å)

Sb - 4 S 2.41, 2.76 K - 6 S 3.17, 3.21

The structure is characterized by SbS_2^- chains built up of $[SbS_4]$ 'trigonal bipyramids' sharing edges. The S-Sb-S angles are 173.7, 102.4, 90.6, and 85.5°. $KFeS_2$ and $KInTe_2$ (1) are similar chain structures. The K^+ ions are in a nearly octahedral environment.

1. Structure Reports, 10, 123; 39A, 100.

ARSENIC CALCIUM

As_3Ca_5

A. HÜTZ and G. NAGORSEN, 1975. Z. Metallk., 66, 314.

Hexagonal, Mn_5Si_3 type (1), $P6_3/mcm$, a = 8.43, c = 6.75 Å, c/a = 0.80, D_x = 3.42, Z = 2. Photographic data.

Atomic positions

	x	y	z
As in 6(g)	-0.392	-0.392	1/4
Ca(1) in 6(g)	-0.750	-0.750	1/4
Ca(2) in 4(d)	1/3	2/3	0

[Interatomic distances (Å)]

As - 9 Ca 2.90 - 3.58

Ca - 5 As 2.90 - 3.58
- 10 Ca 3.63 - 3.98

Ca - 6 As 3.10
- 8 Ca 3.38, 3.63

1. Strukturbericht, 4, 137, 246; Structure Reports, 24, 78.

ARSENIC CALCIUM SILICON

$As_{16}Ca_{10+x}Si_{12-2x}$ (x = 0.7 to 2.5)

M. HAMON, J. GUYADER, P. L'HARIDON and Y. LAURENT, 1975. Acta Cryst., B31, 445-449.

Monoclinic, $P2_1/m$, a = 7.134, b = 17.651, c = 7.267 Å, β = 111.74°, D_m = 3.70 (average), Z = 1. Mo radiation, R = 0.060 for 1167 reflexions, diffractometer data, with allowance for anomalous dispersion. 14 site-sets with occupancy factors are given. There is a pseudo-cubic triclinic cell with a = 12.0, b = 12.0, c = 11.9 Å, α = 90.5, β = 89.5, γ = 84.8°, related by the transformation -1 ½ -1/ 1 ½ 1/1 0 -1. The related phase in the Cu-Si-P system is isotypic.

Interatomic distances (Å)

Ca(1) - 6 As	2.96 - 3.17		Si(1)	- 3 As	2.33 - 2.59
				- 1 Si	2.32
Ca(2) - 6 As	2.92 - 3.11				
			Si(2)	- 3 As	2.35, 2.51
Ca(3) - 6 As	2.94 - 3.13			- 1 Si	2.32
Ca(4) - 6 As	2.82 - 3.07		Si(3)	- 3 As	2.30 - 2.35
				- 1 Si	2.33
			Si(4)	- 3 As	2.36 - 2.42
				- 1 Si	2.33

The arsenic atoms have a slightly distorted cubic close packing. Of the 16 octahedral sites, 10 are occupied by calcium atoms. The other octahedral sites are filled with calcium or silicon-silicon pairs, with each silicon atom bonded to 3 As and 1 Si. The average Si-As is 2.39 Å and Si-Si 2.33 Å.

ARSENIC CHROMIUM COBALT

ARSENIC COBALT IRON

$AsCo_{0\cdot10}Cr_{0\cdot90}$ at 293°K
$AsCo_{0\cdot05}Fe_{0\cdot95}$

K. SELTE, A. KJEKSHUS, S. AABY and A.F. ANDRESEN , 1975. Acta Chem. Scand., A29, 810-816.

Orthorhombic, MnP type (1), Pnma, Z = 4. R = 0.029-0.047, X-ray and neutron powder diffraction data; all atoms in 4(c): x, 1/4, z.

	$AsCo_{0.10}Cr_{0.90}$	$AsCo_{0.05}Fe_{0.95}$
a(Å)	5.604	5.416
b(Å)	3.443	3.379
c(Å)	6.187	6.011
x(T)	0.0077	0.0042
z(T)	0.2011	0.1986
x(As)	0.2004	0.2013
z(As)	0.5771	0.5792

T = $Co_{0.10}Cr_{0.90}$, $Co_{0.05}Fe_{0.95}$. Data for 4.2°, 6°, and 80°K, and for $AsCo_{0.05}$-$Cr_{0.95}$ and details of magnetic structures and phase changes are also given; see also 2 and 3.

[Interatomic distances (Å)]

	$AsCo_{0.10}Cr_{0.90}$	$AsCo_{0.05}Fe_{0.95}$
As - 2 As	2.99	2.92
- 6 T	2.43 - 2.56	2.34 - 2.52
T - 4 T	2.87, 3.03	2.78, 2.93

1. Strukturbericht, 3, 17, 264; Structure Reports, 27, 319.
2. Structure Reports, 37A, 12, 14; 38A, 25; 40A, 18, 101.
3. This volume, following report.

ARSENIC CHROMIUM VANADIUM

$AsCr_{0.5}V_{0.5}$

K. SELTE, H. HJERSING, A. KJEKSHUS and A.F. ANDRESEN, 1975. Acta Chem. Scand., A29, 312-316.

Orthorhombic, MnP type (1), Pnma, Z = 4. R = 0.027-0.056, X-ray and neutron diffraction data; all atoms in 4(c): x, 1/4, z; see also 2; phase changes and magnetic structures are also discussed.

T(°K)	a (Å)	b (Å)	c (Å)	x(T)*	z(T)*	x(As)	z(As)
80	5.707	3.368	6.237	0.0069	0.1960	0.1975	0.5737
293	5.724	3.382	6.250	0.0086	0.1970	0.1986	0.5728

Interatomic distances (Å)

	80°K	293°K
As - 6 T	2.434 - 2.595	2.448 - 3.588
- 2 As	2.96	2.976
T - 6 T	2.931 - 3.368	2.938 - 3.382

* T = $Cr_{0.5}V_{0.5}$. Parameters for the range 0.75-0.95 are also given.

1. Strukturbericht, 3, 17; Structure Reports, 27, 319.
2. Structure Reports, 38A, 25.

ARSENIC COPPER MERCURY SULPHUR THALLIUM ZINC

(GALKHAITE)

$(AsS_3)_8[(Cu,Zn)_{0.24}Hg_{0.76}]_{12}Tl_{0.96}$ (I, II)

$(As_{0.98}Sb_{0.02})_{1.00}[Cu_{0.17}Hg_{0.74}Tl_{0.01}Zn_{0.14}]_{1.06}S_{2.01}$ (III)

I. V. DIVJAKOVIĆ and W. NOWACKI, 1975. Neues Jb. Miner., Mh., 291-293.

II. Idem, 1975. Z. Kristallogr., 142, 262-270.

III. L.N. KAPLUNIK, E.A. POBEDIMSKAJA and N.V. BELOV, 1975. Dokl. Akad. Nauk SSSR, 225, 561-563.

Cubic, I$\bar{4}$3m*, a = 10.379 (I, II), 10.422 (III) Å, A = 44.96 (I, II), 44 (III); Dm = 5.4 (III). Cu radiation, diffractometer data corrected for absorption, R = 0.045 for 145 reflexions; absolute configuration from Friedel pairs (I, II). Mo radiation, diffractometer data, R = 0.052 for 98 reflexions (III). See 1-3 for early reports; I, II give identical data.

* [Misprinted in I]

Atomic positions

site	x	y	z		occupancy
8(c)	0.2456	0.2456	0.2456	I	As
	0.2564	0.2564	0.2564	III	As
12(d)	1/4	1/2	0	I	0.76Hg, 0.12Cu, 0.12Zn
				III	0.74Hg, 0.17Cu, 0.14Zn
2(a)	0	0	0	I	0.48Tl
				III	As [occupancy ∿1/3]
24(g)	0.3884	0.3884	0.1629	I	S
	0.1127	0.1127	0.3382	III	S
8(c)	0.07	0.07	0.07	III	As [occupancy ∿2/3]

The structure is closely related to those of tennantite and sodalite (I, III). The Hg/Cu/Zn atom in 12(d) centres a regular S_4 tetrahedron (Hg-S = 2.503 Å), the As atom is coordinated to 3 S atoms, forming a trigonal pyramid (As-S = 2.265 Å, S-As-S = 93.9°). The Tl atom is coordinated to 12 S atoms, which form a regular Laves polyhedron (Tl-As = 3.863 Å) (I). The structure of III places the small amount of Tl in the 12(d) site and in addition As in an 8(c) site.

1. V.S. GRUZDEV, 1972. Dokl. Akad. Nauk SSSR, 205, 150.
2. T. BOTINELLY, G.J. NEUERBURG and N.M. CONKLIN, 1973. J. Res. U.S. Geol. Surv., 1, 515.
3. G. JUNGLES, 1974. Mineral Record, 5, 299.

ARSENIC IRON MOLYBDENUM

PHOSPHORUS TUNGSTEN

R. GUÉRIN, M. SERGENT and J. PRIGENT, 1975. Mater. Res. Bull., 10, 957-965.

Orthorhombic, MnP-type (1), Pnma, Z = 4. R = 0.055 for 'AsMo', 208 reflexions; R = 0.080 for PW, 387 reflexions. Diffractometer data. There are extensive solid solutions of the type $M_xMo_{1-x}As$ (M = V, Cr, Fe) and $M_xW_{1-x}P$ (M = Cr, Mn); see also 2-9. The space group $Pna2_1$ was tested and rejected.
[There has been considerable discussion concerning 'MoAs'; 2 and 5 assigned an MnP type structure while 9 assigned a defect FeS_2 (marcasite) type structure; 3, 4, and 6 were unable to prepare an MoAs phase in the binary Mo-As system; 7 showed that the data of 9 could be reinterpreted as belonging to an $FeAs_2$ sample or a composition close to that. The present authors observe the MnP type phase only if Fe is present. Both MnP and $FeAs_2$ can be considered as derived from the NiAs structure.]

	a(Å)	b(Å)	c(Å)	D_m	x(M)*	z(M)	x(X)*	z(X)
$AsFe_{0.2}Mo_{0.8}$ ('AsMo')	5.977	3.364	6.400	8.53	0.0101	0.1869	0.1920	0.5700
PW	5.731	3.248	6.227	12.06	0.0147	0.1885	0.1833	0.5652

* M = Mo or W, X = As or P. All atoms in 4(c): x, 1/4, z.

Interatomic distances (Å)

	AsMo	PW
M - 4 M	2.927, 3.096	2.859, 2.966
- 6 X	2.514 - 2.682	2.468 - 2.538
X - 2 X	2.983	2.777

These structures are characterized by M-M chains of two types. Type I, parallel to a, has M-M = 3.096 (Mo) and 2.966 Å (W) and M-M-M angles of 149.8° (Mo) and 150.1° (W), whereas type II has M-M = 2.927 (Mo) and 2.859 Å (W) and angles M-M-M of 70.2° (Mo) and 69.2° (W), respectively, running parallel to b in the (100) plane.

1. Strukturbericht, 3, 17; Structure Reports, 27, 319.
2. Structure Reports, 29, 101
3. Ibid., 31A, 22.
4. Ibid., 30A, 173, ref, 62
5. Ibid., 31A, 24.
6. Ibid., 33A, 63.
7. Ibid., 40A, 87.
8. Ibid., 18, 262.
9. Ibid., 30A, 21.

ARSENIC NICKEL SULPHUR

(GERSDORFFITE)

AsNiS

J.J. STEGER, H. NAHIGIAN, R.J. ARNOTT and A. WOLD, 1974. J. Solid State Chem., 11, 53-59.

Cubic, NiSbS (ullmannite) type (1), $P2_13$, a = 5.576 Å, D_m = 6.35, Z = 4. Powder diffractometer data, Mo radiation, R(I) = 0.054 for 18 peaks, with allowance for anomalous dispersion; see also 2.

Atomic positions

		x	y	z
4 Ni	in 4(a)	0.006	0.006	0.006
As/S*	in 4(a)	0.610	0.610	0.610
S/As†	in 4(a)	0.381	0.381	0.381

* 3.65 As + 0.35 S   † 0.35 As + 3.65 S

1. Structure Reports, 21, 34
2. Ibid., 32A, 26.

ARSENIC NIOBIUM

NIOBIUM SILICON

Nb_3X (X = As, Si)

R.M. WATERSTRAT, K. YVON, H.D. FLACK and E. PARTHÉ, 1975. Acta Cryst., B31, 2765-2769.

Tetragonal, Ti_3P type (1), $P4_2/n$, Z = 8. Mo radiation, photographic and diffractometer data, with allowance for anomalous dispersion; absorption and extinction corrections for Nb_3Si only; see also 2, 3.

Nb_3X	a(Å)	c(Å)	c/a	R	Reflexions
$AsNb_3$	10.294	5.199	0.51	0.043	964
Nb_3Si	10.224	5.189	0.51	0.032	786

Atomic positions

All atoms in 8(g)

$AsNb_3$	x	y	z
Nb(1)	0.1652	0.6554	0.7272
Nb(2)	0.1021	0.2617	0.5181
Nb(3)	0.0562	0.5364	0.2411
As	0.0430	0.2714	0.0203
Nb_3Si			
Nb(1)	0.1653	0.6525	0.7185
Nb(2)	0.1043	0.2665	0.5230
Nb(3)	0.0603	0.5364	0.2370
Si	0.0442	0.2782	0.0293

Interatomic distances (Å)

		$AsNb_3$	Nb_3Si
Nb(1)	- 2 X	2.62, 2.65	2.61, 2.65
	- 12 Nb	2.62 - 3.25	2.64 - 3.22
Nb(2)	- 4 X	2.61 - 2.68	2.59 - 2.70
	- 11 Nb	2.89 - 3.54	2.89 - 3.59
Nb(3)	- 4 X	2.61 - 3.67	2.58 - 3.78
	- 11 Nb	2.86 - 3.54	2.85 - 3.59
X	- 10 Nb	2.61 - 3.67	2.58 - 3.79
	- 2 X	3.86	3.78

The Cr_3Si type structure is replaced by the Ti_3P or Ni_3P types as the electronegativity difference between the component atoms becomes larger.

1. Structure Reports, 32A, 111.
2. Ibid., 30A, 157; 34A, 138, 140.
3. E. GANGLBERGER, H. NOWOTNY and F. BENESOVSKY, 1966. Mh. Chem., 97, 1696.

ARSENIC NITROGEN URANIUM

NITROGEN PHOSPHORUS URANIUM

NITROGEN SELENIUM URANIUM

NITROGEN SULPHUR URANIUM

U_2N_2M (M = As, P, S, Se)

J. LECIEJEWICZ, Z. ŻOŁNIEREK, S. LIGENZA, R. TROĆ and H. PTASIEWICZ, 1975. J. Phys., C, Solid State Phys., 8, 1697-1704.

Trigonal, Ce_2O_2S type (1), P$\bar{3}$m1, U in 2(d): 1/3, 2/3, z_1; N in 2(d): 1/3, 2/3, z_2; M in 1(a): 0, 0, 0. Neutron diffractometer data. Lattice parameters, R, and z values at 4° and 300°K are given in the following table; magnetic structures are given in the paper; see also 2.

U_2N_2M	(Å)	4°K	300°K
U_2N_2S	a = 3.818	z_1 = 0.2798	z_1 = 0.2798
	c = 6.610	z_2 = 0.6264	z_2 = 0.6274
		R = 0.0622	R = 0.0538
U_2N_2Se	a = 3.863	z_1 = 0.3054	z_1 = 0.2965
	c = 6.867	z_2 = 0.6306	z_2 = 0.6244
		R = 0.0855	R = 0.0258
U_2N_2P	a = 3.805	z_1 = 0.2770	z_1 = 0.2747
	c = 6.596	z_2 = 0.6290	z_2 = 0.6280
		R = 0.0608	R = 0.0480
U_2N_2As	a = 3.830	z_1 = 0.2789	z_1 = 0.2802
	c = 6.739	z_2 = 0.6190	z_2 = 0.6203
		R = 0.0390	R = 0.0722

[Interatomic distances (Å)]

		U - 9 U	U - 4 N	U - 3 M	M - 6 U
N_2SU_2	4°K	3.65, 3.82	2.29	2.88	2.88
	300	3.65, 3.82	2.30, 2.29	2.88	2.88
N_2SeU_2	4	3.48, 3.86	2.23, 2.27	3.06	3.06
	300	3.58, 3.86	2.25, 2.30	3.02	3.02
N_2PU_2	4	3.67, 3.81	2.28, 2.32	2.86	2.86
	300	3.70, 3.81	2.29, 2.33	2.85	2.85
AsN_2U_2	4	3.71, 3.83	2.29, 2.32	2.90	2.90
	300	3.70, 3.83	2.29, 2.31	2.91	2.91

1. Structure Reports, 12, 174.
2. Ibid., 34A, 135.

ARSENIC SELENIUM URANIUM

ARSENIC SULPHUR URANIUM

ARSENIC TELLURIUM URANIUM

AsXU (X = Se, S, Te)

D. PIETRASZKO and K. LUKASZEWICZ, 1975. Bull. Acad. Polon. Sci., Ser. Sci. Chim., 23, 337-340.

AsSeU, tetragonal, PbFCl type (1), P4/nmm, a = 3.981, c = 8.371 Å, c/a = 2.102, Z = 2, D_x = 9.831. R = 0.138 for 214 reflexions. See also 3, 5.

AsSU, tetragonal, PbFCl type (1), P4/nmm, a = 3.878, c = 8.164 Å, c/a = 2.105, Z = 2, D_x = 9.326. R = 0.077 for 192 reflexions. See also 3, 4, 5.

AsTeU, tetragonal, UGeTe type (2), I4/mmm, a = 4.151, c = 17.270 Å, c/a = 4.161, Z = 4, D_x = 9.836. R = 0.115 for 313 reflexions. See also 3, 4, 5.

Mo radiation, diffractometer data corrected for absorption.

Atomic positions

AsSeU*	x	y	z
U in 2(c)	3/4	3/4	0.2696
Se in 2(c)	3/4	3/4	0.6315
As in 2(a)	3/4	1/4	0
AsSU*			
U in 2(c)	3/4	3/4	0.2888
S in 2(c)	3/4	3/4	0.6334
As in 2(a)	3/4	1/4	0
AsTeU			
U in 4(e)	0	0	0.1199
Te in 4(e)	0	0	0.3122
As in 4(c)	0	1/2	0

* [Transformed to setting with origin on a centre.]

Interatomic distances (Å)

	AsSeU	AsSU	AsTeU
U - 4 X	2.934	2.822 [2.815]	3.159
- 1 X	3.030	2.847 [2.813]	3.321
- 4 As	3.010	3.028 [3.053]	2.932
As - 4 X	3.671	3.566	3.850
- 4 As	2.815	2.742	2.935
X - 4 X	3.574	3.503	3.638

X = Se, S, or Te

1. Strukturbericht, 2, 45; 3, 64, 369.
2. Structure Reports, 34A, 92.
3. F. HULLIGER, 1968. J. Less-Common Metals, 16, 113.
4. Structure Reports, 40A, 24, 25.
5. Ibid., 38A, 20, 158.

BARIUM IRON SULPHUR

$Ba_5Fe_9S_{18}$

I.E. GREY, 1975. Acta Cryst., B31, 45-48.

Tetragonal, P4/ncc, a = 7.776, c = 49.86 Å, Z = 4, D_x = 3.89. Cu radiation, R = 0.092 for 218 reflexions, photographic data corrected for absorption, powder diffractometer data, 15 site-sets. See also 1, 2.

Interatomic distances (Å)

The Fe atoms are coordinated to 5 atoms, 4 S (2.20-2.32 Å) plus 1 Fe (2.66-2.88 Å), and the Ba atoms are coordinated to 8 S (3.11-3.48 Å). The S-Fe-S angles range from 102 to 116°. Mean distances are Fe-S = 2.26, Ba-S = 3.28 Å.

The structure consists of chains of edge-shared $[FeS_4]$ tetrahedra, lying parallel to [001] (Fig. 1), with Ba atoms packed between the chains. The structure may be considered as a superstructure based on the NH_4CuMoS_4 (3) structure type. In the latter compound the ammonium ions lie in the (001) planes of sulphur atoms and their repeat distance along [001] equals the S-S repeat distance of 5.4 Å. However, in $Ba_5Fe_9S_{18}$ the Ba atoms pack more densely than ammonium, so that the Ba-Ba repeat distance along [001], ∿ 5.0 Å, is considerably shorter than the S-S repeat, ∿ 5.5 Å. The structure is repeated along [001] after every tenth barium atom or after every ninth sulphur or iron layer (containing 4 S and 2 Fe, respectively).

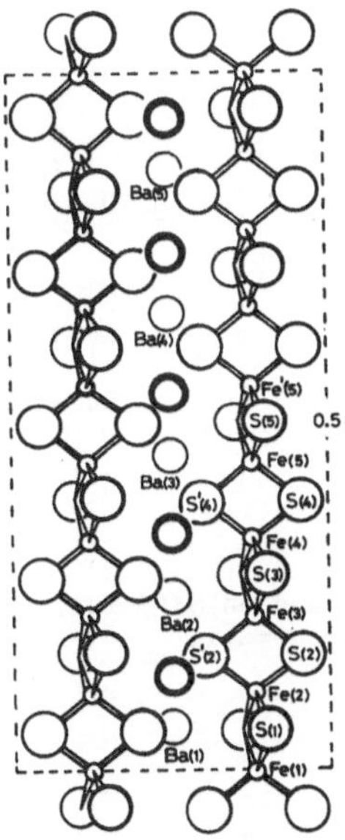

Fig. 1. The infinite chains of edge-shared $[FeS_4]$ tetrahedra and Ba ions in the tetragonal $Ba_5Fe_9S_{18}$ structure, as viewed along [$\bar{1}\bar{1}$0].

1. I.E. GREY, 1974. J. Solid State Chem., 11, 128.
2. Structure Reports, 37A, 17.
3. Ibid., 35A, 144.

BARIUM SULPHUR

I. S. YAMAOKA, J.T. LEMLEY, J.M. JENKS and H. STEINFINK, 1975. Inorg. Chem., 14, 129-131.

Ba_2S_3, tetragonal, $I4_1md$, a = 6.112, c = 15.950 Å, c/a = 2.61, Z = 4, D_x = 4.13. R = 0.06 for 174 reflexions.

BaS_3, tetragonal, $P\bar{4}2_1m$, a = 6.871, c = 4.1681 Å, c/a = 0.61, Z = 2, D_x = 3.94. R = 0.03 for 151 reflexions. See also 1.

Mo radiation, diffractometer data (stationary-crystal, stationary-counter method), corrected for absorption.

Atomic positions

Ba_2S_3			x	y	z
Ba(1)	in	4(a)	0	0	-0.2031
Ba(2)	in	4(a)	0	0	0.2051
S(1)	in	4(a)	0	0	0
S(2)	in	8(b)	0	0.3101	0.6426
BaS_3					
Ba	in	2(a)	0	0	0
S(1)	in	2(c)	0	1/2	0.2060
S(2)	in	4(e)	0.1797	0.6797	0.4760

Interatomic distances (Å)

Ba_2S_3		BaS_3	
Ba(1) - 9 S	3.11 - 3.91	Ba - 12 S	3.20 - 3.54
Ba(2) - 9 S	3.14 - 3.42	S(1) - S	2.07
S(2) - S(2)	2.32		

Ba_2S_3 contains a sulphide ion and an S_2^{2-} ion; in BaS_3 the anion is S_3^{2-}, with S-S = 2.07 Å and the S-S-S angle = 115°

BaS_2

II. I. KAWADA, K. KATO and S. YAMAOKA, 1975. Acta Cryst., B31, 2905-2906.

Monoclinic, C2/c [YbS_2 type (2)], a = 9.299, b = 4.736, c = 8.993 Å, β = 118.37°, Z = 4, D_x = 3.84. Mo radiation, R = 0.056 for 642 observed reflexions, diffractometer data. [Similar data given in 3.]

Atomic positions

	x	y	z
Ba in 4(e)	0	0.1446	1/4
S in 8(f)	0.1603	0.3545	0.0206

In the structure of BaS_2, arrays of disulphide ions S_2^{2-}, with an S-S bond length of 2.118 Å, run parallel to <110> at approximately z = 0 and z = 1/2. The S...S distance of 3.143 Å between neighbouring disulphide ions is rather short, possibly suggesting the existence of some weak bonding between them. The S-S-S angle in these arrays is 165.08°. The Ba ion is located between the arrays of disulphide ions and is surrounded by eight S atoms; the Ba-S distances range from 3.151 to 3.223 Å.

1. Strukturbericht, 4, 123.
2. Structure Reports, 40A, 97.
3. Ibid., 40A, 35.

BARIUM SULPHUR TIN

SODIUM SULPHUR TIN

J.-C. JUMAS, E. PHILIPPOT, F.V.-G. DANIEL, M. RIBES and M. MAURIN, 1975. J. Solid State Chem., 14, 319-327.

α-Ba_2S_4Sn, monoclinic, modified β-K_2SO_4 type (1), $P2_1/c$, a = 8.481, b = 8.526, c = 12.280 Å, β = 112.97°, D_m = 4.09, Z = 4. Mo radiation, R = 0.036 for 2081 reflexions (diffractometer data). See also 2.

Na_4S_4Sn, tetragonal, $P\bar{4}2_1c$, a = 7.837, c = 6.950 Å, D_m = 2.64, Z = 2. Cu radiation, R = 0.088 for 242 reflexions. Photographic data corrected for absorption. See also 3.

Atomic positions

α-Ba_2S_4Sn: all atoms in 4(e)

	x	y	z
Ba(1)	0.95870	0.19389	0.64948
Ba(2)	0.57694	0.12422	0.83611
Sn	0.29995	0.18794	0.49054
S(1)	0.7537	0.0154	0.3922
S(2)	0.2709	0.1016	0.2953
S(3)	0.0870	0.1176	0.9447
S(4)	0.5721	0.1922	0.0981

Na_4S_4Sn

	x	y	z
Na in 8(e)	0.1997	0.5910	0.0313
S in 8(e)	0.0922	0.2376	0.1883
Sn in 2(a)	0	0	0

Interatomic distances (Å)

α-Ba_2S_4Sn			Na_4S_4Sn		
Sn	- 4 S	2.35 - 2.43	Sn	- 4 S	2.39
Ba(1)	- 8 S	3.09 - 3.77	Na	- 5 S	2.79 - 3.09
Ba(2)	- 8 S	3.10 - 3.69			

Both structures are characterized by [SnS_4] tetrahedra with S-Sn-S angles of 108 and 114° (Na_4SnS_4) and 102 to 114° (Ba_2SnS_4). The tetrahedra are linked by the coordination polyhedra of the Na or Ba atoms.

1. Strukturbericht, 2, 86; Structure Reports, 16, 284; 22, 447.
2. Structure Reports, 37A, 152.
3. Ibid., 38A, 169.

BERYLLIUM LITHIUM PHOSPHORUS

BeLiP

A. EL MASLOUT, J.-P. MOTTE, A. COURTOIS and C. GLEITZER, 1975. J. Solid State Chem., 15, 213-217.

Tetragonal, anti-PbFCl type (1), P4/nmm, a = 3.617, c = 6.032 Å, c/a = 1.66, D_m = 1.98, Z = 2. Mo radiation, R(I) = 0.06 for 14 reflexions, X-ray and powder neutron diffraction data; see also 2.

Atomic positions [transformed to centre as origin]

	x	y	z
Li in 2(c)	3/4	3/4	0.341
Be in 2(a)	3/4	1/4	0
P in 2(c)	3/4	3/4	0.781

Interatomic distances (Å)

Li - 4 Be	2.74	Be - 4 P	2.24	P - 4 Be	2.24
- 5 P	2.65, 2.66	- 4 Be	2.56	- 5 Li	2.65, 2.66

The structure belongs to those of the PbFCl (Cu_2Sb) type with c/a ∿ 1.5, and is characterized by shorter distances, here Be-P (2.24 Å) and Li-P (2.65 Å).

1. Strukturbericht, 2, 45; 3, 370; Structure Reports, 29, 38, 342.
2. Structure Reports, 39A, 108.

BERYLLIUM PHOSPHORUS

Be_3P_2

A. EL MASLOUT, J.-P. MOTTE, A. COURTOIS, J. PROTAS and C. GLEITZER, 1975. J. Solid State Chem., 15, 223-228.

Tetragonal, $I4_1/acd$, a = 10.22, c = 20.39 Å, c/a = 1.995, D_m = 2.23, Z = 32. Mo radiation, R(I) = 0.04 for 79 reflexions; single-crystal X-ray for cell and neutron powder data for structure; contrary to 1.

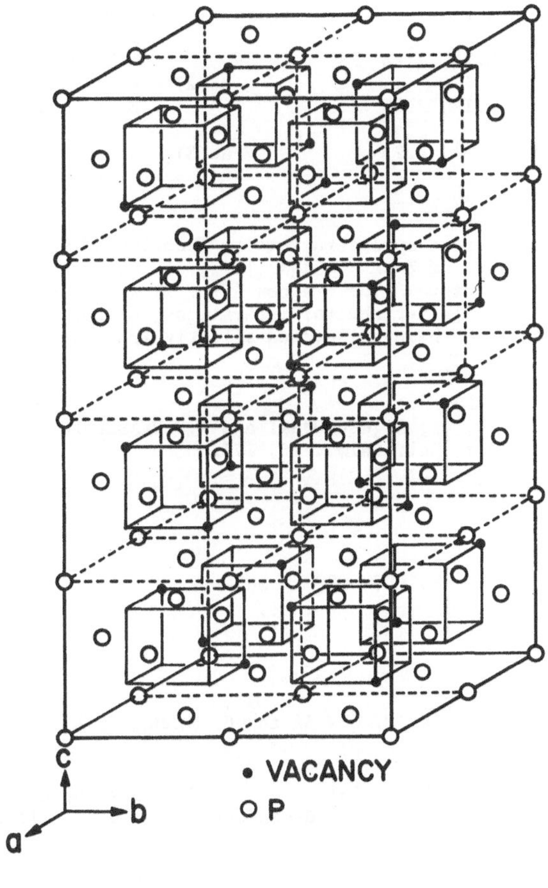

Fig. 1. A perspective view of the tetragonal Be_3P_2 structure. The cubes defining the Be sites are outlined.

Atomic positions

	x	y	z
P(1) in 8(a)	0	1/4	3/8
P(2) in 8(b)	0	1/4	1/8
P(3) in 16(c)	0	0	0
P(4) in 16(e)	-0.002	0	1/4
P(5) in 16(f)	0.254	0.504	1/8
Be(1) in 32(g)	0.126	0.381	0.063
Be(2) in 32(g)	0.380	0.369	0.065
Be(3) in 32(g)	0.133	0.129	0.059

[Interatomic distances (Å)]

P(1) - 4 Be	2.12	Be(1) - 4 P	2.19 - 2.25
		- 5 Be	2.53 - 2.65
P(2) - 8 Be	2.25, 2.28		
		Be(2) - 4 P	2.12 - 2.25
P(3) - 6 Be	2.19, 2.25	- 3 Be	2.60 - 2.66
P(4) - 6 Be	2.11, 2.23	Be(3) - 4 P	2.11 - 2.28
		- 4 Be	2.58 - 2.70
P(5) - 6 Be	2.19 - 2.25		

The structure is a defect superstructure of fluorite (CaF_2) with only a formal relationship to that of Zn_3As_2 (Cd_3As_2) (2, 3), which has a somewhat different defect superstructure. The P atoms (Fig. 1) are in cubic close-packing with the Be atoms, in tetrahedral interstices, occupying 6 of the 8 corners of a distorted cube, being displaced slightly from the ideal positions.

1. Strukturbericht, 3, 353.
2. Structure Reports, 20, 41; S. WEGLOWSKI and K. LUKASZEWICZ, 1968. Bull. Acad. Polon. Sci., Ser. Sci. Chim., 16, 177.
3. Structure Reports, 33A, 25.

BISMUTH COPPER SULPHUR

$Bi_5Cu_4S_{10}$

I. K. MARIOLACOS, V. KUPČÍK, M. OHMASA and G. MIEHE, 1975. Acta Cryst., B31, 703-708.

Monoclinic, C2/m, a = 17.539, b = 3.931, c = 12.847 Å, β = 108.0°, Z = 2, D_x = 6.39. Cu radiation, R = 0.12 for 796 independent reflexions, diffractometer data; see also 1.

Atomic positions

	x	y	z
Bi(1) in 2(a)	0	0	0
Bi(2) in 4(i)	0.1963	0	0.3427
Bi(3) in 4(i)	0.4127	0	0.2240
Cu(1) in 4(h)	0.2076	1/2	0.0608
Cu(2) in 4(h)	0.4018	1/2	0.4736
S(1) in 4(h)	0.0443	1/2	0.1684
S(2) in 4(h)	0.2586	1/2	0.2452
S(3) in 4(i)	0.3364	0	0.4962
S(4) in 4(h)	0.3463	1/2	0.0268
S(5) in 4(h)	0.4818	1/2	0.3613

Interatomic distances (Å)

Bi(1) - 6 S	2.821, 2.846	Cu(1) - 4 S	2.258 - 2.602
Bi(2) - 5 S	2.630 - 3.035	Cu(2) - 4 S	2.304 - 2.446
Bi(3) - 5 S	2.619 - 3.131		

Bismuth atoms have both octahedral and tetragonal pyramidal coordination, and Cu atoms are surrounded by four S atoms. Two of the cell constants of $Cu_4Bi_5S_{10}$ are similar to those of hodrushite, $MBi_5Cu_4S_{11}$ (M = Bi, Fe, Pb (2)), and of cuprobismutite (3). These three substances are layer structures. Both $Cu_4Bi_5S_{10}$ and cuprobismutite have single-layer structures, and hodrushite has a double-layer structure containing one layer of $Cu_4Bi_5S_{10}$ and one layer of cuprobismutite. A structure for cuprobismutite is proposed and verified from powder data.

$Bi_4Cu_4S_9$

II. Y. TAKÉUCHI and T. OZAWA, 1975. Z. Kristallogr., 141, 217-232.

Orthorhombic, Pbnm, a = 11.589, b = 32.05, c = 3.951 Å, Z = 4, D_x = 6.238. Cu radiation, R = 0.08 for 816 reflexions, photographic data corrected for absorption, 17 site-sets; the crystal used was multiple; see also 4, 5.

Interatomic distances (Å)

Cu(1) - 4 S	2.23, 2.40	Bi(1) - 6 S	2.69 - 3.05
Cu(2) - 3 S	2.23, 2.26	Bi(2) - 7 S	2.63 - 3.15
Cu(3) - 4 S	2.32 - 2.37	Bi(3) - 7 S	2.60 - 3.64
Cu(4) - 5 S	2.25 - 3.20	Bi(4) - 7 S	2.63 - 3.50

The structure (Figs. 1 and 2) can be described as an alternate stacking of Cu-S sheets and Bi-S slabs, parallel to (010). The Cu-S sheet is corrugated and contains four- and three-coordinated copper atoms, as in the covellite structure, which are joined together by S-S bonds. The three-coordinated copper atom is not in the plane formed by three sulphur atoms but displaced from the plane by 0.48(1) Å; mean Cu-S distance is 2.25 Å. The Bi-S slabs consist of Bi-S polyhedra which are in turn closely related to those of the bismuthinite structure. The structural formula apparently does not obey the valence rule, but this is explained by the presence of S-S bonds (2.10 Å). The mean Cu-S distances are 2.36 Å (C.N. 4) and

2.25 Å (C.N. 3), and the corresponding S-Cu-S angles are 109 and 117°, respectively. The Bi-S polyhedra are characterized by 3 short Bi-S distances (average = 2.72 Å), with average S-Bi-S angle of 90°.

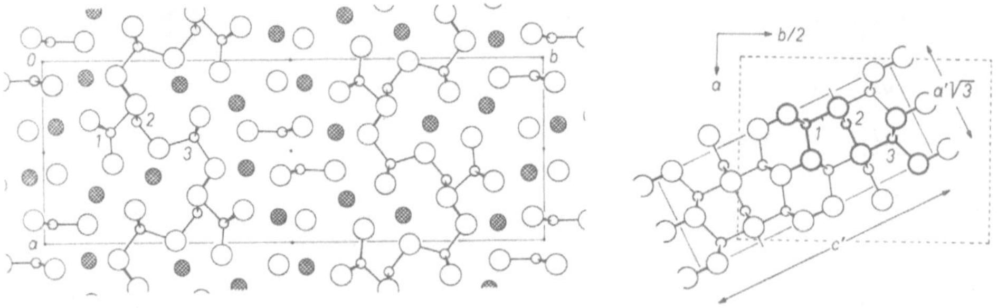

Fig. 1. The (001) projection of the orthorhombic $Cu_4Bi_4S_9$ structure (left), compared with the (001) projection of the covellite structure (large open circles, S; small open circles, Cu; hatched circles, Bi). The Cu-S linkages are emphasized.

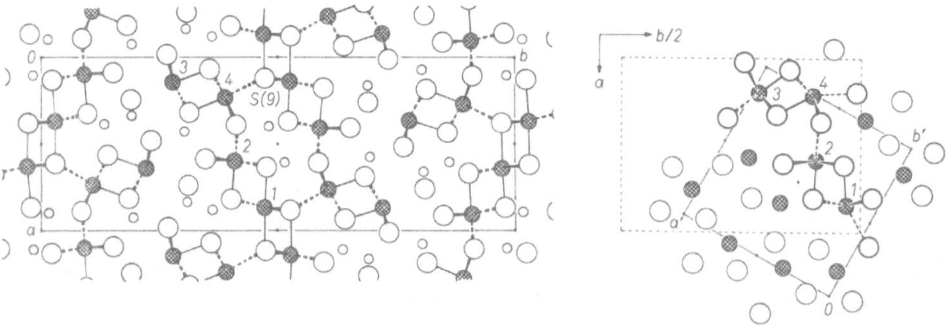

Fig. 1. The (001) projection of the orthorhombic $Cu_4Bi_4S_9$ structure (left), compared with the bismuthinite (001) projection. The Bi-S linkages are emphasized.

$BiCuS_2$ (emplectite)

III. J.C. PORTHEINE and W. NOWACKI, 1975. Z. Kristallogr., 141, 387-402.

Orthorhombic, Pnma, a = 6.1426, b = 3.9189, c = 14.5282 Å, Z = 4, D_x = 6.393. Cu radiation, R = 0.063 for 375 reflexions, diffractometer data corrected for absorption, with allowance for anomalous dispersion; in agreement with 6, 7.

Atomic positions

All atoms in 4(c)

	x	y	z
Bi	0.23156	1/4	0.06304
Cu	0.7509	3/4	0.1719
S(1)	0.6362	1/4	0.0980
S(2)	0.1258	3/4	0.1777

Interatomic distances (Å)

Bi - 3 S	2.54, 2.65	S(1) - 2 Cu	2.34
- 2 S	3.16	- 1 Bi	2.54
Cu - 4 S	2.30 - 2.34	S(2) - 2 Cu	2.30, 2.32
		- 2 Bi	2.65

The three smallest Bi-S distances (2.536 and two of 2.653 Å) span a trigonal pyramid with Bi at the vertex. The BiS_3 pyramids are coupled by corner sharing to form endless chains with composition BiS_2 and period b. Cu is coordinated by four S atoms (at distances 2.304, 2.317, and two of 2.343 Å) in a nearly regular tetrahedron. The BiS_2 chains join with chains of CuS_4 tetrahedra to form sheets parallel to (001). $BiCuS_2$ has structural units which can be described as double chains of BiS_5 square pyramids. Such a pyramid is formed by the five nearest S neighbours of the Bi atom, while the Bi itself lies near the centre of the basal plane. Each BiS_5 pyramid shares two opposite edges of the basal plane with adjacent pyramids in one endless chain; two side edges are shared with pyramids belonging to the second chain, which is parallel and equivalent to the first chain. The double chain has a composition of Bi_2S_4 and a period of 4 Å corresponding to the shortest lattice dimension, b.

1. M. KODERA, V. KUPČÍK and E. MAKOVICKÝ, 1970. Miner. Mag., 37, 641.
2. Structure Reports, 33A, 41.
3. Ibid., 16, 275.
4. A. SUGAKI and H. SHIMA, 1971. Proc. IMA-IAGOD Meetings, 1970, IMA Vol., 270-271 (abstract).
5. T. OZAWA and Y. TAKÉUCHI, 1972. Acta Cryst., A28, S70.
6. Strukturbericht, 3, 393.
7. V. KUPČÍK, 1965. Programme and Abstracts, Meeting of Crystallographic Section of German Miner. Soc., Oct. 13-16, 1965, Marburg, 16-17.

BISMUTH PHOSPHORUS SULPHUR

$BiPS_4$

H. ZIMMERMANN, C.D. CARPENTER and R. NITSCHE, 1975. Acta Cryst., B31, 2003-2006.

Orthorhombic, Ibca, a = 10.601, b = 11.112, c = 19.661 Å, D_m = 4.18, Z = 16. Mo radiation, R = 0.047 for 881 independent reflexions. Diffractometer data, corrected for absorption. See also 1.

Atomic positions

		x	y	z
Bi(1)	in 8(e)	0	1/4	0.3986
Bi(2)	in 8(e)	0	1/4	0.1414
P(1)	in 8(d)	1/4	0.2210	0
P(2)	in 8(c)	0.0347	0	1/4
S(1)	in 16(f)	0.0979	0.1094	0.0030
S(2)	in 16(f)	0.2603	0.3274	0.0860
S(3)	in 16(f)	0.0709	0.5032	0.1657
S(4)	in 16(f)	0.1395	0.1583	0.2429

Interatomic distances (Å)

Bi(1) - 6 S	2.70 - 3.11	P(1) - 4 S	2.03, 2.07
Bi(2) - 8 S	2.68 - 3.30	P(2) - 4 S	2.00, 2.08

The structure (Fig. 1) consists of a network of S tetrahedra, with every second one occupied by a P atom. Chains of edge-shared tetrahedra, running parallel to a and b alternately, form layers perpendicular to c. Six- and eightfold coördination of S around Bi occurs. The [PS_4] tetrahedra are nearly regular, with angles ranging from 105 to 116° and edge lengths between 3.26 and 3.39 Å. The unoccupied tetrahedra are less regular, with edge lengths between 3.26 and 3.83 Å.

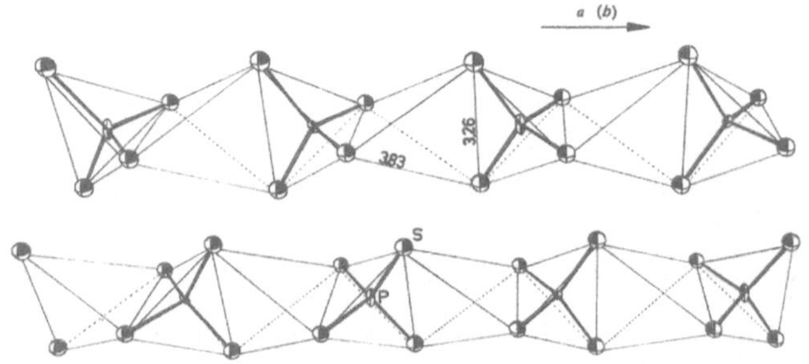

Fig. 1. The chains of [S_4] and [PS_4] tetrahedra in the orthorhombic $BiPS_4$ structure. The chains extend parallel to a and b in alternate layers.

1. Structure Reports, 38A, 161.

BISMUTH RUBIDIUM SULPHUR

Bi_3RbS_5

D. SCHMITZ and W. BRONGER, 1974. Z. Naturforsch., 29B, 438-439.

Orthorhombic, Pmnn, a = 4.161, b = 12.902, c = 18.478 Å, Z = 4. R = 0.10, photographic data; see also 1.

Atomic positions [note non-standard setting]

S(5) in 2(a), S(6) in 2(d), all others in 4(g)

	x	y	z
Bi(1)	0	0.053	0.156
Bi(2)	0	0.268	0.803
Bi(3)	0	0.338	0.476
Rb	0	0.416	0.120
S(1)	0	0.704	0.050
S(2)	0	0.303	0.068
S(3)	0	0.390	0.320
S(4)	0	0.091	0.301
S(5)	0	0	0
S(6)	1/2	1/2	0

Interatomic distances (Å)

Bi(1) - 6 S	2.72 - 2.99	Rb - 9 S	3.23 - 3.93*
Bi(2) - 6 S	2.74 - 3.26		
Bi(3) - 6 S	2.75 - 2.98		

* [3.400 misprinted as 3.414]

The structure (Fig. 1) can be described as a block structure with octahedrally coordinated Bi atoms.

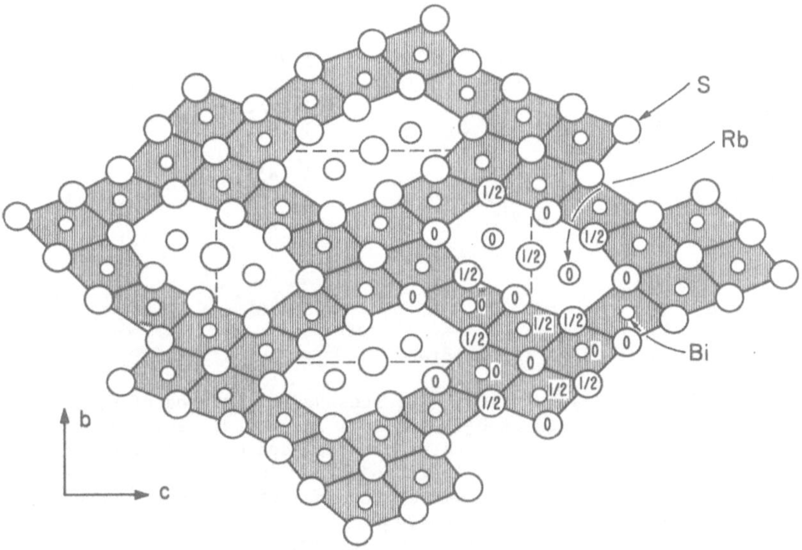

Fig. 1. The orthorhombic Bi_3RbS_5 structure projected onto (100).

1. Structure Reports, 38A, 37.

BORON CARBON

$B_{13.8}C_{1.5}$

H.L. YAKEL, 1975. Acta Cryst., B31, 1797-1806.

Rhombohedral, $R\bar{3}m$, a = 5.2065 Å, α = 66.010°, D_m = 2.456, Z = 1 (hexagonal cell, a = 5.6720, c = 12.1428 Å, c/a = 2.1408, Z = 3). Mo radiation, R = 0.085 for 971 independent reflexions, diffractometer data corrected for absorption; boron-rich material containing 8 ± 1 at. % C. For earlier work on B_4C see 1, and on boron-rich material see 2-4.

Atomic positions (hexagonal coordinates)

		x	y	z
B(1)	in 18(h)	0.83622	-0.83622	0.35934
B(2)	in 18(h)	0.89291	-0.89291	0.11354
0.742 B(3)	in 3(b)	0	0	1/2
0.259 B(4)	in 6(c)	0	0	0.4146
0.043 B(5)	in 36(i)	0.264	0.051	0.501
0.742 C	in 6(c)	0	0	0.3816

Interatomic distances (Å)

	Boron-rich	Carbon-rich(4)	Number
B(1)-B(1)	1.781	1.7618	6
B(1)-B(2)	1.802	1.7860	6
B(1)-B(2)	1.805	1.8008	12
B(2)-B(2)	1.822	1.8062	6
Mean	1.803	1.7911	
O*-B(1)	1.695	1.6813	6
O -B(2)	1.734	1.7252	6
Mean	1.715	1.7032	
Mean angle			
B-O-B	63.43°	63.436°	

* O is the centre of an icosahedron.

The structure (Fig. 1) contains $[B_{12}]$ icosahedra composed of B(1) and B(2) joined by [CBC] chains composed of B(3) and C as in earlier descriptions (1, 4), but about one quarter of the [CBC] chains are replaced by $[B_4]$ groups of a type not previously observed. Each atom in these planar $[B_4]$ groups (composed of B(4) and B(5)) is linked to two other B atoms in the $[B_4]$ group and to one B atom in each of three $[B_{12}]$ icosahedra. The C atoms of the [CBC] chains are similarly linked to three icosahedra. The distances are summarized above and compared with those for a carbon-rich "B_4C" (4). The boron-rich icosahedron is slightly larger (2%) in the boron-rich case.

The composition from the structure analysis is 9.72 ± 6 at.% C, significantly above that from ion-microprobe analysis of the actual crystal used, 8 ± 1 at.% C.

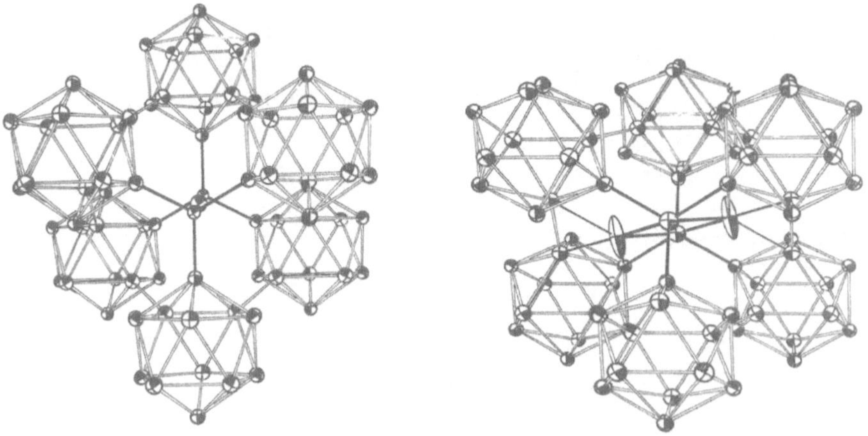

Fig. 1. A perspective view of the two types of structural units in boron-rich "B_4C".

1. Structure Reports, 9, 154.
2. J.L. HOARD and R.E. HUGHES, 1967. "Chemistry of Boron and its Compounds", Ed. E.L. MUETTERTIES, pp. 25-154. John Wiley, New York.
3. K.L. WALTERS and J.L. GREEN, 1970. Quart. Stat. Report on the Advanced Plutonium Fuels Program, Oct. 1, 1970 - Dec. 31, 1970, pp. 14-16. Rept. LA-4595-MS, Los Alamos Scient. Lab., Los Alamos, N.M., U.S.A.
4. A.C. LARSON and D.T. CROMER, 1972. Acta Cryst., A28, 553.

BORON IRON PHOSPHORUS

B_2Fe_5P

L. HÄGGSTRÖM, R. WÄPPLING, T. ERICSSON, Y. ANDERSSON and S. RUNDQVIST, 1975. J. Solid State Chem., 13, 84-91.

Tetragonal, Cr_5B_3 type (1), I4/mcm, a = 5.482, c = 10.332 Å, c/a = 1.88, Z = 4. Cr radiation, R = 0.07 for 58 observed (h0ℓ) reflexions. X-ray powder diffractometer and single-crystal data; see also 2.

Atomic positions

			x	y	z
Fe(1)	in	16(ℓ)	0.1695	0.6695	0.1400
Fe(2)	in	4(c)	0	0	0
B	in	8(h)	0.384	0.884	0
P/B*	in	4(a)	0	0	1/4

* 3.64 P + 0.36 B.

Interatomic distances (Å)

Fe(1)	- 3 B	2.15, 2.20		B	- 1 B	1.80
	- 2 P/B	2.33			- 8 Fe	2.15, 2.20
	- 11 Fe	2.50 - 2.94				
				B/P	- 10 Fe	2.33, 2.58
Fe(2)	- 4 B	2.20				
	- 2 P/B	2.58				
	- 8 Fe	2.50				

1. Structure Reports, 17, 67.
2. Ibid., 27, 95.

BORON TITANIUM

$(B_{12})_4B_2Ti_{1.3-2.0}$

E. AMBERGER and K. POLBORN, 1975. Acta Cryst., B31, 949-953.

Tetragonal, I-tetragonal boron type (1), $P4_2/nnm$, a = 8.830, c = 5.072 Å, c/a = 0.57, D_m = 2.645, Z = 1. Cu radiation, R = 0.07 for 146 reflexions, photographic data.

Atomic positions [transformed to origin at centre]

			x	y	z
Ti*	in	2(a)	3/4	1/4	3/4
B(1)	in	16(n)	0.0776	0.3358	0.1631
B(2)	in	16(n)	0.9847	0.3334	0.8481
B(3)	in	8(m)	0.8753	0.3753	0.135
B(4)	in	8(m)	0.9933	0.4933	0.3366
B(5)	in	2(b)	3/4	1/4	1/4

* Occupied by 1.87 Ti

Ti-B distances range from 2.26-2.54 Å (C.N. 14), B-B distances 1.66-1.98 Å.

The structure (Fig. 1) can be described in terms of four icosahedra (containing 48 of the boron atoms), which are arranged in a flat tetrahedron. At its centre (position 2(a)) lies a titanium atom, with an environment of a 14-corner polyhedron. The single boron atoms occupy position 2(b) and have a distorted tetrahedral environment with respect to the boron atoms of the icosahedra. Each titanium atom is equally bonded to two single boron atoms (and vice versa) thus forming linear chains TiB... parallel to c. The intra-icosahedral B-B distances range from 1.772 to 1.978 Å, with a mean of 1.816 Å; inter-icosahedral values range from 1.661 to 1.777 Å.

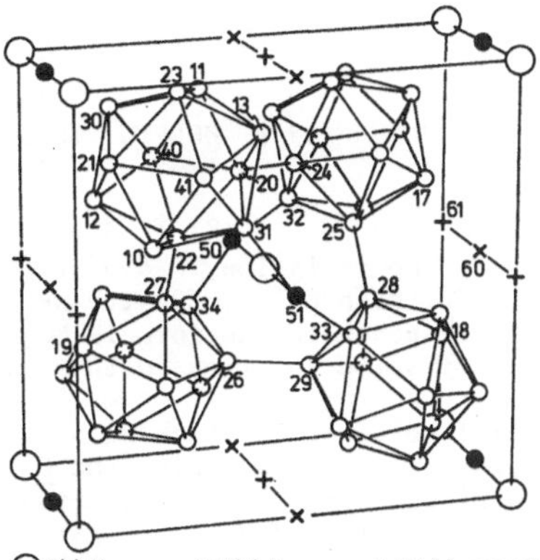

Fig. 1. A perspective view of the tetragonal $(B_{12})_4B_2Ti_{1.87}$ structure. The numbers identify the atoms, the first digit being the order in the atom list, the second digit being the consecutive numbering derived from the first 10, 11....

1. Structure Reports, 22, 211; 38A, 39; 40A, 38.

CAESIUM GALLIUM SULPHUR

$CsGaS_2$

D. SCHMITZ and W. BRONGER, 1975. Z. Naturforsch., 30B, 491-493.

Monoclinic, $RbFeS_2$ type (1), C2/c, a = 7.425, b = 12.21, c = 5.907 Å, β = 113.1°, Z = 4. R = 0.05 for 134 reflexions (diffractometer data).

Atomic positions

	x	y	z
Cs in 4(e)	0	-0.355*	1/4
Ga in 4(e)	0	0.003	1/4
S in 8(f)	0.179	-0.100	0.092

* [Change to -0.355 from 0.355 required for satisfactory distances.]

Interatomic distances (Å)

Cs - 8 S 3.64 - 3.72	Ga - 2 Ga 2.95
	Ga - 4 S 2.27, 2.28

1. Structure Reports, 33A, 3.

CARBON MANGANESE SILICON

CMn_8Si_2

P. SPINAT, C. BROUTY, A. WHULER and P. HERPIN, 1975. Acta Cryst., B31, 541-547.

Triclinic, P1, a = 6.4492, b = 6.5187, c = 9.944 Å, α = 84.709, β = 99.588, γ = 119.979°, D_m = 6.71, A = 32. Mo radiation, R = 0.12 for 2027 reflexions, diffractometer data, corrected for absorption; 32 site-sets; see 1, 2 for earlier work; the structure corresponds to the formula $Mn_{22.6}Si_{5.4}C_4$. This structure allows the formation of solid solutions with Mo and Fe.

The structure consists of layers with a stacking sequence ABA'CD... where the A, A', and D layers are composed of Mn atoms only, the C layer of Si, while the B layer contains Mn, C, and a mixed Mn/Si site. The arrangement is similar to that of Mn_5SiC and belongs to the family of Frank-Kasper structures (Fig. 1). The C atoms occupy trigonal prisms of Mn atoms (Mn-C 1.83-2.26 Å), with 3 extra atoms opposite the faces making a tetrakaidecahedral coordination; Si atoms are 10 or 12 coordinated while Mn atoms are 11, 12, or 14 coordinated. The unique feature of this phase is the environment of the mixed Mn/Si site which has 8 Mn neighbours at the corners of a square prism. Mo atoms substitute only for the Mn atoms with 14 neighbours, but Fe atoms occupy first the 8, then the 11, and finally the 12 C.N. sites. The resulting contractions effectively convert the 14 C.N. sites to 11 or 12 C.N.

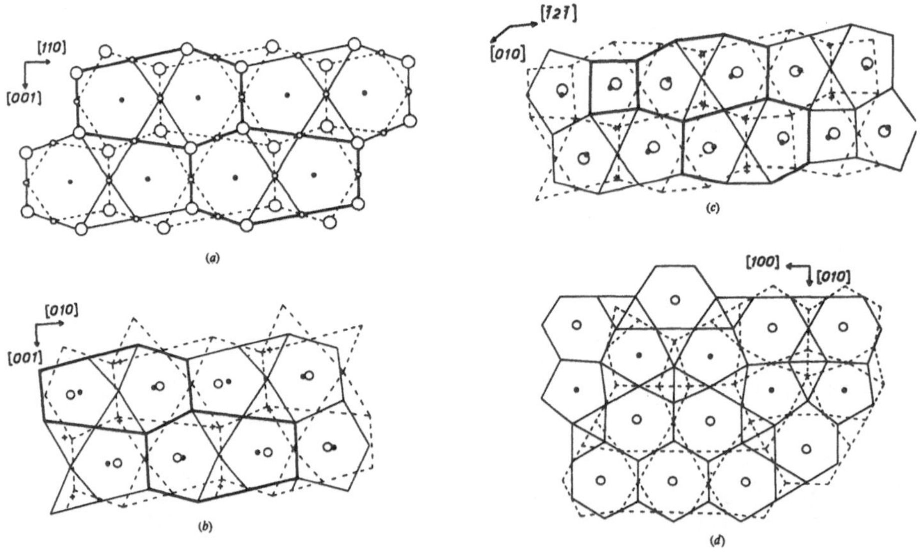

Fig. 1. The triclinic 'Mn_8Si_2C' structure (c) compared with (a) $MnZn_2$, (b) Mn_5SiC, and (d) Mn_5Si. Solid lines represent the main structural layers. Isolated circles or crosses represent intermediate layers.

1. J.P. SÉNATEUR, P. SPINAT and R. FRUCHART, 1965. Colloque Internat. du CNRS, Orsay, 127-134.
2. Structure Reports, 38A, 161.

CARBON TITANIUM

NITROGEN TITANIUM

NITROGEN ZIRCONIUM

A.N. CHRISTENSEN, 1975. Acta Chem. Scand., A29, 563-568.

Cubic, NaCl type, Fm3m, Z = 4. Neutron single crystal data, measured data reduced to 16 independant hkℓ for each phase.

AB	a(Å)	D_x	R	hkℓ measured	B occupancy
$CTi_{0.98}$	4.328	4.87	0.038	96	0.98(3)
$NTi_{0.76}$	4.235	5.02	0.037	445	0.76(4)
NZr	4.585	7.29	0.039	337	1.00(1)

[Interatomic distances (Å)]

AB	A - 6 B B - 6 A	A - 12 A B - 12 B
CTi	2.16	3.06
NTi	2.12	2.99
NZr	2.29	3.24

CERIUM LITHIUM PHOSPHORUS

$CeLi_2P_2$

A. EL MASLOUT, J.-P. MOTTE, A. COURTOIS and C. GLEITZER, 1975.
C.R. Acad. Sci., Paris, C, 280, 21-23.

Hexagonal, anti-Ce_2O_2S (La_2O_3) type (1), P$\bar{3}$m1, a = 4.189, c = 6.834 Å, c/a = 1.631, D_m = 3.41, Z = 1. Diffractometer data, R = 0.15; Li_2P_2Pr is isotypic.

Atomic positions

	x	y	z
Ce in 1(a)	0	0	0
Li in 2(d)	1/3	2/3	0.682
P in 2(d)	1/3	2/3	0.273

Interatomic distances (Å)

Ce - 6 P 3.05	Li - 4 P 2.44, 2.80	P - 4 Li 2.44, 2.80

1. Strukturbericht, 1, 744; Structure Reports, 12, 174.

CERIUM LUTETIUM SULPHUR

$Ce_4Lu_{11}S_{22}$

N. RODIER and V. TIEN, 1975. Bull. Soc. Fr. Minér. Crist., 98, 30-35.

Monoclinic, B2/m, a = 38.60, b = 11.21, c = 3.910 Å, γ = 91.3°, Z = 2. Mo radiation, R = 0.032 for 1520 reflexions, diffractometer data, corrected for absorption; 19 site-sets.

Interatomic distances (Å)

		Average			Average
Ce(1) - 8 S	2.90 - 3.03	2.97	Lu(3) - 6 S	2.60 - 2.74	2.67
Ce(2) - 8 S	2.89 - 3.37	2.99	Lu(4) - 6 S	2.61 - 2.73	2.69
Lu(1) - 7 S	2.66 - 2.84	2.77	Lu(5) - 6 S	2.60 - 2.74	2.68
Lu(2) - 6 S	2.60 - 2.74	2.68	Lu(6) - 6 S	2.69, 2.72	2.70

Both Ce atoms are coordinated to 8 S atoms; one Lu has 7-fold coordination (not previously observed for Lu), and the remaining Lu atoms are octahedrally coordinated. $Ce_4Lu_{11}S_{22}$ is closely related to $CeYb_3S_6$ (1), with Yb(1) = Lu(1) or (2), Yb(2) = Lu(4) or (6), and Yb(3) = Lu(3) or (5).

1. Structure Reports, 40A, 50.

CERIUM NICKEL SILICON

$Ce_7Ni_2Si_5$

M.G. MYSKIV, 1974. Vest. L'vov, Univ. Ser. Chem., 15, 17-21.

Orthorhombic, new type, Pnma, a = 23.31, b = 4.299, c = 13.90 Å, D_m = 6.01, Z = 4. Cu and Cr radiations, R = 0.136 for 190 reflexions. Diffractometer data, 14 site-sets, all 4(c): x, 1/4, z.

Interatomic distances (Å)

Ce(1)	- 10 Ce	3.60-4.30		Ni(1)	- 3 Si	2.24, 2.52
	- 2 Ni	3.02				
	- 5 Si	3.00-3.37		Ni(2)	- 3 Si	2.27 - 2.72
Ce(2)	- 9 Ce	3.60-4.30		Si(1)	- 7 Ce	2.97 - 3.28
	- 2 Ni	3.10			- 2 Ni	2.52
	- 6 Si	3.12-3.30				
				Si(2)	- 7 Ce	3.00 - 3.37
Ce(3)	- 9 Ce	3.67-4.30			- 1 Ni	2.24
	- 2 Ni	3.13			- 1 Si	2.75
	- 6 Si	2.97-3.29				
				Si(3)	- 7 Ce	3.03 - 3.28
Ce(4)	- 10 Ce	3.70-4.30			- 1 Ni	2.27
	- 2 Ni	2.99			- 1 Si	2.75
	- 5 Si	2.94-3.36				
				Si(4)	- 8 Ce	3.16 - 3.27
Ce(5)	- 11 Ce	3.77-4.46			- 1 Ni	2.42
	- 1 Ni	3.07				
	- 5 Si	3.09-3.21		Si(5)	- 8 Ce	3.07 - 3.37
					- 2 Ni	2.71
Ce(6)	- 11 Ce	3.78-4.46				
	- 1 Ni	2.97				
	- 5 Si	3.07-3.37				
Ce(7)	- 10 Ce	3.65-4.30				
	- 2 Ni	3.00				
	- 5 Si	3.18-3.30				

This structure is based on trigonal prismatic coordination for the Ni and Si atoms, and is very similar to that of Mo_4P_3 (1) and $Ce_6Ni_2Si_3$ (2).

1. Structure Reports, 40A, 49
2. Ibid., 30A, 67.

CHROMIUM GALLIUM

GALLIUM IRON

M_3Ga_4 (M = Cr, Fe)

M.J. PHILIPPE, B. MALAMAN, B. ROQUES, A. COURTOIS and J. PROTAS, 1975. Acta Cryst., B31, 477-482.

Monoclinic, C2/m, Z = 6. Mo radiation, diffractometer data. Cr_3Ga_4 is isotypic with Fe_3Ga_4; see also 1-4.

M_3Ga_4	a(Å)	b(Å)	c(Å)	β(°)	D_m	R	hkℓ
Cr_3Ga_4	10.135	7.845	7.986	*105.58	7.01	0.078	699
Fe_3Ga_4	10.091	7.666	7.866	106.67	7.58	0.095	661

* [Angle given 105°60'; private communication from authors]

Atomic positions

Cr_3Ga_4	x	y	z
Cr(1) in 2(a)	0	0	0
Cr(2) in 4(i)	0.5018	0	0.3246
Cr(3) in 4(i)	0.2295	0	0.6531
Cr(4) in 8(j)	0.1408	0.1924	-0.1493
Ga(1) in 4(i)	0.2803	0	0.0895
Ga(2) in 4(i)	0.0475	0	0.3529
Ga(3) in 8(j)	0.6042	0.2097	0.1526
Ga(4) in 8(j)	0.3487	0.1917	0.4447
Fe_3Ga_4			
Fe(1) in 2(a)	0	0	0
Fe(2) in 4(i)	0.4954	0	0.3106
Fe(3) in 4(i)	0.2237	0	0.6374
Fe(4) in 8(j)	0.1358	0.2016	-0.1521
Ga(1) in 4(i)	0.2689	0	0.0844
Ga(2) in 4(i)	0.0419	0	0.3482
Ga(3) in 8(j)	0.6082	0.2034	0.1509
Ga(4) in 8(j)	0.3520	0.1857	0.4501

Interatomic distances (Å)

	Cr_3Ga_4	Fe_3Ga_4
M(1) - 4 M	2.57	2.58
- 8 Ga	2.67-2.74	2.60-2.65
M(2) - 4 M	2.68-2.97	2.75-2.95
- 6 Ga	2.52-2.55	2.46-2.51
M(3) - 3 M	2.51-2.68	2.60-2.75
- 9 Ga	2.57-3.39	2.48-3.41
M(4) - 5 M	2.51-3.02	2.58-3.09
- 7 Ga	2.55-2.65	2.48-2.58
Ga(1) - 7 M	2.52-3.39	2.46-3.41
- 6 Ga	3.01-3.12	2.93-3.10
Ga(2) - 5 M	2.62-2.80	2.48-2.72
- 9 Ga	2.76-3.35	2.75-3.32
Ga(3) - 6 M	2.53-2.67	2.48-2.65
- 7 Ga	2.77-3.29	2.73-3.12
Ga(4) - 5 M	2.52-2.75	2.50-2.64
- 9 Ga	2.55-3.35	2.60-3.32

The structure is a layer structure (Fig. 1), with layers consisting of linked pentagons and hexagons.

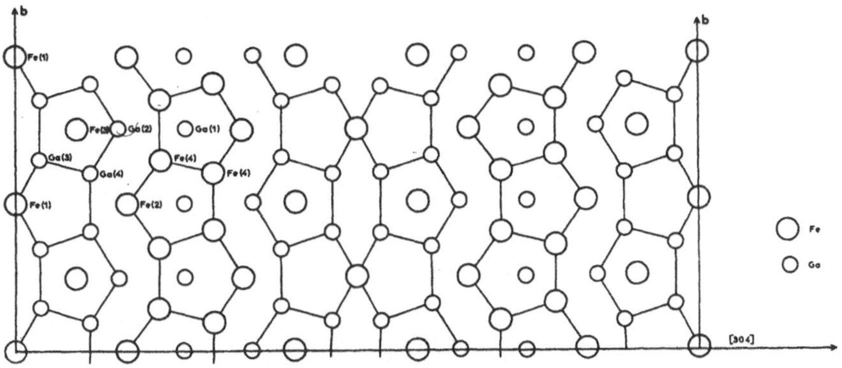

Fig. 1. The monoclinic Fe_3Ga_4 structure projected as indicated (equivalent to a projection onto (100) for a cell transformed $\bar{1}0\bar{1}/0\bar{1}0/304$).

1. Structure Reports, 30A, 125, 134.
2. E. WACHTEL and J. MAYER, 1967. Z. Metallk., 58, 761.
3. J.D. BORNAND and P. FESCHOTTE, 1972. J. Less-Common Metals, 29, 84.
4. Structure Reports, 40A, 107.

CHROMIUM NIOBIUM SILICON

CrNbSi

J. STEINMETZ, B. MALAMAN, J.M. ALBRECHT and B. ROQUES, 1975. Mater. Res. Bull., 10, 571-576.

Hexagonal, Fe_2P type (1), $P\bar{6}2m$, a = 6.598, c = 3.359 Å, c/a = 0.51, D_m = 6.60, Z = 3. Cu radiation, R = 0.10, diffractometer data; see also 2.

Atomic positions

			x	y	z
Cr	in	3(f)	0.247	0	0
Nb	in	3(g)	0.589	0	1/2
Si(1)	in	2(c)	1/3	2/3	0
Si(2)	in	1(b)	0	0	1/2

Interatomic distances (Å)

Cr	- 2 Cr	2.82	Si(1)	- 3 Cr	2.53
	- 6 Nb	2.81, 2.90		- 6 Nb	2.60
	- 4 Si	2.34, 2.53			
			Si(2)	- 6 Cr	2.34
Nb	- 4 Nb	3.45		- 3 Nb	2.71
	- 6 Cr	2.81, 2.90			
	- 5 Si	2.60, 2.71			

1. Structure Reports, 23, 68.
2. Ibid., 26, 113.

CHROMIUM PHOSPHORUS

CrP

K. SELTE, A. KJEKSHUS and A.F. ANDRESEN, 1972. Acta Chem. Scand., 26, 4188-4190.

Orthorhombic, MnP type structure (1), Pnma, D_m = 5.46, Z = 4. R ranges from 0.016 to 0.035, powder neutron diffraction data taken at 17°, 81°, and 293°K; in agreement with 2. All atoms in 4(c): x, 1/4, z; magnetic data also given.

		17°K	81°K	293°K
a(Å)		5.347	5.348	5.356
b(Å)		3.108	3.109	3.117
c(Å)		5.996	5.997	6.007
Cr	x	0.0089	0.0077	0.0074
	z	0.1952	0.1942	0.1954
P	x	0.1848	0.1856	0.1853
	z	0.5641	0.5647	0.5644

Interatomic distances (Å)

	17°K	81°K	293°K
Cr - 4 Cr	2.75, 2.81	2.76, 2.80	2.76, 2.82
- 6 P	2.33 - 2.40	2.32 - 2.42	2.33 - 2.41
P - 6 Cr	2.33 - 2.40	2.32 - 2.42	2.33 - 2.41
- 2 P	2.63	2.64	2.64

1. Strukturbericht, 3, 17, 263; Structure Reports, 27, 319
2. Structure Reports, 30A, 98.

CHROMIUM SULPHUR URANIUM

CrS_3U

I. H. NOËL, J. PADIOU and J. PRIGENT, 1975. C.R. Acad. Sci., Paris, C, 280, 123-126.

Orthorhombic, $ScYS_3$ type (1), Pnam, a = 7.163, b = 6.095, c = 8.851 Å, D_m = 6.50, Z = 4. Mo radiation, R = 0.047 for 965 reflexions, diffractometer data, corrected for absorption; see also 2.

Atomic positions

			x	y	z
U	in	4(c)	0.38288	0.0518	1/4
Cr	in	4(a)	0	0	0
S(1)	in	4(c)	0.0421	0.8594	1/4
S(2)	in	8(d)	0.3285	0.8320	0.5590

Interatomic distances (Å)

U - 6 S 2.71 - 2.84   Cr - 6 S 2.39 - 2.62
  - 2 S 3.07

The structure exhibits octahedral coordination for Cr and trigonal prismatic coordination (plus 2 extra) for the U atom. The [CrS_6] octahedra share corners, with the Cr-S-Cr angle being 135.5°.

$CrS_{17}U_8$

II. H. NOEL, M. POTEL and J. PADIOU, 1975. Acta Cryst., B31, 2634-2637.

Monoclinic, C2/m, a = 13.290, b = 8.423, c = 10.427 Å, β = 101°35', D_m = 7.20, Z = 2. Mo radiation, R = 0.054 for 1660 independent reflexions, diffractometer data corrected for absorption with allowance for anomalous dispersion; see also 2.

Atomic positions

			x	y	z
U(1)	in	8(j)	0.05888	0.24551	0.70435
U(2)	in	4(i)	0.20474	0	0.4552
U(3)	in	4(i)	0.68206	0	0.02194
Cr	in	2(a)	0	0	0
S(1)	in	8(j)	0.1273	0.3051	0.4670
S(2)	in	8(j)	0.1327	0.1889	0.9735
S(3)	in	4(i)	0.2106	0	0.7219
S(4)	in	4(i)	0.0569	0	0.2259
S(5)	in	4(i)	0.3021	0	0.2433
S(6)	in	4(i)	0.5202	0	0.1667
S(7)	in	2(c)	0	0	1/2

Interatomic distances (Å)

U(1) - 8 S 2.75 - 2.95   Cr - 6 S 2.33, 2.43

U(2) - 8 S 2.76 - 2.85

U(3) - 8 S 2.73 - 2.99

Chromium atoms are octahedrally surrounded by sulphur atoms; uranium atoms occupy three non-equivalent positions forming two kinds of coordination polyhedra, a bicapped trigonal prism and two dodecahedra.

1. Structure Reports, 35A, 101.
2. Ibid., 40A, 106.

CHROMIUM TELLURIUM

Cr_2Te_3

T. HAMASAKI, T. HASHIMOTO, Y. YAMAGUCHI and H. WATANABE, 1975. Solid State Comm., 16, 895-897.

Trigonal, [Cr_2S_3 type (1)], P$\bar{3}$1c, a = 6.814, c = 12.073 Å (at 300°K), a = 6.823, c = 11.800 Å (at 85°K). R = 0.043 (300°K) and 0.040 (85°K); neutron diffraction data; see also 2.

Atomic positions

			x	y	z
Cr(1)	in 2(c)		1/3	2/3	1/4
Cr(2)	in 2(b)		0	0	0
Cr(3)	in 4(f)		1/3	2/3	0
Te	in 12(i)	300°K	0.324	-0.006	0.374
		85°K	0.320	-0.004	0.374

[Interatomic distances (Å)]

		300°K	85°K
Cr(1)	- 2 Cr	3.02	2.95
	- 6 Te	2.71	2.72
Cr(2)	- 6 Te	2.70	2.65
Cr(3)	- 1 Cr	3.02	2.95
	- 6 Te	2.73, 2.78	2.73, 2.77
Te	- 4 Cr	2.70 - 2.78	2.65 - 2.77
	- 8 Te	3.70 - 3.86	3.63 - 3.84

1. Structure Reports, 21, 95.
2. Ibid., 28; 45; 29, 111; 35A, 47.

COBALT GERMANIUM MANGANESE

$CoGeMn_{0.98}$

W. JEITSCHKO, 1975. Acta Cryst., B31, 1187-1190.

Orthorhombic, low-temperature form (296°K), NiSiTi type, Pnma, a = 5.947, b = 3.826, c = 7.051 Å. Hexagonal, high-temperature form (343°K), Ni_2In-type, $P6_3/mmc$, a = 4.087, c = 5.316 Å (a = 5.316, b = 4.087, c = 7.079 Å for corresponding orthorhombic cell). Mo radiation, diffractometer data with allowance for anomalous dispersion. Atomic positions, occupancies, and R values (unique reflexions in parentheses) are given in the following table; coordinates are expressed in terms of Pnma, with all atoms in 4(c).

Temperature		296°K			343°K	
Structure type		TiNiSi			Ni_2In	
Scattering factor	Mn	Co	Ge	Mn	Co	Ge
Occupancy	0.997	0.974	1*	0.997*	0.974*	1*
x	0.0234	0.1581	0.7660	0	1/4	3/4
y	1/4	1/4	1/4	1/4	1/4	1/4
z	0.1899	0.5593	0.6198	1/4	7/12	7/12
B($Å^2$)	0.64	0.53	0.66	0.58	1.12	0.69
R		0.060			0.076	
Reflexions		636			118	

* Held constant

Interatomic distances (Å)

A comparison of interatomic distances in the high- and low-temperature forms is given below. Major differences in C.N. and distances are underlined.

	TiNiSi type	Ni_2In type	Δ
Mn - 1 Ge	2.617	2.708*	-0.091
- 2 Ge	2.620	2.708	-0.088
- 2 Ge	2.650	2.708	-0.058
- 1 Ge	3.396	2.708*	+0.688
- 1 Co	2.725	2.708*	+0.017
- 1 Co	2.794	2.708*	+0.086
- 2 Co	2.820	2.708*	+0.112
- 2 Co	2.845	2.708*	+0.137
- 2 Mn	3.092	2.658	+0.434
C.N.	11	14	+3
Co - 2 Ge	2.336	2.360*	-0.024
- 1 Ge	2.352	2.360	-0.008
- 1 Ge	2.370	2.658	-0.288
- 1 Ge	3.640	2.658	+0.982
- 1 Mn	2.725	2.708*	+0.017
- 1 Mn	2.794	2.708*	+0.086
- 2 Mn	2.820	2.708*	+0.112
- 2 Mn	2.845	2.708*	+0.137
- 2 Co	2.810	3.539	-0.729
C.N.	12	11	-1
Ge - 1 Mn	2.617	2.708*	-0.091
- 2 Mn	2.620	2.708*	-0.088
- 2 Mn	2.651	2.708*	-0.057
- 1 Mn	3.396	2.708*	+0.688
- 2 Co	2.336	2.360*	-0.024
- 1 Co	2.352	2.360*	-0.008
- 1 Co	2.370	2.658	-0.288
- 1 Co	3.640	2.658	+0.982
[- 2 Ge	3.495]		
C.N.	9	11	+2

* [There are minor differences between the values listed and those calculated from the parameters given]

The structures (Fig. 1) differ in the C.N. of all atoms due to large changes in positional parameters. The thermal motion of Co is considerably larger than that of the other two atoms in the high-temperature form, where the largest thermal motions are in directions along which atom movement occurs in producing the low-temperature form. This is to be expected in the 'soft-mode' model of diffusionless phase transitions.

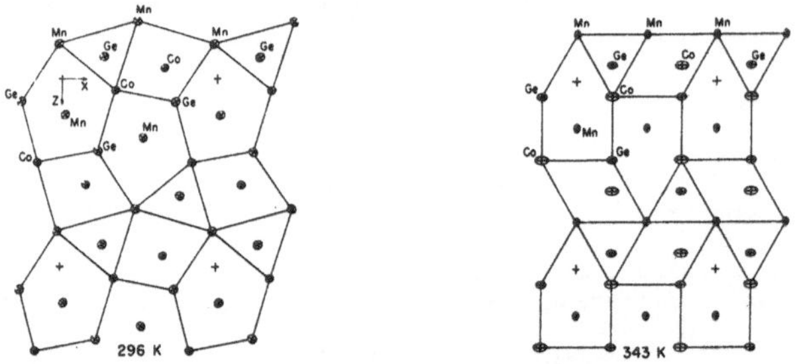

Fig. 1. The low-temperature (TiNiSi type) and high-temperature (Ni_2In type) forms of $CoGeMn_{0.98}$. Atoms at z = 3/4 are connected by lines; the remaining atoms are at z = 1/4.

COBALT INDIUM

$CoIn_2$

H.H. STADELMAIER and H.K. MANAKTALA, 1975. Acta Cryst., B31, 374-378.

Orthorhombic, $CuMg_2$ type (1), Fddd, a = 9.402, b = 17.846, c = 5.282 Å, D_m = 8.68, Z = 16. Mo radiation, R = 0.068 for 399 reflexions, diffractometer data, corrected for absorption; compare the monoclinic description in 2.

Atomic positions

	x	y	z
In(1) in 16(e)	0.9649	1/8	1/8
In(2) in 16(f)	1/8	0.7131	1/8
Co in 16(f)	1/8	0.9971	1/8

Interatomic distances (Å)

In(1)	- 4 Co	2.68, 2.73	Co	- 2 Co	2.70
	- 11 In	3.01 - 3.56		- 8 In	2.68 - 2.76
In(2)	- 4 Co	2.74, 2.76			
	- 11 In	3.00 - 3.56			

The structure is analysed geometrically in terms of the packing of rigid spheres to demonstrate the tetrahedral close-packing characteristic of many metallic structures. In addition the square-antiprismatic coordination of Co and the μ-phase C.N. 15 polyhedra around In are described.

1. Structure Reports, 8, 64; 15, 73.
2. J.D. SCHÖBEL and H.H. STADELMAIER, 1970. Z. Metallk., 61, 342.

COBALT SULPHUR

Co_9S_8

V. RAJAMANI and C.T. PREWITT, 1975. Canad. Miner., 13, 75-78.

Cubic, Fm3m, a = 9.923 Å, Z = 4. R = 0.029 for 97 independent reflexions, diffractometer data, corrected for absorption and extinction; the composition was $Co_{8.85}S_8$, but refinement with vacancies in 32(f) was no better; see 1, 2 for earlier work.

Atomic positions

	x	y	z
Co(1) in 4(b)	1/2	1/2	1/2
Co(2) in 32(f)	0.12623	0.12623	0.12623
S(1) in 8(c)	1/4	1/4	1/4
S(2) in 24(e)	0.2623	0	0

Interatomic distances (Å)

Co(2)	- 4 S	2.127, 2.227	S(2) - 4 S	3.336
	- 3 Co(2)	2.505		
	- 3 Co	3.474	Co(1) - 6 S	2.359
	- 3 Co	3.543		

The calculated valency for Co is 1.78, and the observed Co-6S (2.359 Å) and Co-4S (average = 2.202 Å) distances are somewhat larger than would be expected for Co(II) (2.34 and 2.18 Å, respectively), consistent with this lower valency. There are three metal-metal interactions, (i) Co(2)-Co(2) = 2.505 Å, (ii) Co(2)-S(1)-Co(2), with angle 109.5°, and (iii) Co(2)-S(2)-Co(1), with angle 127°. A broad, partly-filled energy band composed of completely delocalized d-electrons would account for the observed metallic conductivity and Pauli paramagnetism.

1. Structure Reports, 27, 169.
2. Strukturbericht, 4, 137.

COBALT SULPHUR TITANIUM

$Co_{0.25}S_2Ti$

M. DANOT and R. BREC, 1975. Acta Cryst., B31, 1647-1652.

The same data and results were reported in Structure Reports, 40A, 54.

COBALT YTTRIUM

Co_2Y_3

J.M. MOREAU, E. PARTHÉ and D. PACCARD, 1975. Acta Cryst., B31, 747-749.

Orthorhombic, Pnnm, a = 12.248, b = 9.389, c = 3.975 Å, Z = 4, D_x = 5.58. Mo radiation, R = 0.10 for 190 observed reflexions, single-crystal data; see also 1, 2.

Atomic positions

All atoms in 4(g)

	x	y	z
Y(1)	0.128	0.193	0
Y(2)	0.387	0.373	0
Y(3)	0.137	0.574	0
Co(1)	0.269	0.860	0
Co(2)	0.462	0.883	0

Interatomic distances (Å)

Y(1)	- 5 Co	2.83 - 3.57
	- 8 Y	3.57 - 3.67
Y(2)	- 3 Co	2.76, 3.03
	- 9 Y	3.45 - 3.65
Y(3)	- 7 Co	2.94 - 3.14
	- 7 Y	3.45 - 3.67
Co(1)	- 1 Co	2.37
	- 8 Y	2.76 - 3.57
Co(2)	- 2 Co	2.37, 2.39
	- 7 Y	2.89 - 3.03

The structure (Fig. 1) is composed of ribbons of 4 trigonal prisms and is a shift variant of the Dy_3Ni_2 (3) structure.

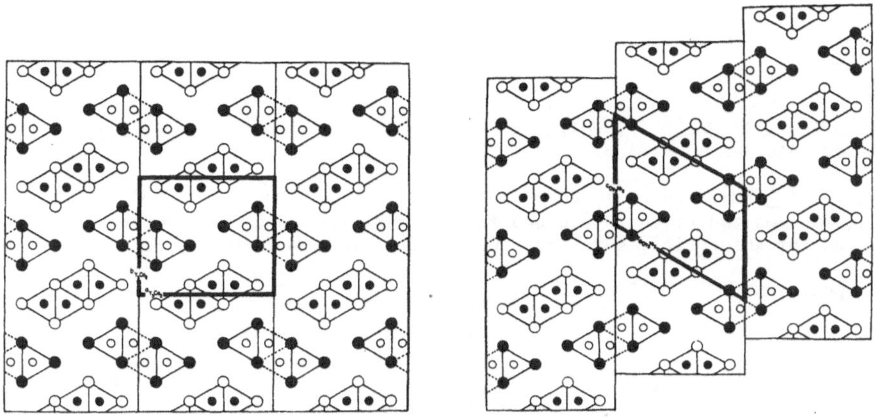

Y_3Co_2 Dy_3Ni_2

Fig. 1. Projections of the orthorhombic Co_2Y_3 and monoclinic Dy_3Ni_2 structures showing their relationship. Large circles represent Dy or Y, small circles Co or Ni; open circles are at 1/2, full circles at 0 height.

1. K.H.J. BUSCHOW, 1971. Philips Res. Rep., 26, 49.
2. A.E. RAY, 1974. Cobalt, 1, 13.
3. Structure Reports, 40A, 62.

COPPER GALLIUM GOLD

AuCu-III ($Au_{0.5}Cu_{0.45}Ga_{0.05}$)

M. DIRAND, A. COURTOIS, J. HERTZ and J. PROTAS, 1975. C.R. Acad. Sci., Paris, C, 280, 559-561.

Orthorhombic, Au_2CuZn type (1), Pbam, a = 8.92, b = 4.56, c = 2.83 Å, Z = 4. Co radiation, R(I) = 0.12 for 31 reflexions, diffractometer data; see also 1. This phase was obtained from an alloy of composition $Au_{0.5}Cu_{0.45}Ga_{0.05}$.

Atomic positions

			x	y	z
Au	in	4(g)	0.141	0.083	0
Cu/Ga	in	4(h)	0.101	0.662	1/2

[Interatomic distances (Å)]

Au	- 5 Au	2.63 - 3.00	Cu/Ga	- 3 Cu/Ga	2.33, 2.83
	- 8 Cu/Ga	2.41 - 3.02		- 8 Au	2.41 - 3.02

1. Structure Reports, 22, 107.

COPPER HAFNIUM

$Cu_{51}Hf_{14}$

J.-P. GABATHULER, P. WHITE and E. PARTHÉ, 1975. Acta Cryst., B31, 608-610.

Hexagonal, $Ag_{51}Gd_{14}$ type (1), P6/m, a = 11.18, c = 8.235 Å, c/a = 0.74, Z = 1, D_x = 10.69. Mo radiation, R = 0.085 for 568 reflexions, diffractometer data, corrected for absorption; some ordering was detected; see also 2, 3.

Atomic positions

			x	y	z
Hf(1)	in	2(e)	0	0	0.3114
Hf(2)	in	6(j)	0.1138	0.3893	0
Hf(3)	in	6(k)	0.4712	0.1417	1/2
Cu(1)	in	2(c)	1/3	2/3	0
Cu(2)	in	4(h)	1/3	2/3	0.2931
Cu(3)	in	6(k)	0.0611	0.2403	1/2
Cu(4)	in	12(ℓ)	0.1914	0.2651	0.2365
Cu(5)	in	12(ℓ)	0.4942	0.1163	0.1520
Cu(6)	in	12(ℓ)	0.1044	0.4373	0.3296
Cu(7)*	in	6(j)	0.1146	0.1352	0

* occupancy = 47%

[Interatomic distances (Å)]

Hf(1)	- 1 Hf	3.11	
	- 18 Cu	2.72 - 2.93	
Hf(2)	- 15 Cu	2.66 - 2.91	
Hf(3)	- 14 Cu	2.74 - 2.97	
Cu(1)	- 3 Hf	2.83	
	- 8 Cu	2.41, 2.55	
Cu(2)	- 3 Hf	2.75	
	- 7 Cu	2.41 - 2.58	
Cu(3)	- 4 Hf	2.81 - 2.90	
	- 8 Cu	2.42 - 2.55	
Cu(4)	- 4 Hf	2.72 - 2.97	
	- 10 Cu	2.32 - 3.15	
Cu(5)	- 4 Hf	2.81 - 2.90	
	- 8 Cu	2.46 - 3.15	
Cu(6)	- 4 Hf	2.74 - 2.80	
	- 9 Cu	2.45 - 3.27	
Cu(7)	- 4 Hf	2.66 - 2.93	
	- 11 Cu	1.41 - 3.15	

This structure occurs with rare-earth - gold and rare-earth - silver alloys; all have radius ratios between 1.22 and 1.31. With smaller values of this ratio the $MoNi_4$ or $ZrAu_4$ structures occur.

1. Structure Reports, 37A, 84.
2. A.J. PERRY, 1974. Mater. Sci. Eng., 13, 57.
3. A.J. PERRY and W. HUGI, 1972. J. Inst. Met., 100, 378.

COPPER HAFNIUM SILICON

$CuHfSi_2$

L.S. ANDRUKHIV, L.A. LYSENKO, Ja. P. JARMOLJUK and E.I. GLADYŠEVSKIJ, 1975. Dop. Akad. Nauk Ukr., No. 7, 645-648.

Tetragonal, [AsCuSiZr type, (1)], P4/nmm, a = 3.732, c = 8.99 Å, c/a = 2.41, Z = 2. Cr radiation, R = 0.145 for 35 reflexions, photographic data; see also 2.

Atomic positions

			x	y	z
Hf	in	2(c)	1/4	1/4	0.741
Cu	in	2(a)	3/4	1/4	0
Si(1)	in	2(c)	1/4	1/4	0.189
Si(2)	in	2(b)	3/4	1/4	1/2

Interatomic distances (Å)

Hf	- 4 Cu	2.98	Si(1)	- 4 Hf	2.71
	- 8 Si	2.71, 2.86		- 4 Cu	2.52
Cu	- 4 Hf	2.98	Si(2)	- 4 Hf	2.86
	- 4 Cu	2.64		- 4 Si	2.64
	- 4 Si	2.52			

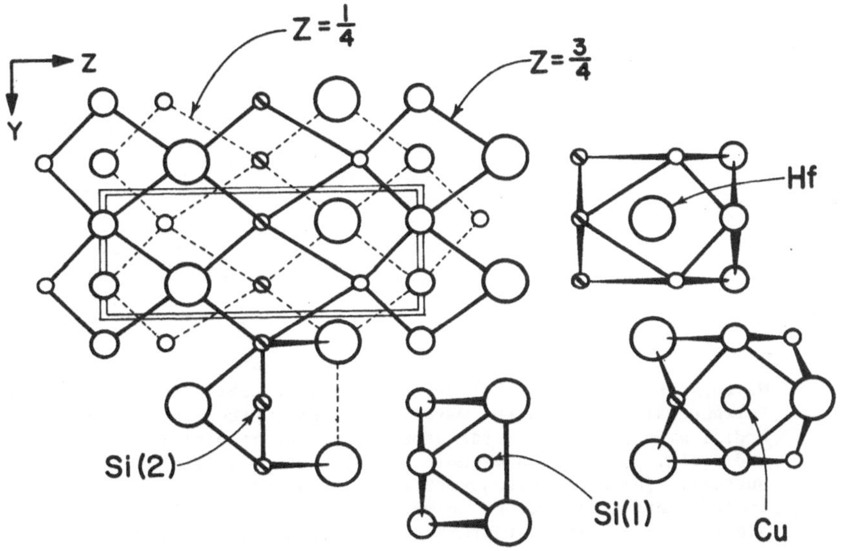

Fig. 1. A projection of the tetragonal $CuHfSi_2$ structure onto (100) with the coordination polyhedra outlined.

The structure (Fig. 1), with the metal atoms 12-coordinated and the Si atoms 8-coordinated, is related to those of Al_2CeGa_2 (3), and PbFCl (4) [This structure was recently reported for ZrCuSiAs (1) where it was related to the NaCl, MgAgAs, $MnCu_2Al$, TiCu, and PbFCl types.]

1. Structure Reports, 40A, 17.
2. Ibid., 40A, 60.
3. Ibid., 29, 97.
4. Strukturbericht, 2, 45; 3, 370; Structure Reports, 29, 38, 342.

COPPER IRON SULPHUR

$Cu_4Fe_5S_8$ (haycockite)

I. J. F. ROWLAND and S.R. HALL, 1975. Acta Cryst., B31, 2105-2112.

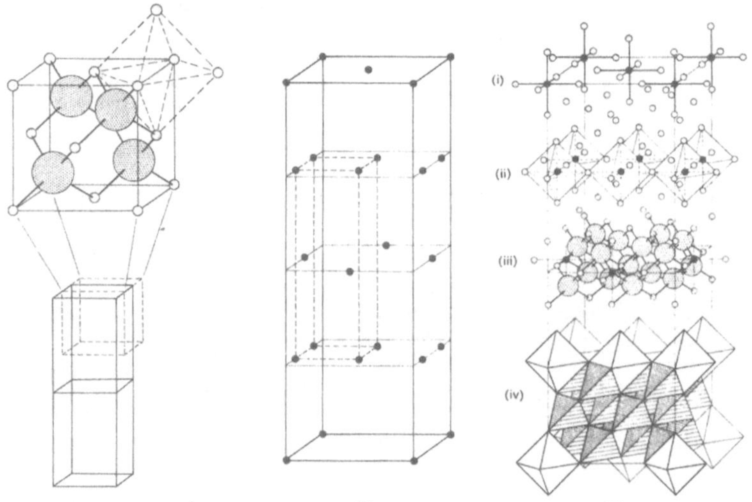

Fig. 1. Haycockite. (a) The sphalerite-like pseudo-cubic subcell containing tetrahedrally coordinated metal and sulphur atoms, with the octahedral arrangement of metal atoms about the interstitial sites indicated. Three pseudo-cubic subcells comprise the pseudo-tetragonal subcell. (b) Eight pseudo-tetragonal subcells, offset one-half cell in the a and c directions to permit correlation of the extra metal atoms, comprise the orthorhombic unit cell. (c) Atomic coordination in the orthorhombic unit cell: (i) octahedral coordination of extra metal atoms at 1(a) and 1(e) at z = 1; (ii) octahedra about four extra metal atoms at 4(u) at z $\sim$ 0.75; (iii) tetrahedral coordination of sulphur to metal atoms including extra at 1(f) and 1(g) at z = 1/2, (iv) metal tetrahedra about sulphur atoms, and metal octahedra about extra metal atoms from z = 0 to z $\sim$ 0.25.

Orthorhombic, P222, a = 10.705, b = 10.734, c = 31.630 Å, Z = 12, D_x = 4.33, R = 0.14 for 4129 reflexions; there is a pseudo-cubic f.c. subcell ($F\bar{4}3m$) with a ∿ 5.3 Å and also a pseudo-tetragonal subcell, P222, a = 5.352, b = 5.367, c = 15.815 Å, R = 0.18 for 2035 reflexions. Mo radiation, diffractometer data corrected for absorption; the orthorhombic cell has 64 site-sets; natural material from Lydenburg, Transvaal, yielded only one fragment suitable for data collection for which a special strategy was used to allow for the subcells and the presence of three twin-components; the Fe atoms were identified from their lower temperature factors. See also 1.

The structure (Fig. 1) is based on the tetrahedral arrangement found in other chalcopyrite-like structures (2) for $Cu_{48}Fe_{48}S_{96}$ plus an octahedral arrangement for the 12 interstitial Fe atoms, thus yielding the cell contents of $Cu_{48}Fe_{60}S_{96}$. The average metal-metal distance for adjacent interstitial atoms is 2.68 Å; for isolated interstitial metal atoms they are 2.67, 2.69, and 2.72 Å (along a, b, and c respectively). Average Cu-S = 2.32, Fe-S = 2.30, and M(interstitial)-S = 2.31 Å.

Cu_5FeS_4 (bornite, low-temperature form)

II. K. KOTO and N. MORIMOTO, 1975. Acta Cryst., B31, 2268-2273.

Orthorhombic, Pbca, a = 10.950, b = 21.862, c = 10.950 Å, Z = 16; this cell is 2 x 4 x 2 times the cubic subcell of the high-temperature form (3, 4). Mo radiation, R = 0.101 for 54 substructure reflexions and 0.148 for 1008 superstructure reflexions; diffractometer data, 20 site-sets given; a redetermination with symmetry differing from 5-7; modified partial Patterson functions were used to solve the superstructure.

Interatomic distances (Å)

		mean			mean
M(1) - 4 S	2.271 - 2.520	2.352	M(7) - 4 S	2.255 - 2.541	2.389
M(2) - 4 S	2.253 - 2.502	2.368	M(8) - 4 S	2.307 - 2.530	2.365
M(3) - 4 S	2.265 - 2.487	2.391	M(9) - 4 S	2.288 - 2.737	2.409
M(4) - 4 S	2.271 - 2.322	2.298	M(10) - 4 S	2.261 - 2.979	2.453
M(5) - 4 S	2.256 - 2.344	2.307	M(11) - 4 S	2.321 - 2.774	2.439
M(6) - 4 S	2.253 - 2.481	2.380	M(12) - 4 S	2.269 - 2.843	2.428

M = (5Cu + Fe)/6

In the structure of bornite (Fig. 2) sulphur atoms are arranged in cubic closest packing. Ordering of metal atoms in the tetrahedral or triangular interstices of sulphur atoms results in two structural units, antifluorite-type and sphalerite-type cubes. Vacancies for metal atoms cluster in the sphalerite-type cube. The two different cubes alternate three-dimensionally, resulting in a characteristic mosaic pattern structure. In the antifluorite-type cube the eight metal atoms are displaced towards one of the triangles or one of the edges of the S tetrahedra, giving rise to 3 short and 1 longer or 2 short and 2 long distances. In the sphalerite-type cubes the four metal atoms are markedly displaced from the centres to one of the triangles of the S tetrahedra, giving rise to almost trigonal coordination (3 short and 1 long). The most probable sites for Fe atoms are M(4) and M(5) with the shortest mean distances and most regular tetrahedral environments (8).

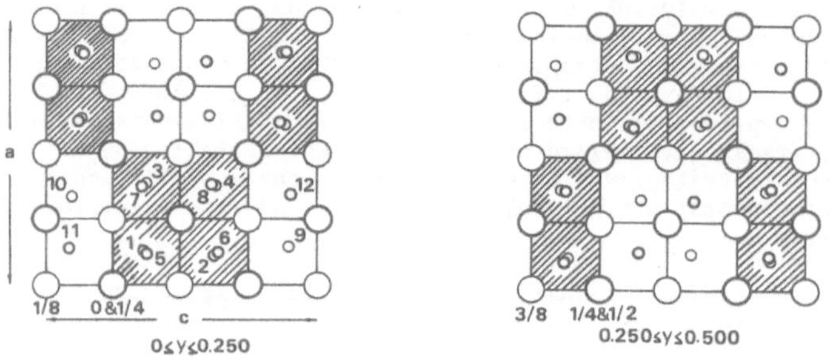

Fig. 2. The orthorhombic structure of bornite, the low form of Cu_5FeS_4; projections of two slabs onto (010) are given with S atoms represented by large circles and marked with their approximate (ideal) y coordinates; the metal atoms are numbered as in the list of interatomic distances.

1. Structure Reports, 38A, 164.
2. Ibid., 39A, 51-53.
3. Ibid., 26, 137.
4. Ibid., 29, 43.
5. G. ALLAIS, 1968. Bull. Soc. Fr. Minér. Crist., 91, 600.
6. Strukturbericht, 4, 149.
7. Structure Reports, 12, 215.
8. Ibid., 13, 280.

COPPER SULPHUR

CuS (covellite)

R. KALBSKOPF, F. PERTLIK and J. ZEMANN, 1975. Miner. (Tschermaks) Petrogr. Mitt., 22, 242-249.

Hexagonal, $P6_3/mmc$, a = 3.796, c = 16.360 Å, c/a = 4.31, Z = 6. Mo radiation, R = 0.056 for 59 independent reflexions, photographic data; see also 1, 2.

Atomic positions

	x	y	z
2 Cu(1) in 4(f)	1/3	2/3	0.7576*
Cu(2) in 4(f)	1/3	2/3	0.1073
S(1) in 2(c)	1/3	2/3	1/4
S(2) in 4(e)	0	0	0.0622

* (Model II - Model I has 2 Cu in 2(d), z = 3/4)

Interatomic distances (Å) (σ 0.004 to 0.011 Å)

Cu(1) - 3 S 2.195	S(1) - 3 S 3.773
Cu(2) - 4 S 2.312, 2.334	S(2) - 3 S 3.796
	S(2) - S(2) 2.037

S-Cu-S angles 108.6 and 110.3°

The 3-coordinated Cu(1) either occupies the position 2(d) (Model I) and has a strongly anisotropic temperature factor, or its position is split parallel to [00.1] by 0.25(6) Å (Model II), in which case its temperature factor is approximately isotropic. The interatomic distance within the S_2-group is 2.04(1) Å. Model II is preferred though the R-value is the same for both models.

1. Strukturbericht, 2, 230.
2. Structure Reports, 18, 380; 33A, 79.

COPPER SULPHUR TITANIUM

$Cu_{0.70}S_2Ti$

I. N. LE NAGARD, O. GOROCHOV and G. COLLIN, 1975. Mater. Res. Bull., 10, 1287-1296.

Rhombohedral, $AgCrSe_2$ (h.t.) type (1), R$\bar{3}$m, a = 3.4385, c = 18.901 Å, c/a = 5.50, D_m = 4.04, Z = 3 (rhombohedral cell, a = 6.604 Å, α = 30°17', Z = 1). Mo radiation, R = 0.037 for 111 independent reflexions, diffractometer data corrected for absorption.

Atomic positions (hexagonal axes)

	x	y	z
Ti in 3(a)	0	0	0
2.1 Cu in 6(c)	0	0	0.1440
S in 6(c)	0	0	0.2567

Interatomic distances (Å)

Ti - 2 Cu 2.72	S - 3 Ti 2.46
- 6 S 2.46	- 4 Cu 2.13, 2.36
Cu - 1 Ti 2.72	
- 3 Cu 2.16	
- 4 S 2.13, 2.36	

$Cu_{0.92}S_4Ti_2$

II. N. LE NAGARD, G. COLLIN and O. GOROCHOV, 1975. Mater. Res. Bull., 10, 1279-1286.

Cubic, spinel type, Fd3m, a = 9.985 Å, D_m = 3.78, Z = 8. R = 0.038 for 67 reflexions, diffractometer data corrected for absorption. Ti in 16(d);
7.4 Cu in 8(a) (origin); S in 32(c): x = 0.3805. Ti-6S = 2.44, Cu-4S = 2.26, S-S = 3.68 Å. See also 2.

1. Structure Reports, 39A, 101.
2. Ibid., 21, 115.

COPPER TIN

(ζ-BRONZE)

$Cu_{20}Sn_6$

J.K. BRANDON, W.B. PEARSON and D.J.N. TOZER, 1975. Acta Cryst., B31, 774-779.

Hexagonal, new structure, $P6_3$, a = 7.330, c = 7.864 Å, c/a = 1.07, D_m = 8.95, Z = 1. Mo radiation, R_w = 0.062 for 434 reflexions, diffractometer data with allowance for anomalous dispersion; contrary to 1.

Atomic positions

			x	y	z
Cu(1)	in	2(a)	0	0	-0.0223
Cu(2)	in	2(b)	2/3	1/3	-0.0900
Cu(3)	in	6(c)	0.6472	-0.0179	-0.0845
Cu(4)	in	2(b)	1/3	2/3	0.0737
Cu(5)	in	6(c)	0.3574	0.0350	0.0831
Cu(6)	in	2(b)	2/3	1/3	0.2450
Sn	in	6(c)	0.6800	-0.0200	0.25*

* fixed to define origin

Interatomic distances (Å)

Cu(1)	- 6 Cu	2.57, 2.64	Cu(5)	- 8 Cu	2.57 - 2.92
	- 6 Sn	2.90, 3.12		- 4 Sn	2.63 - 2.90
Cu(2)	- 8 Cu	2.51 - 2.65	Cu(6)	- 8 Cu	2.57 - 2.67
	- 3 Sn	2.73		- 3 Sn	2.64
Cu(3)	- 8 Cu	2.51 - 2.92	Sn	- 13 Cu	2.63 - 3.12
	- 4 Sn	2.64 - 2.86			
Cu(4)	- 8 Cu	2.59 - 2.65			
	- 3 Sn	2.80			

The previously reported model (1) for the structure could be described in terms of nine layers of atoms perpendicular to c with two columns of tin atoms parallel to c. In contrast, the new structure (Fig. 1) has six main layers of atoms perpendicular to c with all the tin atoms in two of these layers separated by c/2. Rather than being related to γ-brass, the new $Cu_{20}Sn_6$ model can be visualized as a superstructure based on ζ Ag-Zn (2). Average interatomic distances for Cu-Cu and Cu-Sn are 2.65 Å and 2.79 Å, respectively, with no first neighbour Sn-Sn contacts present in the structure. The coordination shells around the various Cu atoms have 11 or 12 nearest neighbours whereas those around Sn atoms have 13 neighbours.

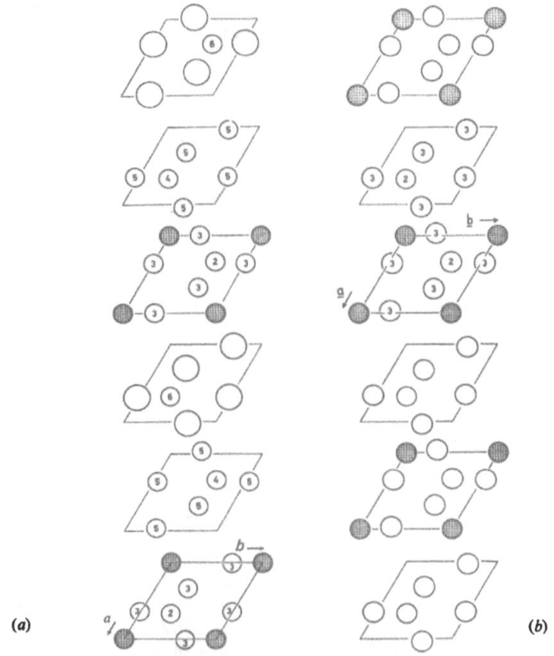

Fig. 1. The main layers of the hexagonal $Cu_{20}Sn_6$ structure (a), compared with the trigonal ζ Ag-Zn structure (b); small circles are Cu or Ag atoms, hatched circles represent Cu(1) atoms lying between the layer and the one above; the layers drawn have z values, reading from the botton to the top of -0.08, 0.08, 0.25, 0.42, 0.58, 0.75, respectively.

1. Strukturbericht, 2, 716.
2. Structure Reports, 15, 120.

DYSPROSIUM GERMANIUM SULPHUR

GERMANIUM LANTHANUM SULPHUR

A. MICHELET, A. MAZURIER, G. COLLIN, P. LARUELLE and J. FLAHAUT, 1975. J. Solid State Chem., 13, 65-76.

$Dy_6Ge_{2.5}S_{14}$, hexagonal $Al_{3.33}Ce_6S_{14}$ type (1), $P6_3$, a = 9.73, c = 5.82 Å, c/a = 0.60, D_m = 5.54, Z = 1. R = 0.067 for 247 independent reflexions, photographic data. A hexagonal supercell exists, with a' = a√3, c' = 2c, space group P6.

$GeLa_2S_5$, monoclinic, Ce_2SiS_5 type (2), $P2_1/a$, a = 7.887, b = 7.675, c = 12.720 Å, γ = 101.40°, Z = 4, D_x = 4.55. Cu radiation, R = 0.038 for 745 independent reflexions, photographic data; see also 3.

$Ge_3La_4S_{12}$, rhombohedral, $Ge_3La_4S_{12}$ type, R3c, a = 11.52 Å, α = 114°41', D_m = 4.29, Z = 2 (hexagonal cell, a = 19.40, c = 8.10 Å, c/a = 0.417, Z = 6). Cu radiation, R = 0.065 for 894 independent reflexions, photographic data; contrary to 4, 5, see also 6.

Atomic positions

$Dy_6Ge_{2.5}S_{14}$

	x	y	z
Dy in 6(c)	0.359	0.140	0.250
S(1) in 6(c)	0.251	0.100	0.818
S(2) in 6(c)	0.523	0.427	0.504
S(3) in 2(b)	1/3	2/3	0.531
Ge(1) in 2(b)	1/3	2/3	0.175
Ge(2)* in 2(a)	0	0	0.047

* occupancy = 0.25

$GeLa_2S_5$

All atoms in 4(e)

La(1)	0.0419	0.2319	0.0935
La(2)	0.3649	0.8401	0.1667
Ge	0.5912	0.3380	0.1165
S(1)	0.3815	0.1706	0.0259
S(2)	0.5004	0.8613	0.3790
S(3)	0.2163	0.5784	0.0061
S(4)	0.6977	0.1287	0.2051
S(5)	0.5474	0.5349	0.2326

$Ge_3La_4S_{12}$

La(1) in 6(a), all others in 18(b)

La(1)	0	0	0
La(2)	0.0030	0.2307	0.2028
Ge	0.2000	0.1875	0.1523
S(1)	0.1549	0.3789	0.1618
S(2)	0.1246	0.0643	0.2511
S(3)	0.1145	0.2005	0.9974
S(4)	0.3960	0.0593	0.1817

Interatomic distances (Å)

$Dy_6Ge_{2.5}S_{14}$

Ge - 4 S 2.07, 2.22
Ge - 6 S 2.59, 2.63

Dy - 7S 2.72 - 3.06, average 2.86

$GeLa_2S_5$

Ge - 4 S	2.17 - 2.25	La(1) - 6 S	2.91 - 3.15
		- 2 S	2.83, 3.03
		La(2) - 6 S	2.90 - 3.12
		- 3 S	3.17 - 3.32

$Ge_3La_4S_{12}$

Ge - 4 S	2.19 - 2.23	La(1) - 6 S	2.90, 2.91
		- 3 S	3.37
		La(2) - 6 S	2.86 - 3.08
		- 3 S	3.00 - 3.73

$Ge_3La_4S_{12}$ contains discrete [GeS_4] tetrahedra; La(1) centres a tri-capped trigonal prism; La(2) has a similar though less-regular environment.

$GeLa_2S_5$ also contains discrete [GeS_4] tetrahedra, and La in trigonal prisms plus 2 or 3 extra S neighbours.

$Dy_6Ge_{2.5}S_{14}$ contains [GeS_4] tetrahedra and [GeS_6] octahedra, and the Dy atoms centre a mono-capped trigonal prism. The structure as described refers to a disordered sub-structure; long exposures show super-structure reflexions corresponding to a cell with a' = a$\sqrt{3}$ and c' = 2c; locating 3 Ge atoms at 0,0,0,; 1/3,2/3,1/2; 2/3,1/3,1/2 in P6 gives reasonable agreement with the observations (4) [see also 5 for a related structure].

1. Structure Reports, 33A, 4.
2. Ibid., 38A, 162.
3. Ibid., 39A, 64.
4. E.S. SARKISOV, R.A. LIDIN and V.V. ŠUM, 1970. Izv. Akad. Nauk SSSR, Neorg. Mater., 6, 2054.
5. E.S. SARKISOV, R.A. LIDIN and Ju.M. KHOŽAINOV, 1968. Ibid., 4, 2033.
6. Structure Reports, 34A, 146.

GALLIUM GERMANIUM LITHIUM

GaGeLi

W. BOCKELMANN and H.-U. SCHUSTER, 1974. Z. anorg. Chem., 410, 233-240.

Hexagonal, $P6_3mc$, a = 4.175, c = 6.783 Å, c/a = 1.625, D_m = 4.546, Z = 2. Diffractometer data, R = 0.097, a redetermination in agreement with 1.

Atomic positions

	x	y	z
Ga in 2(b)	1/3	2/3	0.059
Ge in 2(b)	1/3	2/3	0.444
Li in 2(a)	0	0	0.247

[Interatomic distances (Å)]*

Ge	- 4 Ga	2.53, 2.61	Li	- 6 Ga	2.73, 3.21
	- 6 Li	2.76, 3.17		- 6 Ge	2.76, 3.17
				- 2 Li	3.39
Ga	- 6 Li	2.73, 3.21			
	- 4 Ge	2.53, 2.61			

* [Values calculated from the parameters listed above; distances originally given differ by small amounts.]

1. Structure Reports, 35A, 61.

GALLIUM GERMANIUM PALLADIUM

$\sim Ga_2Ge_5Pd_{12}$

S. HEINRICH and K. SCHUBERT, 1975. Z. Metallk., 66, 353-355.

Hexagonal, Th_7S_{12} type (1), $P6_3/m$, a = 9.448, c = 3.684 Å, c/a = 0.39, Z = 1. Cu radiation, R = 0.13 for 177 reflexions, photographic data, corrected for absorption; an ordered form with c' = 7c was observed in the Ge-rich region of the homogeneity range.

Atomic positions

		x	y	z
Pd(1)	in 6(h)	0.385	0.514	1/4
Pd(2)	in 6(h)	0.015	0.248	1/4
6 Ge/Ga(1)*	in 6(h)	0.717	0.172	1/4
1 Ge/Ga(2)*	in 2(a)	0	0	1/4

* Ratio Ge/Ga = 2.0 for sample used

[Interatomic distances (Å)]

Pd(1)	- 8 Pd	2.95 - 3.11	Ge/Ga(1)	- 8 Pd	2.53 - 2.59
	- 5 Ge/Ga	2.55, 2.59		- 2 Ge/Ga	3.13
Pd(2)	- 8 Pd	2.93 - 3.11	Ge/Ga(2)	- 9 Pd	2.28, 2.93
	- 3 Ge/Ga	2.53, 2.55		- 2 Ge/Ga	1.84
	- 3 Ge/Ga	2.28, 2.93			

1. Structure Reports, 12, 184.

GALLIUM MOLYBDENUM SULPHUR

$GaMo_4S_8$

C. PERRIN, R. CHEVREL and M. SERGENT, 1975. C.R. Acad. Sci., Paris, C, 280, 949-951.

Cubic, $F\bar{4}3m$, a = 9.73 Å, Z = 4. R = 0.035 for 196 reflexions, diffractometer data.

Atomic positions

Ga in 4(a), others in 16(e)

	x	y	z
Mo	0.3974	0.3974	0.3974
S(1)	0.6343	0.6343	0.6343
S(2)	0.1350	0.1350	0.1350
Ga	0	0	0

Interatomic distances (Å)

Mo - 3 Mo	2.82	S(1) - 3 Mo	2.35
- 6 S	2.35, 2.59	- 3 S	3.18
Ga - 4 S	2.28	S(2) - 3 Mo	2.59
		- 3 S	3.16
		- 1 Ga	2.28

The structure contains $[GaS_4]$ tetrahedra and $[Mo_4]$ tetrahedral clusters; the $[Mo_4]$ clusters lie inside the empty corners of a larger cube outlined by 4 S atoms, the $[Mo_4S_4]$ group forming a distorted cube.

GALLIUM SELENIUM

δ-GaSe

I. A. KUHN, R. CHEVALIER and A. RIMSKY, 1975. Acta Cryst., B31, 2841-2842.

II. A. KUHN, A. CHEVY and R. CHEVALIER, 1975. Phys. Status Solidi, 31, 469-475.

Hexagonal, $P6_3mc$, a = 3.755, c = 31.990 Å, c/a = 8.52, Z = 8. Mo radiation, R = 0.098 for 298 reflexions, diffractometer data corrected for absorption (spherical approximation) with allowance for anomalous dispersion.

Atomic positions*

			x	y	z
Ga(1)	in	2(a)	0	0	-0.0380
Ga(2)	in	2(a)	0	0	0.0394
Ga(3)	in	2(b)	1/3	2/3	0.2119
Ga(4)	in	2(b)	1/3	2/3	0.2880
Se(1)	in	2(b)	2/3	1/3	-0.0745
Se(2)	in	2(b)	2/3	1/3	0.0752
Se(3)	in	2(a)	0	0	0.1781
Se(4)	in	2(a)	0	0	0.3274

* I; slightly different values given in II.

Each Ga is surrounded by 1 Ga (2.43 - 2.48 Å) and 3 Se (2.42 - 2.51 Å); each Se by 3 Ga (2.42 - 2.51 Å); Se-Ga-Se angles are 57, 59, 97, 99, 100, and 102°. The Se-Se interlayer distance is 3.880 Å.

GALLIUM VANADIUM

$Ga_{41}V_8$

K. GIRGIS, W. PETTER and G. PUPP, 1975. Acta Cryst., B31, 113-116.

Rhombohedral, R$\bar{3}$, a = 9.4560 Å, α = 94.958°, D_m = 6.49, Z = 1 for $Ga(VGa_5)_8$ (hexagonal cell has a = 13.9382, c = 14.8924 Å, c/a = 1.07). Mo radiation, R = 0.044 for 2457 reflexions, diffractometer data corrected for absorption, 11 site-sets; see also 1.

Interatomic distances (Å) [summarized]

V(1)	- 10 Ga	2.52 - 2.77	
V(2)	- 10 Ga	2.51 - 2.97 [2.74]	
Ga(1)	- 12 Ga	2.93, 2.96	
Ga(2)	- 2 V	2.76	
	- 6 Ga	2.83	
Ga(3)	- 2 V	2.51	
	- 8 Ga	2.75 - 2.83	
Ga(4)	- 2 V	2.65, 2.68	
	- 11 Ga	2.76 - 3.35	
Ga(5)	- 2 V	2.73, 2.77	
	- 11 Ga	2.79 - 3.36	
Ga(6)	- 2 V	2.51, 2.54	
	- 8 Ga	2.74 - 3.12	
Ga(7)	- 2 V	2.58	
	- 8 Ga	2.74 - 3.22	
Ga(8)	- 2 V	2.52, 2.54	
	- 8 Ga	2.70 - 3.36	
Ga(9)	- 2 V	2.57, 2.59	
	- 7 Ga	2.70 - 2.91	

[The V(2)-Ga(9) distance is misprinted as 2.97. There are also minor differences between the values given and those calculated directly from the coordinates.]

The structure (Fig. 1) can be described in terms of linked polyhedra. One Ga atom is in the centre of a cubo-octahedron of Ga atoms, whose triangular faces are each shared by one face of a [VGa_{10}] polyhedron. This [VGa_{10}] polyhedron is formed by 10 triangular and 3 almost square faces, and can be considered as half an icosahedron plus half a cube; the V atom is in the centre.

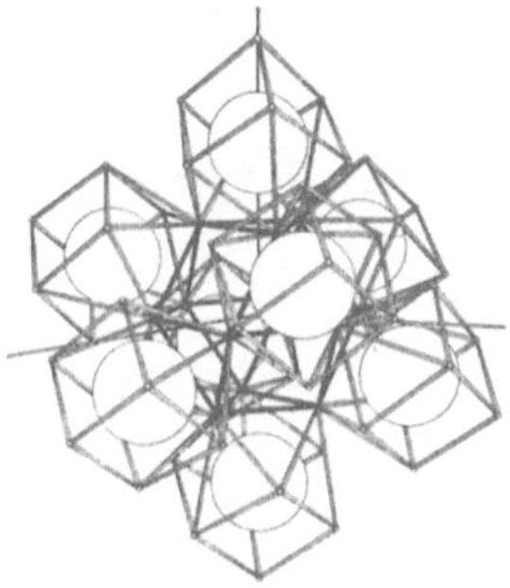

Fig. 1. $Ga_{41}V_8$. View of a [$GaGa_{12}$] cubo-octahedron surrounded by 8 [VGa_{10}] polyhedra; central V atoms represented as spheres.

1. K. GIRGIS, F. LAVES and R. REINMANN, 1966. Naturwissenschaften, 53, 610.

GERMANIUM LITHIUM

Ge_6Li_{11}

I. U. FRANK and W. MÜLLER, 1975. Z. Naturforsch., 30B, 313-315.

Orthorhombic, Cmcm, a = 4.38, b = 24.55, c = 10.64 Å, Z = 4, D_x = 2.97. Cu and Mo radiations, R = 0.069 for 864 reflexions, diffractometer data, 11 site-sets.

Interatomic distances (Å)

Ge(1)	- 2 Ge	2.48, 2.49	Li(3)	- 7 Li	2.77 - 3.20
	- 8 Li	2.62 - 3.00		- 5 Ge	2.73 - 2.81
Ge(2)	- 2 Ge	2.48, 2.49	Li(4)	- 7 Li	2.63 - 3.24
	- 9 Li	2.73 - 3.08		- 5 Ge	2.70, 2.83
Ge(3)	- 2 Ge	2.48	Li(5)	- 5 Li	2.63 - 2.90
	- 8 Li	2.60 - 3.06		- 10 Ge	3.00 - 3.08
Ge(4)	- 11 Li	2.53 - 2.85	Li(6)	- 5 Li	2.56 - 2.96
				- 5 Ge	2.62 - 2.99
Li(1)	- 6 Li	2.73, 2.86			
	- 2 Ge	2.53	Li(7)	- 7 Li	2.56 - 2.92
				- 4 Ge	2.60 - 2.91
Li(2)	- 6 Li	2.73 - 3.24			
	- 4 Ge	2.70 - 2.85			

The structure is characterized by isolated 5-membered Ge rings, with Ge-Ge distances 2.48 - 2.49 Å and angles 106, 109, and 110°.

Ge_2Li_7

II. V. HOPF, W. MÜLLER and H. SCHÄFER, 1972. Z. Naturforsch., 27B, 1157-1160.

Orthorhombic, Cmmm, a = 9.24, b = 13.21, c = 4.63 Å, D_m = 2.25, Z = 4. Cu radiation, R = 0.113, photographic data.

Atomic positions

	x	y	z
Ge(1) in 4(i)	0	0.314	0
Ge(2) in 4(h)	0.142	0	1/2
Li(1) in 2(a)	0	0	0
Li(2) in 2(c)	1/2	0	1/2
Li(3) in 4(g)	0.340	0	0
Li(4) in 4(j)	0	0.180	1/2
Li(5) in 8(p)	0.181	0.159	0
Li(6) in 8(q)	0.319	0.149	1/2

Interatomic distances (Å) *

Ge(1)	- 14 Li	2.64 - 3.38	Li(3)	- 4 Ge	2.87, 2.95
				- 10 Li	2.56 - 3.14
Ge(2)	- 1 Ge	2.62			
	- 13 Li	2.56 - 3.30	Li(4)	- 4 Ge	2.72, 2.91
				- 10 Li	2.81, 3.32
Li(1)	- 4 Ge	2.66			
	- 10 Li	2.66 - 3.32	Li(5)	- 4 Ge	2.64 - 3.15
				- 10 Li	2.56 - 3.43
Li(2)	- 6 Ge	3.31, 3.38			
	- 8 Li	2.58, 2.75	Li(6)	- 3 Ge	2.56 - 2.90
				- 11 Li	2.58 - 3.43

* Some misprints have been corrected.

This structure is related to, but not isotypic with Li_7Si_2 (1).

1. Structure Reports, 30A, 66.

GERMANIUM MANGANESE THORIUM

MANGANESE SILICON THORIUM

Ge_2Mn_2Th
Mn_2Si_2Th

Z. BAN, L. OMĖJEC, A. SZYTULA and Z. TOMKOWICZ, 1975. Phys. Status Solidi, 27, 333-338.

Tetragonal, I4/mmm, Cu_2Si_2Th (Al_2CeGe_2) structure type (1), at 77, 293, 600°K, a = 3.998, 4.019, 4.043, c = 10.445, 10.483, 10.527 Å, Z = 2, z = 0.3786, 0.3775, 0.3781 for Mn_2Si_2Th, a = 4.045, 4.085, 4.103, c = 10.906, 10.924, 10.911 Å, Z = 2, z = 0.3804, 0.3807, 0.3809 for Ge_2Mn_2Th. See also 2. Magnetic structures were also determined. There is a first order phase change at 380°K for Ge_2Mn_2Th, but only a second order one for Mn_2Si_2Th at 485°K.

1. Structure Reports, 29, 91.
2. Ibid., 30A, 153 (Mn_2Si_2Th, ref. 152).

GERMANIUM NIOBIUM

Ge_3Nb_5

S. JAGNER and S.E. RASMUSSEN, 1975. Acta Cryst., B31, 2881-2883.

Tetragonal, W_5Si_3 type (1), I4/mcm, a = 10.146, c = 5.136 Å, c/a = 0.51, Z = 4, D_x = 8.55. Mo radiation, R = 0.017 for 189 reflexions, diffractometer, corrected for absorption, with allowance for anomalous dispersion; see also 2.

Atomic positions

	x	y	z
Nb(1) in 4(b)	0	1/2	1/4
Nb(2) in 16(k)	0.07605	0.22220	0
Ge(1) in 4(a)	0	0	1/4
Ge(2) in 8(h)	0.16546	0.66546	0

Interatomic distances (Å)

Nb(1)	- 4 Ge	2.699	Ge(1)	- 8 Nb	2.707
	- 10 Nb	2.568, 3.192		- 2 Ge	2.568
Nb(2)	- 6 Ge	2.685 - 2.952	Ge(2)	- 10 Nb	2.699 - 2.952
	- 9 Nb	2.895 - 3.370		- 2 Ge	3.533

1. Structure Reports, 19, 277.
2. Ibid., 20, 109.

GERMANIUM PALLADIUM

Ge_9Pd_{25}

W. WOPERSNOW and K. SCHUBERT, 1975. J. Less-Common Metals, 41, 97-103.

Trigonal, $P\bar{3}$, a = 7.351, c = 10.605 Å, c/a = 1.44, Z = 1. Cu radiation, R = 0.12 for 474 reflexions, photographic data. See also 1.

Atomic positions

	x	y	z
Pd(1) in 6(g)	0.308	0.928	0.432
Pd(2) in 6(g)	0.347	0.067	0.922
Pd(3) in 6(g)	0.415	0.089	0.188
Pd(4) in 2(d)	1/3	2/3	0.608
Pd(5) in 2(d)	1/3	2/3	0.242
Pd(6) in 2(c)	0	0	0.251
Pd(7) in 1(b)	0	0	1/2
Ge(1) in 6(g)	0.333	0.993	0.680
Ge(2) in 2(d)	1/3	2/3	0.973
Ge(3) in 1(a)	0	0	0

[Interatomic distances (Å)]

Pd(1)	- 11 Pd	2.75 - 3.50	Pd(6)	- 10 Pd	2.64 - 3.21
	- 4 Ge	2.47 - 3.10		- 4 Ge	2.58 - 2.66
Pd(2)	- 9 Pd	2.78 - 3.31	Pd(7)	- 8 Pd	2.64, 2.67
	- 4 Ge	2.45 - 2.94		- 6 Ge	3.12
Pd(3)	- 10 Pd	2.78 - 3.16	Ge(1)	- 12 Pd	2.52 - 3.43
	- 5 Ge	2.50 - 3.43	Ge(2)	- 10 Pd	2.45 - 2.94
Pd(4)	- 9 Pd	2.75 - 2.86			
	- 3 Ge	2.52	Ge(3)	- 14 Pd	2.48 - 3.42
Pd(5)	- 9 Pd	2.79 - 2.91			
	- 4 Ge	2.61, 2.85			

1. K. KHALAFF and K. SCHUBERT, 1974. Z. Metallk., 65, 379.

GERMANIUM SULPHUR

GeS_2 (h.t.)

G. DITTMAR and H. SCHÄFER, 1975. Acta Cryst., B31, 2060-2064.

Monoclinic, $P2_1/c$, a = 6.720, b = 16.101, c = 11.436 Å, β = 90.88°, D_m = 2.89, Z = 16. Mo radiation, R = 0.09 for 2772 reflexions, diffractometer data, 12 site-sets; see 1-3 for previous work.

Interatomic distances (Å)

Ge(1)	- 4 S	2.22 - 2.23	Ge(3)	- 4 S	2.22 - 2.23
Ge(2)	- 4 S	2.20 - 2.22	Ge(4)	- 4 S	2.19 - 2.22

The structure (Fig. 1) is composed of chains of $[GeS_4]$ tetrahedra which are linked by sharing either corners or edges, and is thus similar to those of SnS_2 and SiS_2.

Fig. 1. Structure of h.t.-GeS_2.

1. Structure Reports, 37A, 158.
2. W. VIAENE and G.H. MOH, 1970. Neues Jb. Miner., Mh., 283.
3. W. PUGH, 1930. J. Chem. Soc., 2369.

GOLD INDIUM

M. PUŠELJ and K. SCHUBERT, 1975. J. Less-Common Metals, 41, 33-44.

Au_9In_4 (h.t.)
Cubic, Cu_9Al_4 type (1), $P\bar{4}3m$, a = 9.843 Å, Z = 4. Cu radiation, R = 0.20 for 92 reflexions, powder photographic data.

Atomic positions

			x	y	z
Au(1)	in	6(f)	0	0	0.366
Au(2)	in	6(g)	1/2	1/2	0.861
Au(3)	in	4(e)	0.835	0.835	0.835
Au(4)	in	4(e)	0.324	0.324	0.324
Au(5)	in	4(e)	0.601	0.601	0.601
Au(6)	in	12(i)	0.314	0.314	0.038
In(1)	in	4(e)	0.113	0.113	0.113
In(2)	in	12(i)	0.807	0.807	0.546

[Interatomic distances (Å)]

Au(1)	- 7 Au	2.64 - 3.16		Au(4)	- 9 Au	2.82 - 3.05
	- 6 In	2.82 - 3.22			- 4 In	2.85, 3.60
Au(2)	- 9 Au	2.74 - 3.12		Au(5)	- 9 Au	2.81, 2.92
	- 4 In	3.10			- 3 In	2.92
Au(3)	- 6 Au	2.88, 3.03		Au(6)	- 8 Au	2.77 - 3.74
	- 6 In	2.83, 2.87			- 5 In	2.89 - 2.99
In(1)	- 10 Au	2.83 - 3.60		In(2)	- 11 Au	2.82 - 3.22
	- 3 In	3.15			- 4 In	3.63, 3.91

Au_7In_3
Trigonal, Au_7In_3 type, $P\bar{3}$, a = 12.215, c = 8.509 Å, c/a = 0.70, D_m = 15.79, Z = 6. Cu radiation, R = 0.20 for 779 reflexions, photographic data, corrected for absorption. 13 site-sets; see also 2, 3.

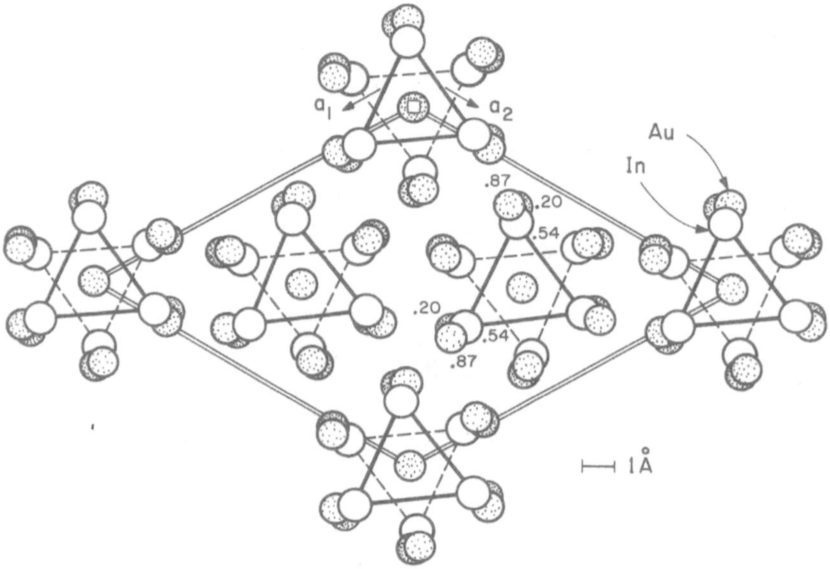

Fig. 1. A projection onto (001) of the trigonal Au_7In_3 structure.

The Au_7In_3 structure (Fig. 1) is related to the γ-brass structure (Cu_5Zn_8). [The Au atoms are 12-coordinated except for one Au which has 13 neighbours, 2 of which are at 3.61 and 3.87 Å. The Au-Au distances range from 2.73 to 3.22 Å, and the Au-In distances range from 2.75 to 3.30 Å, except for the the two distances mentioned above of 3.61 and 3.87 Å. The In atoms are 15- or 16-coordinated, with Au-In distances from 2.75 to 3.30 Å and In-In distances from 3.52 to 3.85 Å.]

1. Strukturbericht, 3, 589; Structure Reports, 30A, 3.
2. Strukturbericht, 6, 163; Structure Reports, 29, 103, 154.
3. Structure Reports, 11, 122.

HAFNIUM IRON SILICON

$Fe_{16}Hf_6Si_7$

L.A. LYSENKO, 1974. Vest. L'vov. Univ. Ser. Chim., 15, 21-23.

Cubic, $Mg_6Cu_{16}Si_7$ ($Mn_{23}Th_6$) type (1), Fm3m, a = 11.48 Å, Z = 4. Cr radiation, R = 0.139 for 41 reflexions, powder data.

Atomic positions

			x	y	z
Hf	in	24(e)	0	0	0.197
Fe(1)	in	32(f)	0.1684	0.1684	0.1684
Fe(2)	in	32(f)	0.377	0.377	0.377
Si(1)	in	4(b)	1/2	1/2	1/2
Si(2)	in	24(d)	1/4	1/4	0

[Interatomic distances (Å)]

Hf	- 4 Hf	3.20
	- 8 Fe	2.75, 2.87
	- 5 Si	2.93, 3.48
Fe(1)	- 3 Hf	2.75
	- 6 Fe	2.51, 2.65
	- 3 Si	2.34
Fe(2)	- 3 Hf	2.87
	- 6 Fe	2.51, 2.82
	- 4 Si	2.45, 2.50
Si(1)	- 6 Hf	3.48
	- 8 Fe	2.45
Si(2)	- 4 Hf	2.93
	- 8 Fe	2.34, 2.50

1. Structure Reports, 16, 84, 112; 20, 95.

HAFNIUM NICKEL SULPHUR

$HfNi_{0.36}S_2$

C. MOREAU, M. SPIESSER and J. ROUXEL, 1975. C.R. Acad. Sci., Paris, C, 280, 1203-1206.

Monoclinic, C2/m, a = 6.27, b = 3.620, c = 5.821 Å, β = 90.0°, D_m = 6.45, Z = 2. R = 0.13 for 456 reflexions, diffractometer data.

Atomic positions

				x	y	z
2	Hf	in	2(a)	0	0	0
4	S	in	4(i)	0.3332	0	0.2514
0.72	Ni	in	4(i)	0.333	0	0.647

Interatomic distances (Å)

Hf - 6 S 2.55 Ni - 4 S 2.16, 2.34 [2.17, 2.30]

The Hf atoms are octahedrally coordinated (S-Hf-S = 89.5 and 90.5°) and the Ni atoms are tetrahedrally coordinated (S-Ni-S = 105 and 114°).

HAFNIUM SULPHUR YTTRIUM

HfS_5Y_2

W. JEITSCHKO and P.C. DONOHUE, 1975. Acta Cryst., B31, 1890-1895.

Orthorhombic, ordered U_3Se_5 type (1), Pnma, a = 11.4585, b = 7.7215, c = 7.2207 Å, Z = 4. Mo radiation, R = 0.072 for 1082 reflexions, diffractometer data with corrections for absorption and extinction, and allowance for anomalous dispersion; see also 2.

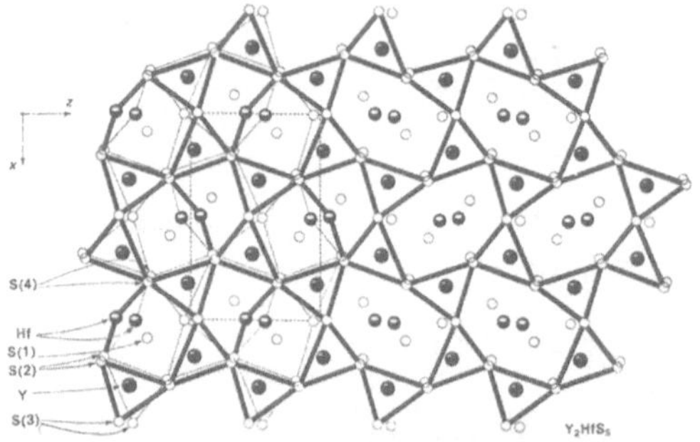

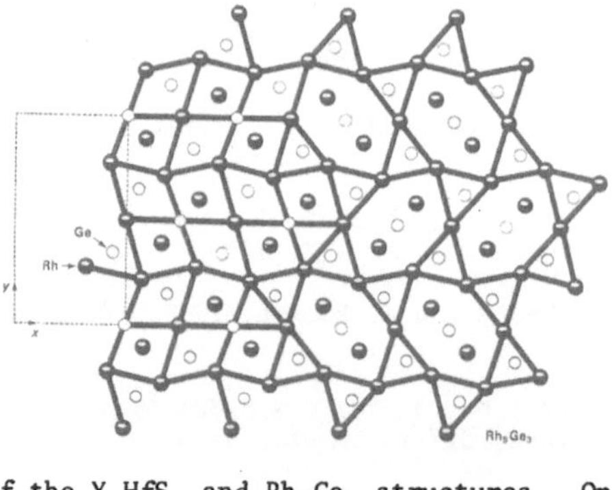

Fig. 1. Comparison of the Y_2HfS_5 and Rh_5Ge_3 structures. On the left hand side of the Y_2HfS_5 drawing, atoms at y = 1/4 and 3/4 are connected by thin and thick lines, respectively. Atoms at y ∿ 0 and ∿ 1/2 are superimposed and not connected. The right hand side of that drawing emphasizes the relation of Y_2HfS_5 to the Rh_5Ge_3 anti-type structure, which can be considered as an undistorted U_3Se_5 (Y_2HfS_5) type structure with half the translation period in the projection direction. In the left hand side of the Rh_5Ge_3 drawing, atoms at z = 0 are connected by lines, atoms at z = 1/2 are unconnected, thus emphasizing the relation of Rh_5Ge_3 to a cubic body-centred structure.

Atomic positions

Y and S(1) in 8(d), others in 4(c)

	x	y	z
Y	0.1778	0.9974	0.0251
Hf	0.0060	1/4	0.5742
S(1)	0.4081	0.0367	0.1630
S(2)	0.1822	1/4	0.3331
S(3)	0.5032	1/4	0.5522
S(4)	0.2921	1/4	0.8125

Interatomic distances (Å)

Y - 8 S	2.81 - 2.96	Hf - 7 S	2.51 - 2.70
- 2 Y	3.82, 3.90	- 2 Hf	4.01

Shortest S-S distances are 3.06 Å.

The structure is an ordered version of the U_3Se_5 (1) type, which can be considered as a distorted anti-type derived from Rh_5Ge_3 (3) (Fig. 1). The Y and Hf atoms are coordinated to eight and seven S atoms, respectively. Four independent S atoms are four- or five-coordinated by the metal atoms. The HfS polyhedra share edges with each other, thus forming a one-dimensionally infinite $[HfS_5]^{6-}$ polyanion.

1. Structure Reports, 38A, 143.
2. This volume, p. 86, 104 [$PbSe_5U_2$ and S_5Sm_2U].
3. Structure Reports, 19, 177.

HAFNIUM, TITANIUM, and ZIRCONIUM TRICHALCOGENIDES

S. FURUSETH, L. BRATTÅS and A. KJEKSHUS, 1975. Acta Chem. Scand., A29, 623-631.

Monoclinic, Se_3Zr type (1), $P2_1/m$, Z = 2. Mo radiation, photographic data corrected for absorption. All atoms in 2(e): x, 1/4, z; data are given in the following tables; NbX_3 and TaX_3 (X = S, Se, Te) do not belong to this $ZrSe_3$ type, contrary to 2.

Compound	a(Å)	b(Å)	c(Å)	β(°)	Variant
TiS_3	4.958	3.4006	8.778	97.32	B
ZrS_3	5.1243	3.6244	8.980	97.28	A
$ZrSe_3$	5.4109	3.7488	9.444	97.48	A
$ZrTe_3$	5.8939	3.9259	10.100	97.82	B
HfS_3	5.0923	3.5952	8.967	97.38	A
$HfSe_3$	5.388	3.7216	9.428	97.78	B

Variant		A	A	A	B	B	B
Compound		ZrS_3	$ZrSe_3$ (1)	HfS_3	TiS_3	$ZrTe_3$	$HfSe_3$
T*	x	0.2837	0.285	0.2839	0.7152	0.7069	0.7145
	z	0.6553	0.656	0.6548	0.6528	0.6660	0.6563
X(1)	x	0.7631	0.762	0.7611	0.2392	0.2391	0.2370
	z	0.5543	0.554	0.5546	0.5505	0.5561	0.5533
X(2)	x	0.4725	0.456	0.4642	0.5320	0.5661	0.5455
	z	0.1716	0.174	0.1702	0.1762	0.1676	0.1733
X(3)	x	0.8799	0.888	0.8768	0.1205	0.0963	0.1113
	z	0.1699	0.169	0.1697	0.1737	0.1616	0.1673
R		0.076	0.11	0.066	0.070	0.057	0.073
Reflexions		353	-	425	316†	322	400

* T = Hf, Ti, Zr; X = S, Se, Te
† diffractometer data

Interatomic distances (Å)

	Variant A			Variant B		
	ZrS_3	$ZrSe_3$	HfS_3	TiS_3	$ZrTe_3$	$HfSe_3$
T - 8 X	2.60-2.72	2.72-2.87	2.59-2.70	2.36-2.86	2.77-3.47	2.59-3.10
Average	2.631	2.77	2.622	2.539	3.028	2.784
Average*	2.655	2.79	2.648	2.516	2.999	2.757
X - X	2.09	2.34	2.10	2.04	2.76	2.33

* After transformation to the other variant, i.e. A→B or B→A by the formula $x_A = 1 - x_B$, $y_A = y_B$, and $z_A = z_B$.

The structure has been described (1, 3, 4) and consists of infinite columns of isolated TX_6 trigonal prisms connected into slabs by two additional T-X bonds through the square faces of adjacent prisms; X-X pairs occur.

The TX_3 compounds occur in two variants, A and B, with positional parameters related as noted above. Both forms have the space group $P2_1/m$ and similar cell parameters. Both forms were observed for $HfSe_3$ but data for only one form are given. An alternate description of the transformation is to say that it simulates the action of a mirror plane parallel to (100) at x = 1/2. The difference between A and B forms is mainly due to a greater scatter in the T-X distances in the B form. [The average T-X distance in the B form is ∿0.025 Å larger and the X-X distance ∿0.01 Å smaller than in the A form.]

1. Structure Reports, 22, 190; 30A, 168, ref. 308.
2. Ibid., 31A, 62.
3. Ibid., 38A, 176, ref. 188.
4. F. HULLIGER, 1968. Structure and Bonding, 4, 83-229; p. 120.

INDIUM MAGNESIUM TELLURIUM

In_2MgTe_4

P. DOTZEL, E. FRANKE, H. SCHÄFER and G. SCHÖN, 1975. Z. Naturforsch., 30B, 179-182.

Tetragonal, β-Cu_2HgI_4 type (1), $I\bar{4}2m$, a = 6.15, c = 12.30 Å, c/a = 2.00, D_m = 5.39, Z = 2. Cu and Mo radiations, R = 0.114 for 156 reflexions, photographic data; see also In_2MnTe_4 (2).

Atomic positions

			x	y	z
Te	in	8(i)	0.274	0.274	0.114
*In/Mg(1)	in	2(a)	0	0	0
†In/Mg(2)	in	4(d)	0	1/2	1/4

* In/Mg(1) = 1.33 In, 0.67 Mg
† In/Mg(2) = 2.67 In, 1.33 Mg

Interatomic distances (Å)

Mg/In(1) - 4 Te	2.77	Te - 3 Te	3.93, 3.95
Mg/In(2) - 4 Te	2.75	- 3 Mg/In	2.75, 2.77

1. Structure Reports, 19, 337, 414.
2. Following report.

INDIUM MANGANESE TELLURIUM

In_2MnTe_4

K.-J. RANGE and H.-J. HÜBNER, 1975. Z. Naturforsch., 30B, 145-148.

Tetragonal, $I\bar{4}2m$, a = 6.191, c = 12.382 Å, c/a = 2.00, D_m = 5.47, Z = 2. Mo radiation, R = 0.092 for 1296 reflexions; diffractometer data; see also In_2MgTe_4 (1).

Atomic positions

			x	y	z
Te	in	8(i)	0.2745	0.2745	0.1136
*Mn/In(1)	in	2(a)	0	0	0
†Mn/In(2)	in	4(d)	0	1/2	1/4

* Mn/In(1) = 1.33 In, 0.67 Mn
† Mn/In(2) = 2.67 In, 1.33 Mn

Interatomic distances (Å)

Mn/In(1) - 4 Te	2.78	Te - 3 Te	3.95, 3.96
Mn/In(2) - 4 Te	2.77	- 3 Mn/In	2.77, 2.78

The structure can be derived from the zincblende structure. It is related to the $CdGa_2S_4$ (2) and β-Cu_2HgI_4 (3) type structures with the same ordered arrangement of vacancies. In $MnIn_2Te_4$, however, the different cations are distributed randomly over the occupied tetrahedral sites.

1. Preceding report.
2. Structure Reports, 19, 414.
3. Ibid., 19, 337.

INDIUM SELENIUM

InSe

A. LIKFORMAN, D. CARRÉ, J. ETIENNE and B. BACHET, 1975. Acta Cryst., B31, 1252-1254.

Rhombohedral, R3m, a = 8.76 Å, α = 26.40° (hexagonal axes, a = 4.00, c = 25.32 Å, D_m = 5.53, Z = 6). R = 0.077 for 380 reflexions, diffractometer data, corrected for absorption; in agreement with 1, but contrary to 2.

Atomic positions (hexagonal axes)

All atoms in 3(a)

	x	y	z
In(1)	0	0	0
In(2)	0	0	0.1114
Se(1)	0	0	0.8281
Se(2)	0	0	0.6165

Interatomic distances (Å)

In(1) - 1 In	2.82	Se(1) - 3 Se	3.86
- 3 Se	2.64	- 3 In	2.63
In(2) - 1 In	2.82	Se(2) - 3 Se	3.85
- 3 Se	2.63	- 3 In	2.64

Each In is surrounded tetrahedrally by 3 Se and 1 In, which interpenetrate forming In-In pairs. This arrangement can be alternatively described as an In-In pair in the middle of a trigonal prism of Se atoms; Se(1) and Se(2) lie in the middle of a trigonal prism and a trigonal antiprism, respectively, both composed of 3 In and 3 Se atoms. [This structure shows formal resemblances to that of $CuCrS_2$ (3).]

1. Structure Reports, 18, 176, 389.
2. Ibid., 22, 142.
3. Ibid., 21, 86.

INDIUM TELLURIUM

In_4Te_3

L.I. MAN, R.K. KARAKHANJAN and R.M. IMAMOV, 1974. Kristallografija, 19, 1166-1169 [Soviet Physics - Crystallography, 19, 725-726].

Orthorhombic, In_4Se_3 type (1), Pnnm, a = 15.55, b = 12.70, c = 4.46 Å, Z = 4. R = 0.114 for 27 reflexions, electron-diffraction data; see also 2-6.

Atomic positions

All atoms in 4(g)

	x	y	z
In(1)	0.037	0.356	0
In(2)	0.421	0.393	0
In(3)	0.710	0.342	0
In(4)	0.183	0.484	0
Te(1)	0.098	0.150	0
Te(2)	0.425	0.149	0
Te(3)	0.767	0.136	0

[Interatomic distances (Å)]*

In(1)	- 3 Te	2.78, 2.83
	- 2 In	2.79, 3.83
In(2)	- 7 Te	3.10 - 3.96
	- 3 In	3.66 - 3.93
In(3)	- 3 Te	2.76, 2.82
	- 1 In	2.77
In(4)	- 4 Te	3.00, 3.49
	- 2 In	2.77, 2.79
Te(1)	- 3 In	2.78, 2.83
Te(2)	- 3 In	2.83, 3.10
Te(3)	- 3 In	2.76, 3.29

* The distances calculated from the coordinates differ slightly from those given in the paper.

1. Structure Reports, 38A, 107.
2. Ibid., 30A, 62.
3. A. THIEL, 1904. Z. anorg. Chem., 40, 324; A. THIEL and H. KOELSCH, 1910. Ibid., 66, 288.
4. W. KLEMM and H. FOGEL, 1934. Ibid., 215, 45.
5. Structure Reports, 18, 389.
6. Ibid., 39A, 73.

IRON NICKEL THORIUM

$FeNi_4Th$

J.B.A.A. ELEMANS, K.H.J. BUSCHOW, H.W. ZANDBERGEN and J.P. de JONG, 1975. Phys. Status Solidi, A, 29, 595-600.

Trigonal, $P\bar{3}m1$, a = 4.9789, c = 4.0336, c/a = 0.810, Z = 1. Neutron powder data, R = 0.035. There is a range of composition of $Th(Fe_{1-x}Ni_x)_5$, with x from 0.1 to 0.8. Refinements were carried out for each composition at 300 and 4.2°K; magnetic structures were also determined. Only the results for x = 0.8 at 300°K are given here.

Atomic positions

			x	y	z
Th	in	1(a)	0	0	0
Fe/Ni	in	2(d)	1/3	2/3	0.029
Fe/Ni	in	3(f)	1/2	0	1/2

This structure is a variant of the $CaCu_5$ type. Fe/Ni = $Fe_{0.2}Ni_{0.8}$.

IRON SELENIUM

$FeSe_2$

J. PICKARDT, B. REUTER, E. RIEDEL and J. SÖCHTIG, 1975. J. Solid State Chem., 15, 366-368.

Orthorhombic, FeS_2 (marcasite) type (1), Pnnm, a = 4.804, b = 5.784, c = 3.586 [misprinted as 2.586] Å, Z = 2, D_x = 7.12. Mo radiation, R = 0.064 for 315 reflexions, diffractometer data corrected for absorption; see also 2-4.

Atomic positions

			x	y	z
Fe	in	2(a)	0	0	0
Se	in	4(g)	0.2134	0.3690	0

[Interatomic distances (Å)]

Fe - 6 Se	2.37, 2.38	Se - 1 Se	2.55
		- 3 Fe	2.37, 2.38

1. Strukturbericht, 1, 495; Structure Reports, 35A, 75; 39A, 77.
2. H.C. GRANGER, 1966. U.S. Geol. Surv. Prof. Pap., 550C, 133.
3. G. KULLERUD, 1958. Geochim. Cosmochim. Acta, 15, 73.
4. Structure Reports, 20, 128.

IRON SILICON ZIRCONIUM

Fe_4Si_2Zr

Ja.P. JARMOLJUK, L.A. LYSENKO and E.I. GLADYŠEVSKIJ, 1975. Dop. Akad. Nauk Ukr., No. 3, 279-282.

Tetragonal, $P4_2/mnm$, a = 7.004, c = 3.755 Å, c/a = 0.54, D_m = 6.63, Z = 2. Cr, Cu radiations, R = 0.098 for 66 reflexions, photographic data.

Atomic positions

	x	y	z
Zr in 2(b)	0	0	1/2
Fe in 8(i)	0.0920	0.3468	0
Si in 4(g)	0.2201	-0.2201	0

Interatomic distances (Å)

Zr	- 2 Zr	3.76
	- 12 Fe	3.05, 3.14
	- 6 Si	2.77, 2.88
Fe	- 3 Zr	3.05, 3.14
	- 6 Fe	2.50 - 2.58
	- 3 Si	2.34, 2.36
Si	- 3 Zr	2.77, 2.88
	- 6 Fe	2.34, 2.36

The structure (Fig. 1) is characterized by layers containing hexagons, 'squares', and triangles.

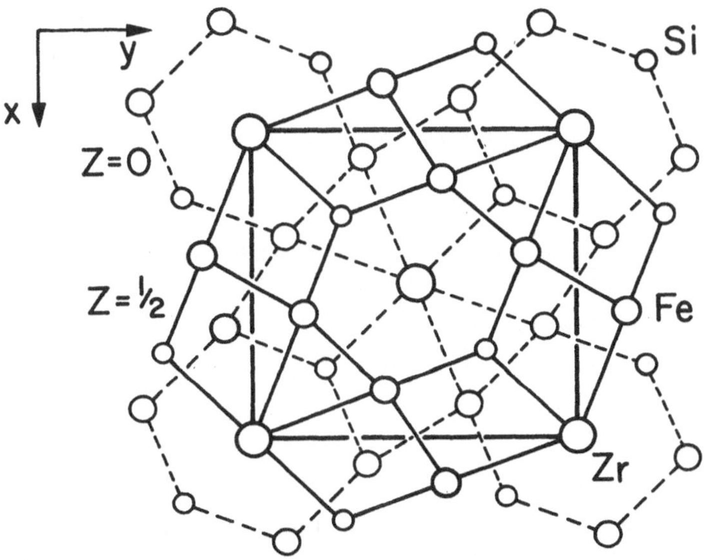

Fig. 1. The tetragonal Fe_4Si_2Zr structure projected onto (001).

IRON TELLURIUM

$Fe_{1.125}Te$

D. FRUCHART, P. CONVERT, P. WOLFERS, R. MADAR, J.P. SENATEUR and R. FRUCHART, 1975. Mater. Res. Bull., 10, 169-174.

Tetragonal, Cu_2Sb (Fe_2As) type (1), P4/nmm, a = 3.8245, c = 6.2818 Å, c/a = 1.64, Z = 2. Neutron powder data, R = 0.033 for 14 reflexions; in agreement with 2, 3; see also 4.

Atomic positions

				x	y	z
2	Fe(1)	in	2(a)	0	0	0
1/4	Fe(2)	in	2(c)	1/2	0	0.561
2	Te	in	2(c)	1/2	0	0.273

1. Strukturbericht, 3, 33.
2. Structure Reports, 18, 197.
3. H. KATSURAKI, 1964. J. Phys. Soc. Japan, 19, 1988.
4. E.F. BERTAUT, P. BURLET and J. CHAPPERT, 1965. J. Solid State Chem., 3, 335.

LANTHANUM PHOSPHORUS

LaP_2

I. H.G. von SCHNERING, W. WICHELHAUS and M. SCHULZE NAHRUP, 1975. Z. anorg. Chem., 412, 193-201.

Monoclinic, h.t.-$LaAs_2$ type (1), Ia, a = 8.883, b = 13.942, c = 8.825 Å, β = 90.15°, D_m = 4.86, Z = 16. Mo radiation, R = 0.063 for 1197 reflexions, diffractometer data. 12 site-sets with isotropic temperature factors; see also 2, 3.

Interatomic distances (Å)

La(1)	- 9 P	2.97 - 3.20	P(3)	- 1 P	2.23
La(2)	- 9 P	2.96 - 3.15	P(4)	- 1 P	2.22
La(3)	- 9 P	2.93 - 3.20	P(5)	- 2 P	2.22
La(4)	- 9 P	2.97 - 3.30	P(6)	- 2 P	2.22, 2.32
P(1)	- 1 P	2.21	P(7)	- 2 P	2.20, 2.32
P(2)	- 2 P	2.21, 2.23	P(8)	- 1 P	2.20

The structure is characterized by P_3^{5-} and P_5^{7-} chains, yielding the formula $La_4P_3P_5$; the P_3 chain has P-P distances of 2.207 and 2.233 Å with angle 107.5°, and the P_5 chain has distances 2.198, 2.315, 2.216, and 2.217 Å and angles of 102.6, 110.3, and 113.6°. La Atoms are coordinated to 9 P at distances ranging from 2.93 to 3.30 Å.

LaP_7

II. W. WICHELHAUS and H.G von SCHNERING, 1975. Naturwissenschaften, 62, 180.

Monoclinic, $P2_1/n$, a = 7.923, b = 11.656, c = 6.989 Å, β = 93.24°, Z = 4, D_x = 3.67. Mo radiation, R = 0.028 for 1271 reflexions, diffractometer data.

Atomic positions

All atoms in 4(e)

	x	y	z
La	0.21745	0.15034	0.10282
P(1)	0.4080	0.4661	0.0968
P(2)	0.7376	0.1144	0.3747
P(3)	0.9531	0.3374	0.2501
P(4)	0.8209	0.0984	0.0848
P(5)	-0.0066	0.5122	0.1552
P(6)	0.5280	0.3105	0.2144
P(7)	0.7516	0.2812	0.0368

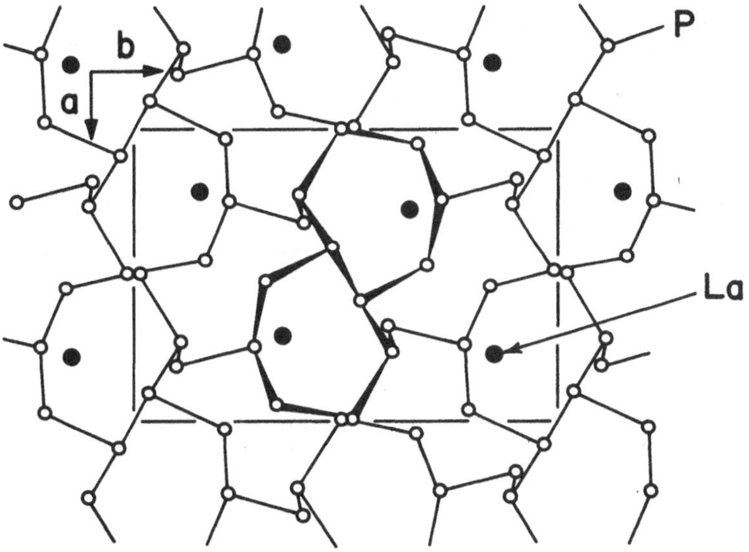

Fig. 1. The monoclinic LaP_7 structure projected onto (001).

La-P distances range from 3.06 to 3.23 Å, and P-P distances from 2.17 to 2.25 Å, with angles from 88 to 114°.

The structure (Fig. 1) is characterized by 7-membered rings, condensed to form 12-membered double rings, and can be formulated to $La^{3+}P_7{}^{3-}$.

1. Structure Reports, 35A, 117.
2. R. SCHMID and H. HAHN, 1970. Z. anorg. Chem., 372, 106.
3. H. HAYAKAWA, K. NOMURA and S. ONO, 1976. J. Less-Common Metals, 44, 327.

LANTHANUM SULPHUR

La_5S_7 ("β'-La_2S_3")

L.F. VEREŠČAGIN, A.A. ELISEEV, G.M. KUZ'MIČEVA, V.V. EVDOKIMOVA, V.I. NOVOKŠONOV and O.P. FIALKOVSKIJ, 1975. Russ. J. Inorg. Chem., 20, 822-824.

Tetragonal, P4bm, a = 11.44, c = 5.28 Å, D_m = 5.0, Z = 2. Mo radiation, photographic data, R = 0.16. This phase is a quenched high pressure form.

Atomic positions

	x	y	z
La(1) in 4(c)	0.668	0.168	0
La(2) in 4(c)	0.136	0.636	0.50
La(3) in 2(a)	0	0	0.172
S(1) in 8(d)	0.713	0.421	0.887
6 S(2) in 8(d)	0.798	0.089	0.508

In this structure La(1) is coordinated to 8 S [3.00, 3.13, 3.19, 3.20 Å], these forming two square pyramids with a common vertex at La(1); La(2) is 6-coordinated, with 2 tetrahedra of S atoms with a common vertex at La(2) [La-S 2.75 - 3.24 Å]; La(3) is 8-coordinated to S atoms which form a square anti-prism [La-S 3.00, 3.09 Å]. S(1) is coordinated to 4 atoms forming a tetrahedron [2 S at 2.76 Å and 2 La at 2.83 and 3.00 Å], and S(2) to 6 atoms, 2 S and 4 La [S-La 3.13 - 3.26 and S-S 2.79 Å].

LEAD SELENIUM URANIUM

$PbSe_5U_2$

M. POTEL, R. BROCHU and J. PADIOU, 1975. Mater. Res. Bull., 10, 205-208.

Monoclinic, $P2_1/c$, a = 8.605, b = 7.788, c = 12.27 Å, β = 90.0°, Z = 4. Mo radiation, R = 0.068 for 1251 reflexions, diffractometer data, corrected for absorption, contrary to 1.

Atomic positions

All atoms in 4(e)*

	x	y	z
Pb	-0.0115	0.4865	0.3203
U(1)	0.5027	0.0215	0.8193
U(2)	0.2523	0.9197	0.4914
Se(1)	0.4522	0.1631	0.5889
Se(2)	0.0407	0.3349	0.0879
Se(3)	0.2762	0.1638	0.3160
Se(4)	0.2805	0.5411	0.4980
Se(5)	0.2668	0.8043	0.7189

* [Misprinted as 4(c)]

Interatomic distances (Å)

Pb - 8 Se 2.95 - 3.50	U(1) - 8 Se 2.92 - 3.11
	U(2) - 7 Se 2.78 - 2.96

The coordination of Pb and U(1) is that of a bi-capped trigonal prism with average Pb-Se = 3.20, U-Se = 2.97; the U(2) coordination is less regular, describable as a pentagonal bipyramid with average U-Se = 2.88 Å. The structure is closely related to those of U_3Se_5 and U_3S_5 (2). [See also HfS_5Y_2 and S_5Sm_2U (3).]

1. Structure Reports, 38A, 170.
2. Ibid., 38A, 143, 148.
3. This volume, p. 76, 104.

LITHIUM SILICON

$Li_{13}Si_4$

U. FRANK, W. MÜLLER and H. SCHÄFER, 1975. Z. Naturforsch., 30B, 10-13.

Orthorhombic, Pbam, a = 7.99, b = 15.21, c = 4.43 Å, D_m = 1.24, Z = 2. Mo radiation, R = 0.058 for 786 reflexions, diffractometer data, contrary to 1.

Atomic positions

	x	y	z
Si(1) in 4(h)	0.4263	0.4321	1/2
Si(2) in 4(g)	0.4165	0.1601	0
Li(1) in 4(g)	0.155	0.026	0
Li(2) in 2(c)	0	1/2	0
Li(3) in 4(g)	0.094	0.195	0
Li(4) in 4(g)	0.270	0.345	0
Li(5) in 4(h)	0.259	0.096	1/2
Li(6) in 4(h)	0.411	0.255	1/2
Li(7) in 4(h)	0.094	0.395	1/2

Interatomic distances (Å)

Li(1) -	6 Li	2.60 - 2.82	Li(6) -	7 Li	2.70 - 3.31
-	5 Si	2.72 - 2.94	-	3 Si	2.64, 2.70
Li(2) -	12 Li	2.79 - 3.28	Li(7) -	8 Li	2.68 - 3.31
-	2 Si	2.53	-	3 Si	2.72, 2.76
Li(3) -	7 Li	2.61 - 2.98	Si(1) -	1 Si	2.38
-	4 Si	2.63, 3.23	-	12 Li	2.69 - 3.23
Li(4) -	8 Li	2.67 - 3.19	Si(2) -	12 Li	2.53 - 3.05
-	4 Si	2.82 - 3.05			
Li(5) -	9 Li	2.60 - 3.29			
-	4 Si	2.69 - 2.90			

The structure contains layers of atoms with isolated pairs of Si atoms (Si-Si = 2.38 Å) in layers with z = 1/2, and isolated Si atoms in the layers with z = 0.

1. Structure Reports, 30A, 66.

LITHIUM TIN

I. U. FRANK, W. MÜLLER and H. SCHÄFER, 1975. Z. Naturforsch., 30B, 1-5

II. Idem, 1975. Ibid., 30B, 6-9.

III. U. FRANK and W. MÜLLER, 1975. Ibid., 30B, 316-322.

I. Li_5Sn_2, rhombohedral, $R\bar{3}m$, a = 4.74, c = 19.83 Å, c/a = 4.18, D_m = 3.56, Z = 3. Mo radiation, R = 0.089 for 270 independent reflexions, diffractometer data, see also 1.

II. Li_7Sn_2, orthorhombic, Ge_2Li_7 type (2), Cmmm, a = 9.80, b = 13.80, c = 4.75 Å, D_m = 2.99, Z = 4. Mo radiation, R = 0.054 for 1055 reflexions, diffractometer data, see also 1.

III. $Li_{13}Sn_5$, trigonal, $P\bar{3}m1$, a = 4.70, c = 17.12 Å, c/a = 3.64, D_m = 3.46, Z = 1. Mo radiation, R = 0.086 for 485 reflexions, diffractometer data, see also 1.

Atomic positions

Li_5Sn_2		x	y	z
Sn	in 6(c)	0	0	0.0727
Li(1)	in 6(c)	0	0	0.352
Li(2)	in 6(c)	0	0	0.210
Li(3)	in 3(b)	0	0	1/2

Li_7Sn_2	x	y	z
Sn(1) in 4(i)	0	0.3127	0
Sn(2) in 4(h)	0.153	0	1/2
Li(1) in 2(a)	0	0	0
Li(2) in 2(c)	1/2	0	1/2
Li(3) in 4(g)	0.359	0	0
Li(4) in 4(j)	0	0.179	1/2
Li(5) in 8(p)	0.187	0.154	0
Li(6) in 8(q)	0.349	0.165	1/2

$Li_{13}Sn_5$			
Sn(1) in 1(a)	0	0	0
Sn(2) in 2(d)	1/3	2/3	0.7764
Sn(3) in 2(d)	1/3	2/3	0.6091
Li(1) in 2(d)	1/3	2/3	0.943
Li(2) in 2(d)	1/3	2/3	0.445
Li(3) in 2(d)	1/3	2/3	0.285
Li(4) in 2(d)	1/3	2/3	0.117
Li(5) in 2(c)	0	0	0.836*
Li(6) in 2(c)†	0	0	0.668
Li(7) in 1(b)	0	0	1/2

* [Not in original; private communication from authors]
† [Misprinted as 2(d)]

Interatomic distances (Å)

Li_5Sn_2

Sn	- 1 Sn	2.88	Li(2)	- 4 Sn	2.72, 2.92
	- 13 Li	2.72 - 3.31*		- 10 Li	2.83 - 3.43
Li(1)	- 6 Sn	2.94, 3.29	Li(3)	- 6 Sn	3.31
	- 8 Li	2.83 - 3.43		- 8 Li	2.87, 2.93

* [Misprinted as 3.11]

Li_7Sn_2

Sn(1)	- 14 Li	2.82 - 3.51	Li(3)	- 4 Sn	2.93, 3.14
				- 10 Li	2.71 - 3.51
Sn(2)	- 1 Sn	3.00			
	- 13 Li	2.81 - 3.40	Li(4)	- 4 Sn	2.89, 3.00
				- 10 Li	2.61 - 3.43
Li(1)	- 4 Sn	2.81			
	- 10 Li	2.80 - 3.51	Li(5)	- 4 Sn	2.86 - 3.20
				- 10 Li	2.71 - 3.66
Li(2)	- 6 Sn	3.40 - 3.51			
	- 8 Li	2.71 - 2.75	Li(6)	- 3 Sn	2.82, 2.98
				- 11 Li	2.61 - 3.47

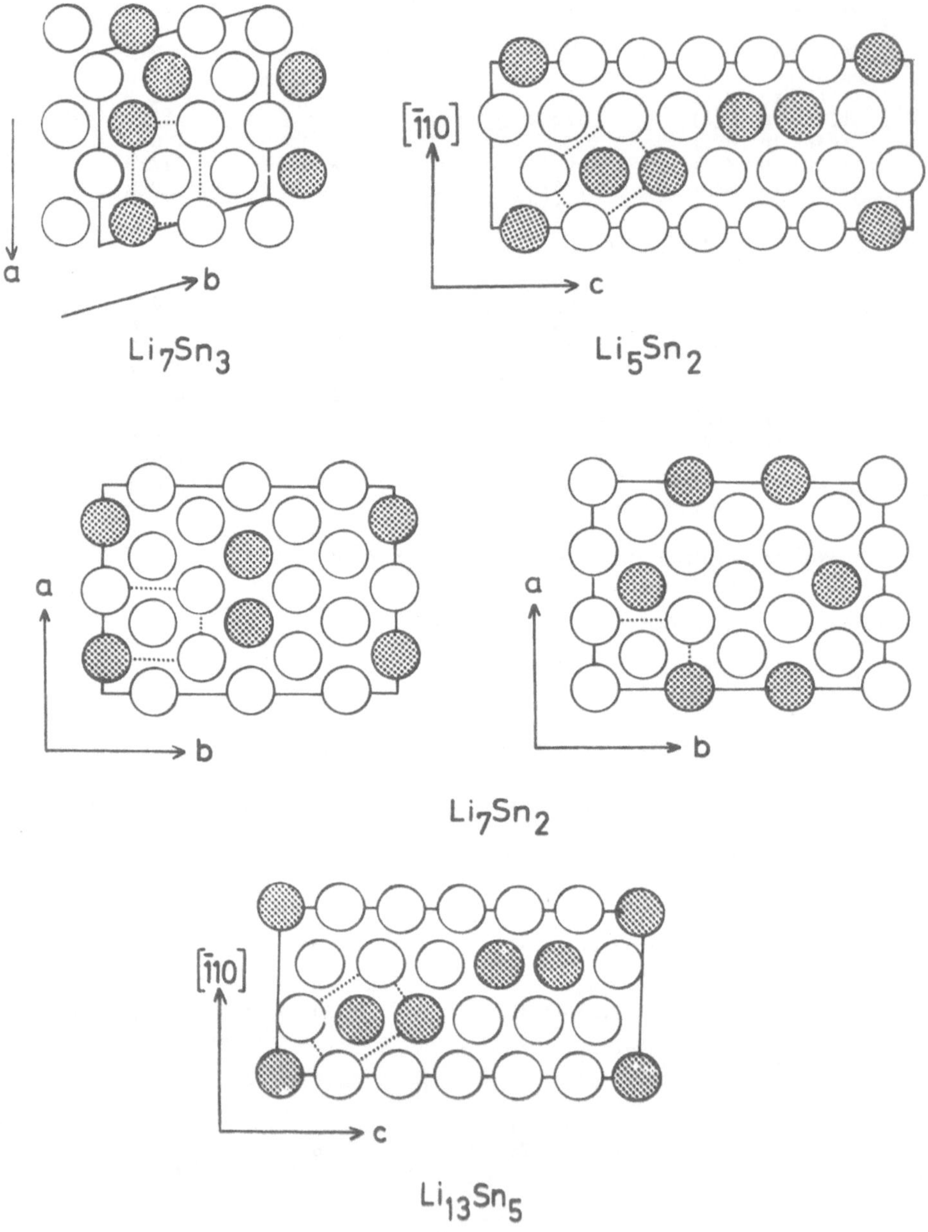

Fig. 1. Projections of Li-Sn structures, showing the atom pairs which occur. The shaded circles represent Sn atoms.

Interatomic distances (Å)

$Li_{13}Sn_5$

Sn(1)	- 14 Li	2.81 - 3.37		Li(3)	- 6 Sn	2.90, 3.27
					- 8 Li	2.76 - 3.41
Sn(2)	- 1 Sn	2.87				
	- 13 Li	2.85 - 3.29		Li(4)	- 6 Sn	3.27, 3.37
					- 8 Li	2.83 - 2.98
Sn(3)	- 13 Li	2.80 - 3.29				
				Li(5)	- 4 Sn	2.81, 2.90
Li(1)	- 4 Sn	2.85, 2.89			- 10 Li	2.83 - 3.41
	- 10 Li	2.90 - 3.35				
				Li(6)	- 6 Sn	2.90, 3.29
Li(2)	- 4 Sn	2.80, 2.87			- 8 Li	2.83 - 3.34
	- 10 Li	2.76 - 3.34				
				Li(7)	- 6 Sn	3.29
					- 8 Li	2.87, 2.88

These structures (Fig. 1) are layer structures based on a b.c.c. packing characterized by 14-fold coordination. Sn-Sn pairs occur frequently and the Sn-Sn distances can be analysed in terms of the valence rule. The Li_5Sn_2 structure (Fig. 1) is characterized by these pairs of atoms (Sn-Sn = 2.88 Å); similar pairs occur in Li_2Si and Li_9Ge_4. Such pairs also occur in alternate layers of the Li_7Sn_2 (Fig. 1) structure (Sn-Sn = 3.00 Å), which resembles the Li_7Pb_2 structure, and in the $Li_{13}Sn_5$ structure.

1. G. GRUBE and E. MEYER, 1934. Z. Elektrochem., 40, 771.
2. This volume, p. 70.

LUTETIUM SULPHUR

Lu_2S_3

K.-J. RANGE and R. LEEB, 1975. Z. Naturforsch., 30B, 637-638.

Rhombohedral, Al_2O_3 (corundum) type (1), $R\bar{3}c$, a = 7.191 Å, α = 55°44', Z = 2 (rhombohedral cell); a = 6.722, c = 18.160 Å, c/a = 2.70, Z = 6 (hexagonal cell), D_x = 6.25. R = 0.08, diffractometer data; see also 2.

Atomic positions (hexagonal axes)

	x	y	z
Lu in 12(c)	0	0	0.3495
S in 18(e)	-0.3044	-0.3044	1/4

Interatomic distances (Å)

Lu	- 6 S	2.64, 2.73	S	- 6 S	3.54 - 4.06.
	- 1 Lu	3.61			

1. Strukturbericht, 1, 240.
2. Structure Reports, 27, 498; 29, 122.

MAGNESIUM PHOSPHORUS

MgP_4

A. EL MASLOUT, M. ZANNE, F. JEANNOT and C. GLEITZER, 1975. J. Solid State Chem., 14, 85-90.

Monoclinic, CdP_4 type (1), $P2_1/c$, a = 5.131, b = 5.090, c = 7.476 Å, β = 81.17°, D_m = 2.56, Z = 2. Co radiation, R = 0.08 for 28 reflexions, powder diffractometer data; the cell was derived from electron diffraction data.

Atomic positions

	x	y	z
Mg in 2(a)	0	0	0
P(1) in 4(e)	0.248	0.294	0.746
P(2) in 4(e)	0.398	0.888	0.402

Interatomic distances (Å)

Mg - 6 P	2.59	P(1) - 2 Mg	2.59, 2.61
		- 2 P	2.18, 2.20
		P(2) - 1 Mg	2.85
		- 3 P	2.20 - 2.85

1. Structure Reports, 20, 59.

MAGNESIUM ZINC

Mg_4Zn_7

Ja. P. JARMOLJUK, P.I. KRIPJAKEVIČ and É.V. MEL'NIK, 1975. Kristallografija, 20, 538-542 [Soviet Physics - Crystallography, 20, 329-331].

Monoclinic, new type, B2/m, a = 25.96, b = 14.28, c = 5.24 Å, γ = 102.5°, D_m = 4.80, Z = 10. Cu radiation, R = 0.088 for 184 hk0 reflexions, photographic data, 24 site-sets.

The Mg_4Zn_7 structure (Fig. 1) shows the distorted tetrahedral packing usual in Frank-Kasper layer structures, resulting from the stacking of two kinds of planar networks: primary (pentagon - triangle) and secondary (3^6) nets. The 14 Zn atoms have C.N. 12, 8 of the Mg atoms have C.N. 16 and the other two C.N. 14 and 15. Interatomic distances range from 2.58 to 2.71 for Zn-Zn, 2.90 to 3.26 for Mg-Zn, and 2.74 to 3.79 Å for Mg-Mg. The structure belongs to a series including Al_3Zr_4, the Laves phases, and R_6X_7, R_2X_3, and $R_{14}X_{23}$ types.

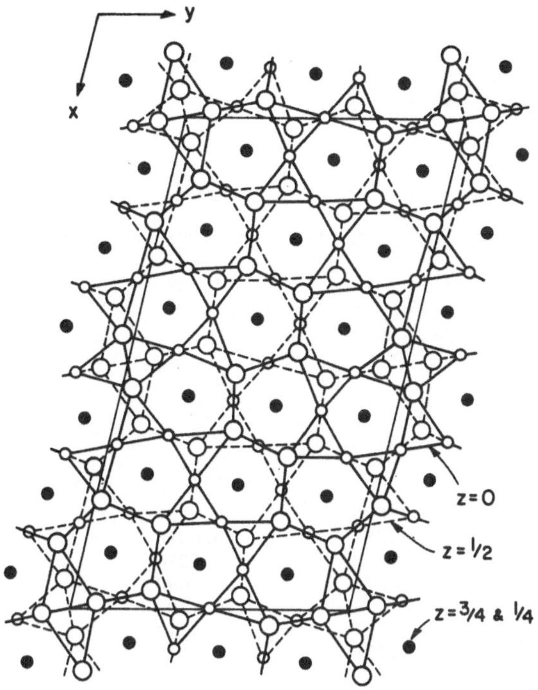

Fig. 1. The monoclinic Mg_4Zn_7 structure projected onto (001).

MANGANESE PHOSPHORUS

$Mn_{1.9}P$

I. R. WÄPPLING, L. HÄGGSTRÖM, T. ERICSSON, S. DEVANARAYANAN, E. KARLSSON, B. CARLSSON and S. RUNDQVIST, 1975. J. Solid State Chem., 13, 258-271.

Hexagonal, P$\bar{6}$2m, a = 6.059, c = 3.440 Å, c/a = 0.57, Z = 3. Mo radiation, R = 0.070 for 58 observed reflexions; photographic data, with allowance for dispersion; see also 1.

Atomic positions

			x	y	z
Mn(1)	in	3(f)	0.2532	0.2532	0*
Mn(2)	in	3(g)	0.5927	0.5927	1/2
P(1)	in	2(c)	1/3	2/3	0
P(2)	in	1(b)	0	0	1/2

* Occupancy = 0.88

[Interatomic distances (Å)]

Mn(1)	- 8 Mn	2.66, 2.76	P(1)	- 9 Mn	2.30, 2.52
	- 4 P	2.30	P(2)	- 9 Mn	2.30, 2.47
Mn(2)	- 10 Mn	2.68, 2.76, 3.18			
	- 5 P	2.47, 2.52			

MnP_4

II. W. JEITSCHKO and P.C. DONOHUE, 1975. Acta Cryst., B31, 574-580.

Monoclinic, C2/c, a = 10.513, b = 5.0944, c = 21.804 Å, β = 94.71°, D_m = 4.09, Z = 16. Mo radiation, R = 0.051 for 1765 reflexions, diffractometer data, with allowance for anomalous dispersion.

Atomic positions

All atoms in 8(f)

	x	y	z
Mn(1)	0.92576	0.1172	0.07002
Mn(2)	0.31551	0.8018	0.18209
P(1)	0.57712	0.3274	0.03698
P(2)	0.76977	0.4090	0.08340
P(3)	0.07973	0.4207	0.07441
P(4)	0.77567	0.8099	0.04916
P(5)	0.47458	0.5175	0.17056
P(6)	0.17128	0.1223	0.21538
P(7)	0.16026	0.5081	0.16917
P(8)	0.46793	0.1058	0.19999

Interatomic distances (Å)

Mn(1)	- 6 P	2.24 - 2.35	P(4)	- 3 P	2.18, 2.25
	- 1 Mn	2.94		- 1 Mn	2.24
Mn(2)	- 6 P	2.22 - 2.41	P(5)	- 2 P	2.20, 2.28
	- 1 Mn	2.94		- 2 Mn	2.24, 2.27
P(1)	- 2 P	2.23	P(6)	- 2 P	2.21, 2.28
	- 2 Mn	2.33, 2.35		- 2 Mn	2.38, 2.41
P(2)	- 2 P	2.18, 2.23	P(7)	- 3 P	2.21, 2.24
	- 2 Mn	2.23, 2.25		- 1 Mn	2.22
P(3)	- 3 P	2.21 - 2.25	P(8)	- 3 P	2.20 - 2.24
	- 1 Mn	2.24		- 1 Mn	2.24

The Mn atoms are in octahedral P coordination, and the P atoms are tetrahedrally coordinated by Mn and P atoms (Fig. 1). Four MnP_6 octahedra share edges to form a linear array of four Mn atoms. The Mn atoms are displaced from the centres of the octahedra toward one another to form pairs with a resulting Mn-Mn bonding distance of 2.941 Å, thus accounting for the diamagnetism. Average Mn-P and P-P distances are 2.282 and 2.225 Å, respectively. MnP_4 can be described as an eight-layer stacking variant of the two-layer CrP_4 structure, although bonding within and between the layers is of equal strength. The continuous Cr-Cr chains in CrP_4 give rise to metallic conductivity, whereas the Mn-Mn bonds in MnP_4 are shorter and stronger, giving rise to a band gap and semi-conductivity; moreover the Mn-Mn bonds are not continuous.

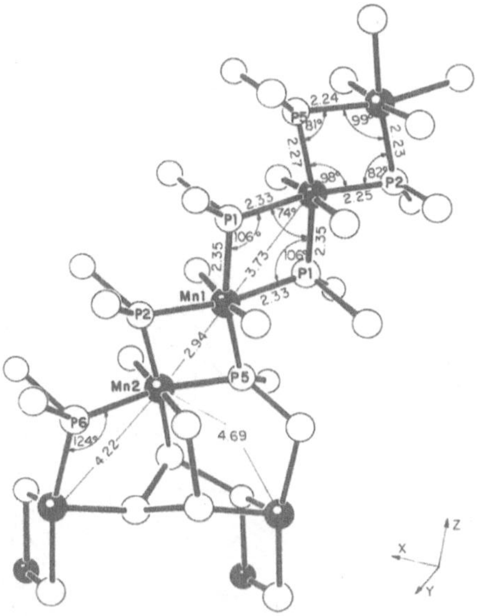

Fig. 1. A perspective view of the immediate environment of the Mn atoms in the monoclinic MnP_4 structure.

1. Structure Reports, 27, 275.

NICKEL SELENIUM

I. E. RØST and K. HAUGSTEN, 1971. Acta Chem. Scand., 25, 3194-3196.

$Ni_{6\pm x}Se_5$, orthorhombic, Cmcm, a = 3.437, b = 11.86, c = 17.06 Å. Mo radiation, R = 0.088 for 360 independent reflexions, photographic date corrected for absorption. The data in this early paper are revised in II below.

II. G. ÅKESSON and E. RØST, 1975. Acta Chem. Scand., A29, 236-240.

Ni_6Se_5, orthorhombic, $Pca2_1$, a = 6.863, b = 17.09, c = 11.821 Å, Z = 8, D_x = 7.16. Mo radiation, R = 0.060 for 1750 observed reflexions, diffractometer data corrected for absorption, 22 site-sets; see also 1-6; the sample used was quenched from 420°C.

Interatomic distances (Å)

Ni(1)	- 5 Se	2.29 - 2.48		Ni(7)	- 5 Se	2.31 - 2.50
	- 3 Ni	2.59 - 2.61			- 2 Ni	2.50, 2.55
Ni(2)	- 5 Se	2.36 - 2.50		Ni(8)	- 5 Se	2.40 - 2.53
	- 2 Ni	2.52, 2.59			- 3 Ni	2.55 - 2.66
Ni(3)	- 5 Se	2.39 - 2.66		Ni(9)	- 5 Se	2.33 - 2.61
	- 4 Ni	2.57 - 2.78			- 2 Ni	2.50, 2.60
Ni(4)	- 4 Se	2.31 - 2.40		Ni(10)	- 4 Se	2.30 - 2.37
	- 3 Ni	2.55 - 2.70			- 3 Ni	2.61 - 2.66
Ni(5)	- 5 Se	2.35 - 2.49		Ni(11)	- 4 Se	2.30 - 2.36
	- 1 Ni	2.55			- 2 Ni	2.62, 2.63
Ni(6)	- 5 Se	2.38 - 2.71		Ni(12)	- 5 Se	2.35 - 2.59
	- 2 Ni	2.66			- 3 Ni	2.52 - 2.78

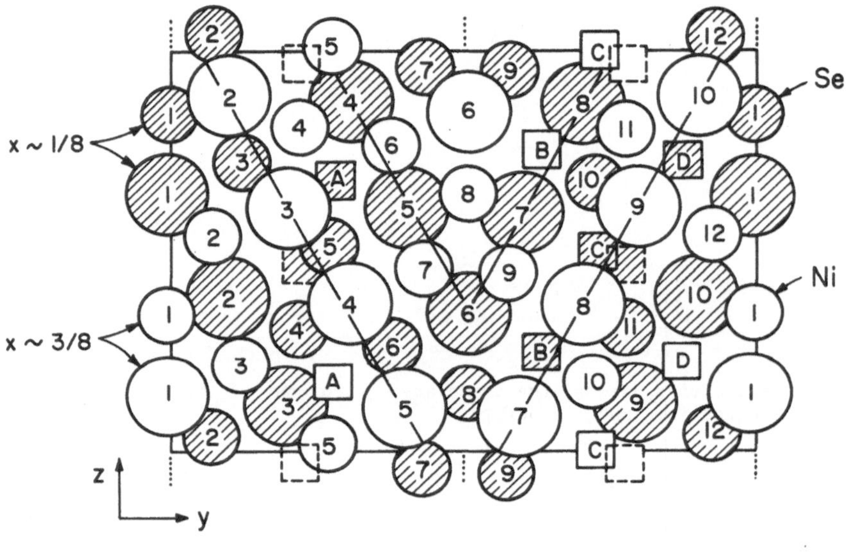

Fig. 1. The orthorhombic, Ni_6Se_5 structure projected onto (100); only atoms with x between ∿ 1/8 and ∿ 3/8 are shown. The zigzag chains are shown; squares mark vacant sites.

An earlier investigation of the subcell was given in I. For the supercell (II) the structure (Fig. 1) has Ni atoms in deformed tetrahedral, octahedral, or square pyramidal positions in the Se lattice. The Se atoms are 11- or 12-coordinated and form zigzag chains. The ordering observed results in a doubling of the a-axis compared with that of the subcell (I); vacant sites (marked by squares in Fig. 1) alternate with filled sites, e.g. A and Ni(10), B and Ni(6), C and Ni(5), and D and Ni(3). The material appears to be isostructural with Ni_7S_4 (7).

1. F. GRØNVOLD, R. MØLLERUD and E. RØST, 1966. Acta Chem. Scand., 20, 1997.
2. Structure Reports, 26, 208.
3. A.L.N. STEVELS, 1969. Diss., Gröningen, Phil. Res. Rep., Suppl., 9.
4. K.L. KOMAREK and K. WESSELY, 1972. Mh. Chem., 103, 923.
5. E. RØST and E. VESTERSJØ, 1968. Acta Chem. Scand., 22, 2118.
6. K. HAUGSTEN and E. RØST, 1968. Ibid., 23, 3599.
7. Structure Reports, 38A, 129.

NICKEL SILICON URANIUM

$Ni_2Si_7U_2$

L.G. AKSELRUD, Ja.P. JARMOLJUK and E.I. GLADYŠEVSKIJ, 1975. Dop. Akad. Nauk Ukr., No. 7, 643-645.

Orthorhombic, Cmmm, a = 3.964, b = 20.85, c = 3.964 Å, D_m = 7.92, Z = 2, D_x= 8.01. Cu radiation, R = 0.078, powder diffractometer data.

Atomic positions

			x	y	z
U	in	4(i)	0	0.170	0
Ni	in	2(a)	0	0	0
*Ni/Si	in	4(j)	0	0.386	1/2
Si(1)	in	4(i)	0	0.437	0†
Si(2)	in	4(j)	0	0.063	1/2
Si(3)	in	4(j)	0	0.722	1/2

* $Ni_{0.5}Si_{0.5}$ † Misprinted as 1/2 in original text.

Interatomic distances (Å)

U	- 10 Si	2.91 - 3.03 [3.00 - 2.98]
	- 4 Ni/Si	3.05
	- 1 Ni	3.56
	- 6 U	3.86, 3.96
Ni	- 8 Si	2.38
Ni/Si	- 5 Si	2.25
	- 4 U	3.05
Si(1)	- 2 Ni/Si	2.25
	- 2 Ni	2.38
	- 5 Si	2.52 - 2.80 [2.63 - 2.80]
	- 2 U	3.03 [2.98]
Si(2)	- 2 Ni/Si	2.25
	- 2 Ni	2.38
	- 5 Si	2.63, 2.80
	- 2 U	2.99
Si(3)	- 1 Ni/Si	2.25
	- 2 Si	2.30
	- 6 U	2.91, 2.99 [3.00]

[The structure closely resembles that of $B_6Cr_2Ni_3$ (1).]

1. Structure Reports, 38A, 44.

NICKEL SULPHUR

(MILLERITE)

β-NiS

I. J.D. GRICE and R.B. FERGUSON, 1974. Canad. Miner., 12, 248-252.

II. V. RAJAMANI and C.T. PREWITT, 1974. Ibid., 12, 253-257.

I. Rhombohedral, NiS (millerite) type (1), R3m, a = 5.6448 Å, α = 116°38', Z = 3 (hexagonal cell, a = 9.6071, c = 3.1434 Å, c/a = 0.33, Z = 9). Mo radiation, R = 0.066 for 155 reflexions, single-crystal diffractometer data, corrected for absorption; crystal had composition $(Ni_{0.981}Fe_{0.016}Co_{0.004})S$, Morbridge Mine, Malarctic, Quebec.

II. Rhombohedral, R3m, a = 5.6519 Å, α = 116.63°, Z = 3 (hexagonal cell, a = 9.6190, c = 3.1499 Å, c/a = 0.33, z = 9). Mo radiation, R = 0.014 for 92 reflexions, single-crystal diffractometer data, corrected for absorption; see also 1, 2. Crystal from Quebec with formula $(Ni_{1.03}Fe_{0.019}Co_{0.004})S$.

Atomic positions (hexagonal axes, all atoms in 9(b))

		x	y	z
I	Ni	0.91225	-0.91225	0.47546
	S	0.11224	-0.11224	0*
II	Ni	-0.08781	0.08781	0.088†
	S	0.1124	-0.1124	0.6164

* Fixed at zero † Value fixed at 0.088, arbitrary origin.

Interatomic distances (Å) (σ = 0.001 to 0.004 Å)

Ni to	1S	2S	2S	2Ni	2Ni
I	2.263	2.264	2.369	2.529	3.143
II	2.260	2.261	2.383	2.540	3.150

The structure is that reported earlier on the basis of powder data (1) but more accurately determined. The results are considered from two points of view. I, on the basis of magnetic susceptibility measurements (by I. Maartense), consider millerite to be diamagnetic (the paramagnetic susceptibility was found to be negligible), thus the reported small value (3) was likely due to traces of Fe. This is consistent with a molecular orbital analysis in which the d electrons from Ni are combined with the p electrons to form 5σ + 3π bonds; the remaining 2 electrons (in $3d_{z^2}$) would be unpaired unless involved in the extra 2 Ni bonds. The dimagnetism is consistent with this interaction. II note that millerite satisfies the general valence rule of Pearson, and accept the reported paramagnetism (3). They assign 6 d electrons of the 3 Ni atoms (in the Ni_3 triangles) to a completely delocalized d band. They also note that a weak interaction is possible parallel to the c axis (Ni-Ni = 3.150 Å). Both I and II agree that there are no localized electrons in anti-bonding molecular orbitals.

1. Strukturbericht, 1, 133; 2, 234.
2. Structure Reports, 27, 293.
3. F. HULLIGER, 1968. Structure and Bonding, 4, 82.

NICKEL ZIRCONIUM

Ni_3Zr

C. BÈCLE, G. DEVELEY, J.-L. GLIMOIS and M. SAILLARD, 1975. C.R. Acad. Sci., Paris, B, 280, 43-44.

Hexagonal, Ni_3Sn type (1), $P6_3/mmc$, a = 5.327, c = 4.321 Å, c/a = 0.81, Z = 2. Cu radiation, R = 0.105 for 24 reflexions, powder diffractometer data; see also 2.

Atomic positions

	x	y	z
Zr in 2(c)	1/3	2/3	1/4
Ni in 6(h)	0.84	-0.84	1/4

Interatomic distances

Zr - 12 Ni	2.66, 2.69	Ni - 8 Ni	2.56 - 2.77
		- 4 Zr	2.66, 2.69

1. Strukturbericht, 5, 7.
2. S.A. POGODIN and V.I. SKOROBOGATOVA, 1957. Izv. Sekt. Fiz.-Khim. Anal. 25, 70.

NIOBIUM SELENIUM

$NbSe_3$

A. MEERSCHAUT and J. ROUXEL, 1975. J. Less-Common Metals, 39, 197-203.

Monoclinic, Pm, a = 10.006, b = 3.478, c = 15.626 Å, β = 109.30°, D_m = 6.406, Z = 6. Cu radiation, R = 0.048 for 433 reflexions, diffractometer data, 24 site-sets. See also 1, 2.

Interatomic distances (Å)

Nb(1) - 8 Se	2.63 - 2.74	Nb(4) - 8 Se	2.55 - 2.82
- 1 Nb	4.31	- 1 Nb	4.25
Nb(2) - 8 Se	2.61 - 2.78	Nb(5) - 8 Se	2.65 - 2.89
- 1 Nb	4.45	- 1 Nb	4.41
Nb(3) - 8 Se	2.62 - 2.94	Nb(6) - 8 Se	2.62 - 2.94
- 1 Nb	4.44		

The structure (Fig. 1) contains isolated chains of [$NbSe_6$] trigonal prisms similar to those in $TaSe_3$ (3) [and $ZrSe_3$ (4)]; the Nb atoms are also coordinated to 2 extra Se atoms opposite the rectangular faces of the prisms, thus forming rumpled slabs parallel to the (010) plane. [These can be considered as less-symmetrical versions of the Re_3B structure (5).]

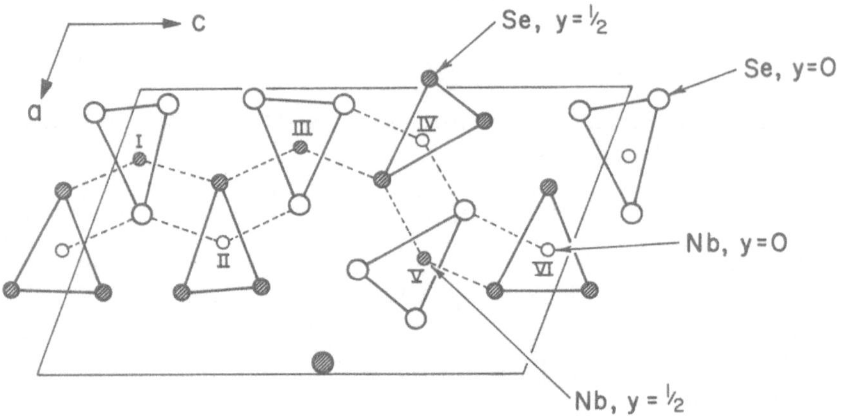

Fig. 1. The (010) projection of the monoclinic $NbSe_3$ structure, showing the rumpled slabs of $[NbSe_6]$ trigonal prisms.

1. E. REVOLINSKY, G.A. SPIERING and D.J. BEERNTSEN, 1965. J. Phys. Chem. Solids, 26, 1029.
2. Structure Reports, 30A, 157.
3. Structure Reports, 31A, 62; L. ASLANOV, Ju.P. SIMANOV, A.V. NOVOSELOVA and Ju. UKRAINSKIJ, 1963. Russ. J. Inorg. Chem., 8, 1381.
4. Structure Reports, 30A, 168.
5. Ibid., 24, 70.

PALLADIUM PLUTONIUM

PdPu

D.T. CROMER, 1975. Acta Cryst., B31, 1760-1761.

Orthorhombic, FeB type (1), Pnma, a = 7.036, b = 4.550, c = 5.663 Å, Z = 4, D_x = 12.65. Mo radiation, R = 0.0495 for 417 reflexions, diffractometer data corrected for absorption, with allowance for anomalous dispersion; see also 2.

Atomic positions

	x	y	z
Pu in 4(c)	0.1795	1/4	0.1413
Pd in 4(c)	0.0447	1/4	0.6510

Interatomic distances (Å)

Pu - 7 Pd	2.93 - 3.06	Pd - 4 Pd	2.92, 3.69
- 8 Pu	3.73 - 3.77	- 7 Pu	2.93 - 3.06

1. Strukturbericht, 2, 7.
2. V.I. KUTAITSEV, N.T. CHEBOTAREV, I.G. LEBEDEV, M.A. ANDRIANOV, V.N. KONEV and T.S. MENSHIKOVA, 1965. Plutonium 1965, 420-449. Ed. A.E. KAY and M.B. WALDRON. Chapman and Hall, London.

PHOSPHORUS SILICON

PSi

T. WADSTEN, 1975. Chem. Scripta, 8, 63-69.

Orthorhombic, $Cmc2_1$, a = 3.5118, b = 20.488, c = 13.607 Å, Z = 24. Cu radiation, R = 0.036 for 450 reflexions, diffractometer data with allowance for anomalous dispersion (δf'). The material is piezoelectric.

Atomic positions

All atoms in 4(a)

	x	y	z
Si(1)	0	0.170695	0.262186
Si(2)	0	0.060176	0.219992
Si(3)	0	0.435930	0.319717
Si(4)	0	0.432388	0.148270
Si(5)	0	0.794999	0.163113
Si(6)	0	0.798080	0.335220
P(1)	0	0.999828	0.362513
P(2)	0	0.230929	0.120265
P(3)	0	0.344322	0.413142
P(4)	0	0.541255	0.117960
P(5)	0	0.689588	0.364728
P(6)	0	0.885976	0.069940

The crystal structure of silicon monophosphide is similar to the SiAs type structure. It is composed of layers of two slightly puckered triangular nets of phosphorus atoms, arranged on top of each other in such a way that the P atoms form somewhat distorted octahedra which are mutually connected by sharing edges. Two silicon atoms are situated in each P_6 octahedron. In this way all silicon atoms possess a tetrahedral environment of one silicon, Si-Si = 2.33-2.34 Å, and three phosphorus atoms atoms, Si-P = 2.26-2.30 Å. The phosphorus atoms have a one-sided environment of three nearest neighbours, all of which are silicon atoms (2.26-2.30 Å). The distances are normal.

PHOSPHORUS SULPHUR

β-P_4S_5

A.M. GRIFFIN and G.M. SHELDRICK, 1975. Acta Cryst., B31, 2738-2740.

Monoclinic, $P2_1/m$, a = 6.389, b = 10.966, c = 6.613 Å, β = 115.65°, Z = 2, D_x = 2.26. Cu radiation, R = 0.029 for 581 reflexions, diffractometer data corrected for absorption.

Atomic positions

	x	y	z
P(1) in 2(e)	0.2851	1/4	0.3388
P(2) in 2(e)	0.8706	1/4	0.7714
P(3) in 4(f)	0.4214	0.1458	0.8575
S(1) in 2(e)	0.6449	1/4	0.4219
S(2) in 4(f)	0.7673	0.1053	0.9179
S(3) in 4(f)	0.2231	0.1053	0.5158

Interatomic distances (Å)

P(1)	- 3 S	2.108, 2.120 (2.117, 2.125)*
P(2)	- 3 S	2.109, 2.129 (2.116, 2.136)*
P(3)	- 1 P	2.284 (2.295)*
	- 2 S	2.105, 2.116 (2.111, 2.116)*

* Distances in brackets are corrected for libration.

S-P-S angles are 97.6, 102.2, 105.5, 107.4, 104.6 and 115.5°.

The molecular geometry is very similar to that of As_4S_5 (1) but with the crystallographic mirror plane at right angles.

1. Structure Reports, 39A, 18.

PHOSPHORUS VANADIUM

P_2V

M. GÖLIN, B. CARLSSON and S. RUNDQVIST, 1975. Acta Chem. Scand., A29, 706-708.

Monoclinic, $As_2Nb(Ge_2Os)$ type (1), C2/m, a = 8.4641, b = 3.1054, c = 7.1698 Å, β = 119.264°, Z = 4. Mo radiation, R = 0.0339 for 516 reflexions, single-crystal diffractometer data, corrected for absorption with allowance for anomalous dispersion; see also 2.

Atomic positions

All atoms in 4(i)

	x	y	z
V	0.84301	0	0.30163
P(1)	0.59962	0	0.39969
P(2)	0.14074	0	0.02893

Interatomic distances (Å)

V	- 8 P	2.43 - 2.49	P(1)	- 9 P	2.70 - 3.24
	- 3 V	2.79, 3.11			
			P(2)	- 5 P	2.21 - 3.11

1. Structure Reports, 24, 144; 29, 18.
2. Ibid., 9, 114; 29, 125.

PLATINUM PLUTONIUM

Pt_3Pu_5

D.T. CROMER and A.C. LARSON, 1975. Acta Cryst., B31, 1758-1759.

Hexagonal, Mn_5Si_3 type (1), $P6_3/mcm$, a = 8.490, c = 6.084 Å, c/a = 0.72, Z = 2, D_x = 15.57. Mo radiation, R = 0.0484 for 167 observed reflexions, diffractometer data corrected for absorption with allowance for anomalous dispersion; see also 2.

Atomic positions

			x	y	z
Pu(1)	in	4(d)	1/3	2/3	0
Pu(2)	in	6(g)	0.2435	0	1/4
Pt	in	6(g)	0.6031	0	1/4

Interatomic distances (Å)

Pu(1)	- 8 Pu	3.042, 3.615	Pt	- 9 Pu	2.944 - 3.309
	- 6 Pt	3.014		- 2 Pt	3.509
Pu(2)	- 10 Pu	3.580 - 3.678			
	- 5 Pt	2.944 - 3.309			

1. Strukturbericht, 4, 24.
2. V.I. KUTAITSEV et al., 1965. Plutonium, Ed. A.E. KAY and M.B. WALDRON. Chapman and Hall, London.

PLUTONIUM RUTHENIUM

Pu_5Ru_3

D.T. CROMER, A.C. LARSON and R.B. ROOF, 1975. Acta Cryst., B31, 1756-1757.

Tetragonal, W_5Si_3 type (1), I4/mcm, a = 10.745, c = 5.719 Å, c/a = 0.53, Z = 4, D_X = 15.07. Mo radiation, R = 0.066 for 265 reflexions, diffractometer data, corrected for absorption and with allowance for anomalous dispersion; see also 2.

Atomic positions

	x	y	z
Pu(1) in 4(b)	0	1/2	1/4
Pu(2) in 16(k)	0.0831	0.2195	0
Ru(1) in 4(a)	0	0	1/4
Ru(2) in 8(h)	0.1596	0.6596	0

Interatomic distances (Å)

Pu(1)	- 10 Pu	2.86, 3.45	Ru(1)	- 2 Ru	2.86
	- 4 Ru	2.82		- 8 Pu	2.90
Pu(2)	- 9 Pu	3.00 - 3.57	Ru(2)	- 2 Ru	3.97
	- 6 Ru	2.84 - 3.25		- 10 Pu	2.82 - 3.25

1. Structure Reports, 19, 276.
2. F.H. ELLINGER, C.C. LAND and K.A. GSCHNEIDNER, 1967. Plutonium Handbook, Vol. 1, p. 219-220. Gordon and Breach, New York.

SAMARIUM SULPHUR URANIUM

S_5Sm_2U

V. TIEN, M. GUITTARD, J. FLAHAUT and N. RODIER, 1975. Mater. Res. Bull., 10, 547-554.

Orthorhombic, U_3S_5 type (1), Pnma, a = 11.80, b = 8.04, c = 7.42 Å, Z = 4. Mo radiation, R = 0.066 for 1294 reflexions, diffractometer data corrected for absorption; see also HfS_5Y_2 and $PbSe_5U_2$ (2).

Atomic positions

Sm and S(1) in 8(d), others in 4(c)

	x	y	z
U	0.50956	1/4	0.9257
Sm	0.67762	0	0.4779
S(1)	0.4021	0.4573	0.1643
S(2)	0.6865	1/4	0.1726
S(3)	0.5009	1/4	0.5402
S(4)	0.2837	1/4	0.8022

1. Structure Reports, 38A, 143, 148.
2. This volume, p. 76, 86.

SELENIUM URANIUM

SULPHUR URANIUM

β-Se_2U
β-S_2U

G.V. ELLERT, G.M. KUZ'MIČEVA, A.A. ELISEEV, V.K. SLOVJANSKIKH and S.P. MOROZOV, 1974. Ž. Neorg. Khim., 19, 2834-2838 [Russ. J. Inorg. Chem., 19, 1548-1551].

Orthorhombic, $PbCl_2$ type (1), Pnam, Z = 4. Mo radiation, R for Se_2U = 0.148, R for S_2U = 0.19, photographic data; see 2 for earlier work. A partial structure for U_3S_5 does not appear to differ from the complete one of 3.

	a(Å)	b(Å)	c(Å)	D_x
β-Se_2U	7.57	8.97	4.26	8.98 [9.09]
β-S_2U	7.06	8.47	4.12	8.02 [8.14]

Atomic positions

Atoms in 4(c)

β-Se_2U	x	y	z
U	0.242	-0.127*	1/4
Se(1)	-0.137*	-0.0714*	1/4
Se(2)	-0.031*	0.335	1/4
β-S_2U			
U	0.250	-0.123*	1/4
S(1)	-0.135*	-0.071*	1/4
S(2)	-0.032*	0.333	1/4

* [Signs changed to give reasonable distances]

Interatomic distances (Å)

β-Se_2U			β-S_2U		
U	- 9 Se	2.86 - 3.25	U	- 9 S	2.72 - 3.13
Se(1)	- 4 U	2.86 - 2.91	S(1)	- 4 U	2.72 - 2.76
	- 6 Se	3.24 - 3.39		- 6 S	3.05 - 3.25
Se(2)	- 5 U	3.07 - 3.25	S(2)	- 5 U	2.89 - 3.13
	- 4 Se	3.40 - 3.43		- 4 S	3.23, 3.25

1. Strukturbericht, 2, 16.
2. Structure Reports, 12, 139, 181; 17, 536; 19, 407; 38A, 147.
3. Ibid., 38A, 148.

SILICON URANIUM

SiU_3

G. KIMMEL and S. NADIV, 1975. Acta Cryst., B31, 1351-1353.

Tetragonal, I4/mcm, a = 6.0328, c = 8.6907 Å, c/a = 1.44, Z = 4. Cu radiation, R = 0.069, diffractometer data; see also 1-5; 1 gave a different arrangement.

Atomic positions

	x	y	z
U(1) in 4(b)	0	1/2	1/4
U(2) in 8(h)	0.226	0.726	0
Si in 4(a)	0	0	1/4

Interatomic distances (Å)

U(1)	- 4 Si	3.02
	- 8 U	2.93, 3.17
U(2)	- 4 Si	3.05
	- 8 U	2.93 - 3.17
Si	- 12 U	3.02 - 3.05

The metallic radii of silicon and uranium in U_3Si are equal (1.53 ± 0.05 Å). The valence of the U and Si ions is approximately the same as that given by 1. The f.c.t. subcell in the U_3Si unit cell is considered to be a distorted $Cu_3\bar{A}u$ type, where the U(2) atoms are not on the (100) plane, but a little outside or inside the (100) faces. There is much less variation in the U-Si distances than in 1.

1. Structure Reports, 11, 284.
2. Ibid., 30A, 169.
3. R.R. BOUCHER, 1971. J. Appl. Cryst., 4. 326.
4. D. de VOOGHT, G. VERNIERS and P. de MEESTER, 1973. J. Nucl. Mater., 46, 303.
5. M. ROSEN, Y. GEFEN, G. KIMMEL and A. HALWANY, 1973. Phil. Mag., 28, 1007.

SODIUM SULPHUR TIN

$Na_4S_8Sn_3$

J.-C. JUMAS, E. PHILIPPOT and M. MAURIN, 1975. J. Solid State Chem., 14, 152-159.

Monoclinic, C2/c, a = 11.247 [also misprinted as 11.427], b = 7.337, c = 17.621 Å, β = 95.27°, D_m = 3.23, Z = 4. Mo radiation, R = 0.043 for 1620 independent reflexions, diffractometer data, corrected for absorption.

Atomic positions

Sn(1) in 4(e), all others in 8(f)

	x	y	z
Sn(1)	0	0.0840	1/4
Sn(2)	0.89857	0.16622	0.03250
S(1)	0.7023	0.0318	0.0159
S(2)	0.1826	0.0703	0.7530
S(3)	-0.0190	0.1510	0.5726
S(4)	0.4798	0.1831	0.6494
Na(1)	0.6399	0.1266	0.1645
Na(2)	0.2625	0.1643	0.1232

Interatomic distances (Å)

Na(1) - 5 S	2.73 - 2.90	Sn(1) - 4 S	2.34 - 2.46
Na(2) - 7 S	2.90 - 3.53	Sn(2) - 5 S	2.40 - 2.60

The structure is characterized by tetrahedrally coordinated Sn(1) atoms and 5-coordinated (trigonal bipyramid) Sn(2) atoms. These polyhedra share corners and edges to form a three-dimensional network of composition $[Sn_3S_8]^{4-}$, with Na^+ cations in the interstices. The trigonal bipyramidal coordination for Sn is unusual.

SODIUM TIN

$Na_{3.7}Sn$

W. MÜLLER and K. VOLK, 1975. Z. Naturforsch., 30B, 494-496.

Orthorhombic, Pnma, a = 9.82, b = 5.57, c = 22.79 Å, Z = 8, D_x = 2.17. Mo radiation, R = 0.063 for 2696 reflexions, diffractometer data corrected for absorption; see also 1-3.

Atomic positions

All atoms in 4(c)

	x	y	z
4 Sn(1)	0.1660	1/4	0.0956
4 Sn(2)	0.3407	1/4	0.8301
3.73 Na(1)	0.3353	1/4	0.3659
3.78 Na(2)	0.4919	1/4	0.1291
3.79 Na(3)	0.1491	1/4	0.2397
3.20 Na(4)	0.9499	1/4	0.9828
3.64 Na(5)	0.6575	1/4	0.9192
3.68 Na(6)	0.9779	1/4	0.8454
4 Na(7)	0.6441	1/4	0.7800
3.76 Na(8)	0.8067	1/4	0.5320

Interatomic distances (Å)

Sn(1) - 11 Na	3.22 - 3.50		Na(5) - 3 Sn	3.30, 3.72	
			- 8 Na	3.18 - 3.81	
Sn(2) - 11 Na	3.16 - 3.72				
			Na(6) - 3 Sn	3.40, 3.58	
Na(1) - 3 Sn	3.36, 3.38		- 8 Na	3.14 - 3.62	
- 8 Na	3.37 - 3.91				
			Na(7) - 2 Sn	3.17, 3.19	
Na(2) - 3 Sn	3.29, 3.37		- 7 Na	3.18 - 3.72	
- 6 Na	3.34 - 3.72				
			Na(8) - 2 Sn	3.16, 3.22	
Na(3) - 3 Sn	3.29, 3.47		- 8 Sn	3.52 - 3.92	
- 7 Na	3.37 - 3.91				
Na(4) - 3 Sn	3.33, 3.50				
- 8 Na	3.06 - 3.92				

The structure is notably less-densely packed than the analogous $Na_{15}Pb_4$; the Sn atoms are isolated and do not occur in pairs.

1. C.H. MATHEWSON, 1905. Z. anorg. Chem., 46, 94.
2. W. HUME-ROTHERY, 1928. J. Chem. Soc., 947.
3. Strukturbericht, 4, 139.

SULPHUR THALLIUM

S_5Tl_2

B. LECLERC and T.S. KABRÉ, 1975. Acta Cryst., B31, 1675-1677.

Orhtorhombic, $P2_12_12_1$, a = 6.660, b = 16.70, c = 6.538 Å, D_m = 5.18, Z = 4. Mo radiation, R = 0.072 for 329 reflexions, photographic data corrected for absorption; see also 1, 2.

Atomic positions

All atoms in 4(a)

	x	y	z
Tl(1)	0.4082	0.5249	0.3440
Tl(2)	0.7768	0.6530	0.0703
S(1)	0.568	0.480	0.924
S(2)	0.645	0.374	0.065
S(3)	0.383	0.298	0.062
S(4)	0.167	0.347	0.247
S(5)	0.257	0.332	0.550

Interatomic distances (Å)

Tl(1) - 6 S	3.04 - 3.49	S(1) - S(2)	2.06	S(2) - S(3)	2.16
		S(3) - S(4)	2.05	S(4) - S(5)	2.08
Tl(2) - 8 S	3.10 - 3.63				
		S-S-S	108, 109, 108°		

This structure is distinguished by the presence of a pentasulphide ion. The two negative charges in this S_5^{2-} ion appear to be localized on the end sulphur atoms. The middle sulphur has no thallium atom at less than 3.57 Å. The thallium is three-coordinated pyramidal; the lone-pair, $6s^2$, of the thallium monovalent ion, which in this case appears to be stereochemically active, occupies the fourth tetrahedral position.

1. E. FRASSON and V. SCATTURIN, 1953. Ann. Chim. Roma, 43, 561.
2. Structure Reports, 20, 192.

SULPHUR THULIUM

S_3Tm_2-III

K.-J. RANGE and R. LEEB, 1975. Z. Naturforsch., 30B, 889-895.

Orthorhombic, Sb_2S_3 type (1), Pnma, a = 10.479, b = 3.805, c = 10.353 Å, D_m = 6.83, Z = 4. R = 0.10 for 150 reflexions, photographic data.

Atomic positions

All atoms in 4(c)

	x	y	z
Tm(1)	0.9888	1/4	0.3127
Tm(2)	0.3078	1/4	0.5049
S(1)	0.047	1/4	0.873
S(2)	0.885	1/4	0.556
S(3)	0.228	1/4	0.193

Interatomic distances (Å)

Tm(1)	- 7 S	2.686 - 2.796
Tm(2)	- 7 S	2.748 - 2.793
S(1)	- 5 Tm	2.731 - 2.807
S(2)	- 5 Tm	2.686 - 2.846
	- 3 S	3.058, 3.282
S(3)	- 5 Tm	2.734 - 2.796
	- 1 S	3.058

1. Strukturbericht, 3, 49; Structure Reports, 21, 35.

SULPHUR TITANIUM

S_2Ti ($S_2Ti_{1.023}$)

I. C. RIEKEL and R. SCHÖLLHORN, 1975. Mater. Res. Bull., 10, 629-634.

II. R.R. CHIANELLI, J.C. SCANLON and A.H. THOMPSON, 1975. Ibid., 10, 1379-1382.

III. A.H. THOMPSON, F.R. GAMBLE and C.R. SYMON, 1975. Mater. Res. Bull., 10, 915-920.

I. Trigonal, CdI_2 type (1), P$\bar{3}$m1, a = 3.409, c = 5.694 Å, c/a = 1.67, Z = 1, D_x = 3.28. Mo radiation, R = 0.043 for 861 reflexions, diffractometer data corrected for absorption and with allowance for anomalous dispersion; chemical analysis gives $Ti_{1.02}S_2$; S atom vacancies are not compatible with the refinement; the excess Ti occurs between layers.

II, III. Trigonal, CdI_2 type (1), P$\bar{3}$m1, a = 3.4073, c = 5.6953, c/a = 1.67, D_m = 3.242, Z = 1. Mo radiation, R = 0.047 for 43 reflexions, diffractometer, corrected for absorption with allowance for dispersion. See also 2.

Atomic positions

	x	y	z
Ti(1) in 1(a)	0	0	0
Ti(2)* in 1(b)	0	0	1/2
S in 2(d)	1/3	2/3	0.24926 (I)
			0.2501 (II, III)

* Occupancy factor = 0.026

Interatomic distances (Å)

	I	II
Ti(1) - 6 S	2.427	2.429
- 2 Ti	2.847	2.846
Ti(2) - 6 S	2.431	-
- 2 Ti	2.847	2.846

	I	II
S - 6 Ti	2.427, 2.431	2.429
S-Ti(1)-S	89.25, 90.75°	89.05°
S-Ti(2)-S	89.02, 90.98°	-

In I, the lattice is nearly undistorted and both Ti atoms are octahedrally coordinated, with the Ti(2) octahedron slightly more distorted. This is contrary to 3 and therefore this distortion cannot be a factor in the $TiS_2 \rightarrow Ti_3S_4$ transition.

In II, III, TiS_2 is shown to be a semi-metal with stoichiometric composition, contrary to 4 who considered it a semiconductor with a 1 e.v. band gap. The structure is very close to the ideal CdI_2 type, in agreement with I.

$S_{12}Ti_8$ (12R)

IV. E. TRONC, R. MORET, J.J. LEGENDRE and M. HUBER, 1975. Acta Cryst., B31, 2800-2804.

Rhombohedral, R$\bar{3}$m, a = 3.420, c = 34.32 Å, c/a = 10.04, Z = 1, D_x = 3.46. Mo radiation, R = 0.028 for 129 reflexions, diffractometer data corrected for absorption; see also 5.

Atomic positions

	x	y	z
Ti(1) in 3(a)	0	0	0
Ti(2) in 3(b)	0	0	1/2
0.1787 Ti(3) in 6(c)	0	0	0.08858
S(1) in 6(c)	0	0	0.37518
S(2) in 6(c)	0	0	0.20858

Interatomic distances (Å)

Ti(1)	- 6 S	2.441	Ti(3)	- 6 S	2.332, 2.544
	- 8 Ti	3.040, 3.420		- 10 Ti	3.040 - 3.420
Ti(2)	- 6 S	2.443	S(1)	- 6 Ti	2.441, 2.544
	- 12 Ti	3.328, 3.420			
			S(2)	- 6 Ti	2.332, 2.443

Every second layer of Ti atoms is partially occupied and these layers are 0.180 Å from the ideal (5); this displacement is not unique to the 12R structure but occurs in other polytypes (6).

1. Strukturbericht, 1, 161, 779.
2. Ibid., 5, 71.
3. J. BENARD and Y. JEANNIN, 1963. Adv. Chem. Series, 39, 191.
4. H.W. MYRON and A.J. FREEMAN, 1974. Phys. Rev., B, 9, 481.
5. Structure Reports, 24, 233; 27, 353.
6. Ibid., 35A, 107.

SULPHUR VANADIUM

I. KAWADA, M. NAKANO-ONODA, M. ISHII, M. SAEKI and M. NAKAHIRA, 1975. J. Solid State Chem., 15, 246-252.

S_4V_3, monoclinic, I2/m, a = 5.831, b = 3.267, c = 11.317 Å, β = 91.78°, D_m = 4.10, Z = 2; see also 1.

S_8V_5, monoclinic, F2/m, a = 11.396, b = 6.645, c = 11.293 Å, β = 91.45°, D_m = 3.90, Z = 4; see also 2, 3.

The space groups are given in terms of the unreduced lattices for a comparison with the basic NiAs structure. In the reduced lattice, S_4V_3 has the space group A2/m, a' = 5.831, b' = 3.267, c' = 12.891 Å, β' = 118.66°, and S_8V_5 has C2/m, a' = 11.396, b' = 6.645, c' = 8.123 Å, β' = 135.98°. Mo radiation, diffractometer data, S_8V_5 corrected for absorption. R for S_4V_3 = 0.052 for 1031 reflexions; R for S_8V_5 = 0.057 for 2362 reflexions.

Atomic positions

S_4V_3		x	y	z
*V(1)	in 2(a)	0	0	0
V(2)	in 4(i)	0.54113	1/2	0.24362
S(1)	in 4(i)	0.33862	0	0.36289
S(2)	in 4(i)	0.66359	0	0.11246
S_8V_5				
V(1)	in 4(a)	0	0	0
V(2)	in 8(g)	1/4	0.28437	1/4
V(3)	in 8(i)	0.51498	0	0.24286
S(1)	in 8(i)	0.16263	0	0.14069
S(2)	in 8(i)	0.16674	1/2	0.11098
S(3)	in 16(j)	0.41329	0.25100	0.12347

* Occupancy = 0.724

Interatomic distances (Å)

	V_3S_4	Av.	V_5S_8	Av.
V(1) - 6 S	2.37 - 2.42	2.406	2.39 - 2.41	2.399
V(2) - 6 S	2.33 - 2.49	2.397	2.31 - 2.45	2.384
V(3) - 6 S	-		2.31 - 2.44	2.390
V - V	2.93, 2.92		2.87, 3.04, 2.91	

The structures are based on the NiAs type (1-3) with metal-deficient and full metal layers alternating between the h.c.p. S layers. The metal-metal bonds form chains in V_3S_4 and clusters in V_5S_8; the inter-chain and inter-cluster distances are 3.3, 3.8 and 3.3, 3.6, 3.8 Å, respectively.

1. Structure Reports, 30A, 165.
2. Ibid., 29, 131,
3. A.B. de Vries, 1972. Dissertation, Groningen.

ADDITIONAL PAPERS

The following reports were omitted from Volume 34A (1969).

ALUMINUM SCANDIUM

$AlSc_2$

S. EYMOND and E. PARTHÉ, 1969. J. Less-Common Metals, 19, 441-443.

Hexagonal, Ni_2In type (1), $P6_3/mmc$, a = 4.888, c = 6.166 Å, D_m not given, Z = 2. Cu radiation, R = 0.10 for powder data. Sc(1) in 2(a); Sc(2) in 2(d); Al in 2(c).

Sc-Sc = 3.08 (x 2), 3.22 (x 6), Al-Sc = 3.22 (x 6), 2.82 (x 3) Å.

1. Structure Reports, 9, 91.

ALUMINUM STRONTIUM

Al_2Sr

G. NAGORSEN, H. POSCH, H. SCHÄFER and A. WEISS, 1969. Z. Naturforsch., 24B, 1191.

Orthorhombic, Imma, a = 7.92, b = 4.84, c = 7.99 Å, D_m = 3.12, Z = 4. R = 0.16 [no details]. Al in 8(i): x = 0.185, z = 0.088; Sr in 4(e): z = 0.700.

ANTIMONIDES

$M_{1+x}Sb$

A. KJEKSHUS and K.P. WALSETH, 1969. Acta Chem. Scand., 23, 2621-2630.

Hexagonal, $P6_3/mmc$, NiAs-type structure for M = Cr, Fe, Co, Ni, Pd, Pt, with additional M atoms in position 2(d) for x > 0.

ANTIMONY IRON

$FeSb_2$

H. HOLSETH and A. KJEKSHUS, 1969. Acta Chem. Scand., 23, 3043-3050.

Orthorhombic, Pnn2, a = 5.833, b = 6.538, c = 3.197 Å, D_m = 8.09, Z = 2. Mo radiation, R = 0.09 for 628 reflexions (film data). Fe in 2(a): z = 0; Sb in 4(c): (0.1881, 0.3565, 0.0097).

The structure is of the marcasite-type, except that it lacks a mirror plane. Fe-Sb = 2.58-2.62 Å.

ARSENIC COPPER LEAD SULPHUR

(SELIGMANNITE)

$CuPbAsS_3$

Y. TAKÉUCHI and N. HAGA, 1969. Z. Kristallogr., 130, 254-260.

Results agree with those in 1 and in a more accurate analysis (2).

1. Structure Reports, 20, 30.
2. Ibid., 35A, 11.

ARSENIC LEAD SULPHUR

(DUFRENOYSITE)

$Pb_2As_2S_5$

B. RIBÁR, C. NICCA and W. NOWACKI, 1969. Z. Kristallogr., 130, 15-40.

Monoclinic, $P2_1$, a = 7.90, b = 25.74, c = 8.37 Å, β = 90°21', Z = 8. Cu radiation, R = 0.091 for 2966 reflexions.

Structure as in 1.

1. Structure Reports, 32A, 227.

BARIUM SULPHUR TANTALUM

$BaTaS_3$

R.A. GARDNER, M. VLASSE and A. WOLD, 1968. Inorg. Chem., 8, 2784-2787.

Hexagonal, $P6_3/mmc$, a = 6.846, c = 5.744 Å, D_m = 5.80, Z = 2. Mo radiation, R = 0.087 for 186 reflexions (films, visual intensities). Ba in 2(d); Ta in 2(a); S in 6(h): x = 0.1686.

Isostructural with $BaVS_3$ (1) and $CsNiCl_3$ (2), as previously reported (3). The structure is based on hexagonal close-packing of BaS_3 layers, with chains of face-shared octahedra along c. Ta ions occupy all S_6 octahedra, Ta-S = 2.46(1) Å. Ba-S = 3.42, 3.47(1) Å.

1. Structure Reports, 34A, 295.
2. Ibid., 19, 332.
3. Ibid., 29, 104.

CHROMIUM GERMANIUM NITROGEN

Cr_3GeN

H. BOLLER, 1969. Mh. Chem., 100, 1471-1476.

Tetragonal, $P\bar{4}2_1m$, a = 5.375, c = 4.012 Å, Z = 2. Cu radiation, R = 0.10.

Atomic positions

			x	y	z
Cr(1)	in	4(e)	0.2041	0.7041	0.0655
Cr(2)	in	2(b)	0	0	1/2
Ge	in	2(c)	0	1/2	0.05480
N	in	2(a)	0	0	0

The structure is closely related to that of filled U_3Si-type and to those of perovskite-type carbides and nitrides. Cr-Cr = 2.60-2.98, Cr-Ge = 2.48-2.73, Cr-N = 1.95, 2.01 Å.

CHROMIUM SELENIUM SULPHUR ZINC

CADMIUM CHROMIUM SELENIUM SULPHUR

$Cr_2ZnS_2Se_2$
$CdCr_2S_{2.4}Se_{1.6}$

E. RIEDEL and E. HORVÁTH, 1969. Z. anorg. Chem., 371, 248-255.

Cubic, Fd3m, a = 10.23, 10.43 Å, Z = 8. Powder data, R = 0.07 and 0.11. Spinels, u = 0.385, 0.390, random arrangement of S and Se.

GERMANIUM LITHIUM

GeLi

E. MENGES, V. HOPF, H. SCHÄFER and A. WEISS, 1969. Z. Naturforsch., 24B, 1351-1352.

Tetragonal, $I4_1/a$, a = 9.75, c = 5.78 Å, D_m = 3.75, Z = 16. R = 0.12 [no details]. Atoms in 16(f), Ge at (0.199, 0.106, 0.269), Li at (0.150, 0.100, 0.820).

IRON PHOSPHORUS

PHOSPHORUS PLATINUM

FeP_2
PtP_2

E. DAHL, 1969. Acta Chem. Scand., 23, 2677-2684.

FeP_2
Orthorhombic, Pnnm, a = 4.973, b = 5.657, c = 2.723 Å, Z = 2. Mo radiation, R = 0.058 for 246 reflexions (films, visual intensities). Marcasite structure, as previously described (1); Fe in 2(a); P in 4(g): x = 0.1683, y = 0.3689. P-P = 2.237(2) Å.

PtP_2
Cubic, Pa3, a = 5.696 Å, Z = 4. Mo radiation, R = 0.16 for 48 reflexions (films, visual intensities). Pyrite structure, as previously described (2); Pt in 4(a); P in 8(c): x = 0.3899. P-P = 2.172(15) Å.

1. Strukturbericht, 3, 22, 310; Structure Reports, 33A, 63.
2. Structure Reports, 24, 47.

NIOBIUM PHOSPHORUS SULPHUR

NIOBIUM PHOSPHORUS SELENIUM

NbPS
NbPSe

P.C. DONOHUE and P.E. BIERSTEDT, 1969. Inorg. Chem., 8, 2690-2694.

Orthorhombic, Immm, a = 3.438, 3.462, b = 11.88, 12.33, c = 4.725, 4.821 Å, D_m = 5.29, not given, Z = 4, for S and Se compounds, respectively. $R(F^2)$ = 0.093 and 0.086 for powder data. Nb in 4(h): y = 0.1232, 0.1188; P in 4(j): z = 0.235, 0.221; S (Se) in 4(g): y = 0.212, 0.211.

Nb has bicapped trigonal prismatic coordination to four P and four S (Se). Nb-Nb = 2.93, 3.44 (3.46), Nb-P = 2.58 (2.63), Nb-S(Se) = 2.59, 2.60 (2.67, 2.71) Å.

PAPERS FROM EARLIER YEARS

Some other papers published in earlier years are reported in this volume. The compounds and abbreviated references are given below

AlGdGe	1974.	Vest. L'vov. Univ., 15, 26
$AsCo_{1-x}Ni_xS$ AsNiS	1974.	J. Solid State Chem., 11, 53
$Bi_{1.3}Ni_9S_8Sb_{0.7}$	1974.	Canad. Miner., 12, 269
Bi_3RbS_5	1974.	Z. Naturforsch., 29B, 438
$CeFe_7Mn_5$	1974.	Vest. L'vov. Univ., 16, 11
$Ce_7Ni_2Si_5$	1974.	Ibid., 15, 17
CeMgSi	1974.	Ibid., 15, 24
CrP	1972.	Acta Chem. Scand., 26, 4188
D_2Ti_3	1974.	Soviet Physics - Crystallography, 19, 468
$Fe_{16}Hf_6Si_7$	1974.	Vest. L'vov. Univ., 15, 21
GaGeLi	1974.	Z. anorg. Chem., 410, 233
Ge_2Li_7	1972.	Z. Naturforsch., 27B, 1157
In_4Te_3	1974.	Soviet Physics - Crystallography, 19, 725
NiS (β)	1974.	Canad. Miner., 12, 248
Ni_6Se_5	1971.	Acta Chem. Scand., 25, 3194
S_2U (β) Se_2U (β)	1974.	Ž. Neorg. Khim., 19, 2834

PAPERS REPORTED ELSEWHERE

Preliminary or duplicate accounts have not been reported here. The compounds and abbreviated references are given below.

Bi_3CuPbS_6 (krupkaite) Bi_5CuPbS_9 (gladite)	Neues Jb. Miner., Mh., 541 (1974)
Bi_4PbS_7 (V phase)	Proc. Japan Acad., 50, 317 (1974) [Same data reported in Structure Reports, 40A, 36]
$Co_{0.25}S_2Ti$	Acta Cryst., B31, 1647 (1975) [Same data reported in Structure Reports, 40A, 54]

TABLE I

This Table lists substances for which structures have been assigned but not refined (usually on the basis of a powder pattern, and assumption of the atomic parameters of the type structure).

Phase	Structure type	Reference
Ag_2CaH	Ag_2CaH	Inorg. Chem., 14, 2910
$AgHg_2Ti$	AuCu	J. Less-Common Metals, 42, 279
$Ag_3Li_8Si_5$	$Ag_3Li_8Si_5$	Z. Naturforsch., 30B, 804
Ag_2MgZn	ClCs	J. Phys. Soc. Japan, 38, 281
$AgPS_2$	$AgPS_2$	Izv. Akad. Nauk SSSR, Neorg. Mater., 11, 1696
$Ag_4P_2S_7$	$Ag_4P_2S_7$	Ibid., 11, 1696
AgS_2Sb (γ)	AgS_2Sb (γ)	Ibid., 11, 1879
$AlAu_2Hf$	$AlCu_2Mn$	J. Less-Common Metals, 39, 341
$AlAu_2Ti$	$AlCu_2Mn$	Ibid., 39, 341
$AlAu_2Ti$	ClCs	Ibid., 39, 341
$(Al,Be)B_{12}$ (β)	AlB_{12} (A form)	Z. anorg. Chem., 414, 203
AlBa	AlBa	J. Less-Common Metals, 39, 1
Al_2Ba	Al_2Ba	Ibid., 39, 1
Al_3Ce	Ni_3Sn	Ibid., 40, 313
AlCuGd	Fe_2P	Ibid., 39, 185
Al_3Dy	Al_3Ho	Ibid., 40, 313
Al_3Dy	$AuCu_3$	Ibid., 40, 313
Al_3Er	Al_3Ho	Ibid., 40, 313
$Al_2Fe_4Ho_3$	Cu_2Mg	Ibid., 40, 285
AlFeHo	$MgZn_2$	Ibid., 40, 285
Al_3FeLu_2	$MgZn_2$	Ibid., 40, 285
$AlFe_2Tb$	$CeNi_3$	Ibid., 40, 207
AlFeY	$MgZn_2$	Ibid., 40, 361
Al_3Gd	$BaPb_3$	Ibid., 40, 313

TABLE I

Phase	Structure type	Reference
AlGdNi	Fe_2P	J. Less-Common Metals, 39, 185
AlGeLi	CaF_2	Z. anorg. Chem., 410, 241
$AlHfPt_2$	$AlHfPt_2$	J. Less-Common Metals, 40, 251
Al_3Ho	Al_3Ho	Ibid., 40, 313
Al_3Ho	$AuCu_3$	Ibid., 40, 313
Al_3Ho	Ni_3Ti	Ibid., 40, 313
Al_3La	Ni_3Sn	Ibid., 40, 313
AlLi	NaTl	Acta Cryst., B31, 1793
AlLiSi	Mg_2Si	Z. anorg. Chem., 410, 241
Al_3Lu	$AuCu_3$	J. Less-Common Metals, 40, 313
$Al_{18}Mg_3Mn_2$	$Al_{18}Cr_2Mg_3$	Z. Metallk., 66, 605
$Al_{0.75}Mo_2S_4$	Al_2MgO_4 (spinel)	J. Solid State Chem., 13, 298
$AlMo_4S_8$	$GaMo_4S_8$	C.R. Acad. Sci., Paris, C, 280, 949
Al_3Nd	Ni_3Sn	J. Less-Common Metals, 40, 313
$AlNi_2V$	$AlCu_2Mn$	Ibid., 39, 341
$AlNi_2V$	ClCs	Ibid., 39, 341
Al_3Pr	Ni_3Sn	Ibid., 40, 313
$AlPt_2Zr$	$AlPt_2Zr$	Ibid., 40, 251
Al_3Sc	$AuCu_3$	Ibid., 40, 313
Al_3Sm	Ni_3Sn	Ibid., 40, 313
Al_2Sr	$CeCu_2$	Ibid., 39, 1
Al_2Sr_3	Ga_2Sr_3	Ibid., 39, 1
Al_3Tb	Al_3Ho	Ibid., 40, 313
Al_3Tb	$BaPb_3$	Ibid., 40, 313
Al_3Y	$BaPb_3$	Ibid., 40, 313
Al_3Yb	$AuCu_3$	Ibid., 40, 313
As_4Ca_2Ga	As_4Ca_2Ga	Rev. Chim. Minér., 12, 433

TABLE I

Phase	Structure type	Reference
$As_5Ca_4Ga_3$	$As_5Ca_4Ga_3$	C.R. Acad. Sci., Paris, C, 281, 457; Rev. Chim. Minér., 12, 433
$As_6Ca_5Ga_2$	$As_6Ca_5Ga_2$	Ibid., C, 281, 457; ibid., 12, 433
$As_7Ca_6Ga_3$	$As_7Ca_6Ga_3$	Rev. Chim. Minér., 12, 433
$As_{16}Ca_{10+x}Si_{12-2x}$	$As_{16}Ca_{10+x}Si_{12-2x}$	Rev. Chim. Minér., 12, 1; Acta Cryst., B31, 445
AsCoRh	Co_2P	Mater. Res. Bull., 10, 603
AsCoS	FeS_2 (pyrite)	Inorg. Chem., 14, 2915
AsCoSe (h.p.)	FeS_2 (pyrite)	Ibid., 14, 2915
As_3Cr_5	β-Sb_3Yb_5	Acta Chem. Scand., A29, 641
$As_4Cu_6S_9$	$As_4Cu_6S_9$	Amer. Min., 60, 998
AsFeMn	$AsFe_2$	Mater. Res. Bull., 10, 603
$As_{0.5}FeMnP_{0.5}$	Fe_2P	Ibid., 10, 603
$AsGa_2InS_3$	SZn (sphalerite)	Ibid., 10, 331
$AsGa_2InSe_3$	SZn (sphalerite)	Ibid., 10, 331
$AsIn_3S_3$	Al_2MgO_4 (spinel)	Ibid., 10, 331
$AsIn_3Se_3$	SZn (sphalerite)	Ibid., 10, 331
As_2Mg_2Zn	La_2O_3 (anti)	Z. Naturforsch., 30B, 132
AsMnRu	Fe_2P	Mater. Res. Bull., 10, 603
As_3V_5 (β)	Bi_3Y_5	Acta Chem. Scand., A29, 641
As_3V_5 (γ)	β-Sb_3Yb_5	Ibid., A29, 641
Au_2CdSi_2	Cr_2Si_2Th	J. Phys. Chem. Solids, 36, 1063
Au_2CeIn	ClCs	Z. Metallk., 66, 110
Au_2CeSi_2	Cr_2Si_2Th	J. Phys. Chem. Solids, 36, 1063
AuCu-III	Au_2CuZn	C.R. Acad. Sci., Paris, C, 280, 559
Au_2DyIn	ClCs	Z. Metallk., 66, 110
Au_2DySi_2	Cr_2Si_2Th	J. Phys. Chem. Solids, 36, 1063
Au_2ErIn	ClCs	Z. Metallk., 66, 110

TABLE I

Phase	Structure type	Reference
$AuErNi_4$	Cu_4MgSn	J. Less-Common Metals, 43, 117
Au_2ErSi_2	Cr_2Si_2Th	J. Phys. Chem. Solids, 36, 1063
Au_2GdIn	ClCs	Z. Metallk., 66, 110
$AuGdNi_4$	Cu_4MgSn	J. Less-Common Metals, 43, 117
Au_2HfIn	$AlCu_2Mn$	Ibid., 39, 341
Au_2HoIn	ClCs	Z. Metallk., 66, 110
$AuHoNi_4$	Cu_4MgSn	J. Less-Common Metals, 43, 117
Au_2HoSi_2	Cr_2Si_2Th	J. Phys. Chem. Solids, 36, 1063
Au_2InLa	ClCs	Z. Metallk., 66, 110
Au_2InNd	ClCs	Ibid., 66, 110
Au_2InPr	ClCs	Ibid., 66, 110
Au_2InSm	ClCs	Ibid., 66, 110
Au_2InTb	ClCs	Ibid., 66, 110
Au_2InTi	$AlCu_2Mn$	J. Less-Common Metals, 39, 341
Au_2InY	ClCs	Z. Metallk., 66, 110
Au_2InYb	ClCs	Ibid., 66, 110
Au_2InZr	$AlCu_2Mn$	J. Less-Common Metals, 39, 341
Au_2InZr	ClCs	Ibid., 39, 341
AuLiSb	AuLiSb	Z. Naturforsch., 30B, 133
$Au_3Li_8Si_5$	$Au_3Li_8Si_5$	Ibid., 30B, 804
$AuLuNi_4$	Cu_4MgSn	J. Less-Common Metals, 43, 117
Au_2NdSi_2	Cr_2Si_2Th	J. Phys. Chem. Solids, 36, 1063
$AuNi_4Dy$	Cu_4MgSn	J. Less-Common Metals, 43, 117
$AuNi_4Sc$	Cu_4MgSn	Ibid., 43, 117
$AuNi_4Tb$	Cu_4MgSn	Ibid., 43, 117
$AuNi_4Tm$	Cu_4MgSn	Ibid., 43, 117
$AuNi_4Y$	Cu_4MgSn	Ibid., 43, 117

TABLE I

Phase	Structure type	Reference
$AuNi_4Yb$	Cu_4MgSn	J. Less-Common Metals, 43, 117
Au_2PrSi_2	Cr_2Si_2Th	J. Phys. Chem. Solids, 36, 1063
Au_2PuSi_2	Cr_2Si_2Th	Ibid., 36, 1063
Au_2Si_2Sn	Cr_2Si_2Th	Ibid., 36, 1063
Au_2Si_2Tb	Cr_2Si_2Th	Ibid., 36, 1063
Au_2Si_2Y	Cr_2Si_2Th	Ibid., 36, 1063
$(C,B)_6Fe_{23}$	C_6Cr_{23}	Russ. J. Phys. Chem., 48, 1525
B_4CoU	B_4CrY	Mh. Chem., 106, 381
B_4CrU	B_4CrY	Ibid., 106, 381
BFe_2	Al_2Cu	J. Phys. Soc. Japan, 39, 1233
B_4FeU	B_4CrY	Mh. Chem., 106, 381
BMgNi	BMgNi	Dop. Akad. Nauk Ukr., 57 (1975)
B_4MnU	B_4CrY	Mh. Chem., 106, 381
B_4MoU	B_4MoTh	Ibid., 106, 381
BNi_4Y	BNi_4Y	Izv. Akad. Nauk SSSR, Neorg. Mater., 11, 1893
BNi_4Y	$BCeCo_4$	Ibid., 11, 1893
B_3Ni_8Y	B_3CeCo_8	Ibid., 11, 1893
$B_6Ni_{12}Y$	$B_6Ni_{12}Y$	Ibid., 11, 1893
B_4ReTm	B_4CrY	Dop. Akad. Nauk Ukr., 58 (1975)
B_4ReU	B_4MoTh	Mh. Chem., 106, 381
B_4UV	B_4CrY	Ibid., 106, 381
B_4UW	B_4MoTh	Ibid., 106, 381
Ba_2CoNbS_6	BaS_3Ti	Mater. Res. Bull., 10, 57
Ba_2CoS_6Ta	$BaNiO_3$	Ibid., 10, 57
Ba_2CoSe_6Ta	$BaNiO_3$	Ibid., 10, 57
Ba_2CrNbS_6	BaS_3Ti	Ibid., 10, 57
Ba_2CrS_5Ta	$BaNiO_3$	Ibid., 10, 57

TABLE I

Phase	Structure type	Reference
Ba_2CrSe_6Ta	$BaNiO_3$	Mater. Res. Bull., 10, 57
Ba_2CuS_6Ta	BaS_3Ti	Ibid., 10, 57
Ba_2FeNbS_6	BaS_3Ti	Ibid., 10, 57
Ba_2FeS_5Ta	$BaNiO_3$	Ibid., 10, 57
Ba_2FeSe_6Ta	$BaNiO_3$	Ibid., 10, 57
Ba_2Hg	Ba_2Hg	J. Less-Common Metals, 39, 271
Ba_2Hg_9	Ba_2Hg_9	Ibid., 39, 271
Ba_2IrS_6Ta	BaS_3Ti	Mater. Res. Bull., 10, 57
Ba_2IrSe_6Ta	$BaNiO_3$	Ibid., 10, 57
Ba_2MnNbS_6	BaS_3Ti	Ibid., 10, 57
$Ba_2Mn_{0.67}Nb_{1.33}S_6$	$BaNiO_3$	Ibid., 10, 57
Ba_2MnS_6Ta	$BaNiO_3$	Ibid., 10, 57
Ba_2NbNiS_6	BaS_3Ti	Ibid., 10, 57
Ba_2NiS_5Ta	$BaNiO_3$	Ibid., 10, 57
Ba_2RhS_6Ta	BaS_3Ti	Ibid., 10, 57
Ba_2S_6TaTi	$BaNiO_3$	Ibid., 10, 57
Ba_2S_6TaV	$BaNiO_3$	Ibid., 10, 57
BaS_3Te	BaS_3Te	C.R. Acad. Sci., Paris, C, 281, 523
Ba_2S_4Te	Ba_2S_4Te	Ibid., C, 281, 523
$Ba_3S_7Te_2$	$Ba_3S_7Te_2$	Ibid., C, 281, 523
$Be_{13}Cm$	$NaZn_{13}$	J. Less-Common Metals, 42, 345
BeP_2	BeP_2	Mater. Res. Bull., 10, 1237
$Be_{13}Pa$	$NaZn_{13}$	J. Less-Common Metals, 42, 345
$BiCe_2$	La_2Sb	Ibid., 41, 329
Bi_2Ce	Bi_2La	Ibid., 41, 329
Bi_3Ce_4	P_4Th_3 (anti)	Ibid., 41, 329
Bi_3Ce_5	Mn_5Si_3	Ibid., 41, 329

TABLE I

Phase	Structure type	Reference
Bi_3Dy_5	Bi_3Y_5	J. Less-Common Metals 41, 329
Bi_3Er_5	Bi_3Y_5	Ibid., 41, 329
BiGd	ClNa	Ibid., 41, 329
Bi_3Gd_5	Bi_3Y_5	Ibid., 41, 329
Bi_3Gd_5	Mn_5Si_3	Ibid., 41, 329
Bi_3Ho_5	Bi_3Y_5	Ibid., 41, 329
Bi_2In (h.p.)	HgSn (γ)	Sov. Phys. Solid State, 17, 1593 Fiz. Tverd. Tela, 17, 2413
Bi_3In (h.p.)	Bi_3Pb (X)	Ibid., 17, 1593; ibid., 17, 2413
Bi_4In (h.p.)	Sn (β)	Ibid., 17, 1593; ibid., 17, 2413
$BiLa_2$	La_2Sb	J. Less-Common Metals, 41, 329
Bi_2La	Bi_2La	Ibid., 41, 329
Bi_3La_4	P_4Th_3 (anti)	Ibid., 41, 329
Bi_3La_5	Mn_5Si_3	Ibid., 41, 329
$BiNd_2$	La_2Sb	Ibid., 31, 329
Bi_2Nd	Bi_2La	Ibid., 41, 329
Bi_3Nd_4	P_4Th_3 (anti)	Ibid., 41, 329
Bi_3Nd_5	Mn_5Si_3	Ibid., 41, 329
$BiPr_2$	La_2Sb	Ibid., 41, 329
Bi_2Pr	Bi_2La	Ibid., 41, 329
Bi_3Pr_4	P_4Th_3 (anti)	Ibid., 41, 329
Bi_3Pr_5	Mn_5Si_3	Ibid., 41, 329
$Bi_{14}S_6Te_{15}$	$Bi_{14}S_6Te_{15}$	Amer. Min., 60, 994
$Bi_{14}S_8Te_{13}$	$Bi_{14}S_8Te_{13}$	Ibid., 60, 994
BiS_2Tl	$FeNaO_2$	Bull. Soc. Chim. Fr., 1037 (1975)
$Bi_2S_5Tl_4$	$Bi_2S_5Tl_4$	Ibid., 1037 (1975)
Bi_3Sm_4	P_4Th_3 (anti)	J. Less-Common Metals, 41, 329

TABLE I

Phase	Structure type	Reference
Bi_3Tb_4	P_4Th_3 (anti)	J. Less-Common Metals, 41, 329
Bi_3Tb_5	Bi_3Y_5	Ibid., 41, 329
Bi_3Tb_5	Mn_5Si_3	Ibid., 41, 329
Bi_3Tm_5	Bi_3Y_5	Ibid., 41, 329
Bi_3Y_5	Bi_3Y_5	Ibid., 41, 329
Bi_3Yb_4	P_4Th_3 (anti)	Ibid., 41, 329
C_6Eu	BaC_6	C.R. Acad. Sci., Paris, C, 281, 929
C_6Sm	BaC_6	Ibid., C, 281, 929
C_6Yb	BaC_6	Ibid., C, 281, 929
$Ca_{10+x}P_{16}Si_{12-2x}$	$As_{16}Ca_{10+x}Si_{12-2x}$	Rev. Chim. Minér., 12, 1; Acta Cryst., B31, 445
$CdHg_2Ti$	AuCu	J. Less-Common Metals, 42, 279
$CdMg_3$	Ni_3Sn	Z. Metallk., 66, 56
$CeFe_2Ge_2$	Cr_2Si_2Th	Solid State Comm., 16, 1005
$CeFe_7Mn_5$	$CeFe_7Mn_5$	Vest. L'vov. Univ. Ser. Chem., 16, 11
$Ce_2Fe_{10}Mn_7$	$Ni_{17}Th_2$	Ibid., 16, 11
$Ce_4Ge_3S_{12}$	$Ge_3La_4S_{12}$	J. Solid State Chem., 13, 65
$Ce_6Ge_{2.5}S_{14}$	$Dy_6Ge_{2.5}S_{14}$	Ibid., 13, 65
Ce_2HfS_5	S_5U_3	Mater. Res. Bull., 9, 1333
Ce_2HfSe_5	S_5U_3	Ibid., 9, 1333
CeMgSi	CeMgSi	Vest. L'vov. Univ. Ser. Chem., 15, 24
CeP_3Si	CeP_3Si (F)	J. Less-Common Metals, 41, 197
CeP_3Si	CeP_3Si (C)	Ibid., 41, 197
Ce_2S_5Si	Ce_2S_5Si	J. Solid State Chem., 13, 65
Ce_2S_5U	S_5U_3	Mater. Res. Bull., 10, 547
$CoFeGe_2Ti_2$	FeSiTi	J. Less-Common Metals, 39, 347
$Co_{0.9}Fe_{0.1}GeTi$	Fe_2P	Ibid., 39, 347
CoGeMn	Co_2P	Inorg. Chem., 14, 1117

TABLE I

Phase	Structure type	Reference
CoGeTi	Fe_2P	J. Less-Commom Metals, 39, 347
CoHfSb	AgAsMg	Ibid., 39, 341
CoPS (h.p.)	FeS_2 (pyrite)	Inorg. Chem., 14, 2915
CoPSe (h.p.)	FeS_2 (pyrite)	Ibid., 14, 2915
CoSSb (h.p.)	AsNiS (ullmannite)	Ibid., 14, 2915
CoSbZr	AgAsMg	J. Less-Commom Metals, 39, 341
Cr_2FeTe_4	Cr_2FeTe_4	J. Solid State Chem., 15, 178
$Cr_2Ga_{0.67}S_4$	Al_2MgO_4 (spinel)	Ibid., 13, 298
Cr_2Gd	Cu_2Mg	Fiz. Metal. Metalloved., 38, 102
CrS_6Sr_2Ta	$BaNiO_3$	Mater. Res. Bull., 10, 57
CrS_2Tl	CrS_2Tl	Naturwissenschaften, 62, 528
Cr_3S_5Tl	Cr_3S_5Tl	Ibid., 62, 528
Cr_5S_8Tl	Cr_5S_8Tl	Ibid., 62, 528
$CrSe_2Tl$	CrS_2Tl	Ibid., 62, 528
$CrSe_3Tl_3$	$CrSe_3Tl_3$	Ibid., 62, 528
Cr_3Se_5Tl	Cr_3S_5Tl	Ibid., 62, 528
$CuGa_5Zr_2$	$AuCu_3$	Russ. Metallurgy (Metally), 4, 151
$Cu_{12.9}Ga_{10.1}Zr_6$	$Mn_{23}Th_6$	Ibid., 4, 151
Cu_2Gd	Cu_2Mg	Fiz. Metal. Metalloved., 38, 102
$CuGe_2Hf$	AsCuSiZr	Dop. Akad. Nauk Ukr., 645 (1975)
$CuGe_2Zr$	AsCuSiZr	Ibid., 645 (1975)
$Cu_{0.59}Hf_{0.41}$	$Ni_{10}Zr_7$	J. Less-Common Metals, 40, 365
$Cu_{0.72}Hf_{0.28}$	$Cu_{0.72}Hf_{0.28}$	Ibid., 40, 365
$Cu_{0.78}Hg_{0.22}$	$Ag_{3.6}Gd$	Ibid., 40, 365
CuLiSi	CuLiSi	Z. Naturforsch., 30B, 804
Cu_4NiY	$CaCu_5$	J. Less-Common Metals, 43, 117
$Cu_{1.970}S$	$Cu_{1.96}S$-III	Izv. Akad. Nauk SSSR, Neorg. Mater., 11, 2129

TABLE I

Phase	Structure type	Reference
$Cu_2S_9Sn_4$	$Cu_2S_9Sn_4$	Bull. Soc. Chim. Fr., 2670 (1975)
Cu_4S_4Sn	Cu_4S_4Sn	Ibid., 2670 (1975)
$Cu_4S_8Sn_3$	$Cu_4S_8Sn_3$	Ibid., 2670 (1975)
$CuSi_2Zr$	AsCuSiZr	Dop. Akad. Nauk Ukr., 645 (1975)
Cu_3Sn	18R	Trans. Jap. Inst. Metals, 16, 581
$Cu_{0.59}Zr_{0.41}$	$Ni_{10}Zr_7$	J. Less-Common Metals, 40, 365
$Cu_{0.72}Zr_{0.28}$	$Cu_{0.72}Zr_{0.28}$	Ibid., 40, 365
$Cu_{0.78}Zr_{0.22}$	$Ag_{3.6}Gd$	Ibid., 40, 365
D_2Ti_3 (γ)	HZr (γ)	Soviet Physics - Crystallography, 19, 468
$DyFe_2Si_2$	Cr_2Si_2Th	Solid State Comm., 16, 1005
$Dy_6S_{14}Si_{2.5}$	$Al_{3.3}Ce_6S_{14}$	J. Solid State Chem., 13, 65
Er_2HfS_5	S_5U_3	Mater. Res. Bull., 9, 1333
Er_2S_5Zr	S_5U_3	Ibid., 9, 1333
Eu_2H_6Ru	H_6RuSr_2	Inorg. Chem., 14, 1866
Eu_3Pd_2	Eu_3Ni_2	J. Less-Common Metals, 40, 263
Eu_3S_4	P_4Th_3	Izv. Akad. Nauk SSSR, Neorg. Mater., 10, 2134
$EuSe_3$	$EuSe_3$	Ibid., 11, 424
Eu_2Se_3	S_3Sc_2	Ibid., 11, 424
Eu_3Se_4	S_4Sc_3	Ibid., 11, 424
Eu_3Se_7	Eu_3Se_7	Ibid., 11, 424
Fe_2GdGe_2	Cr_2Si_2Th	Solid State Comm., 16, 1005
Fe_2GdSi_2	Cr_2Si_2Th	Ibid., 16, 1005
Fe_2Ge_2Nd	Cr_2Si_2Th	Ibid., 16, 1005
FeGeTi	FeSiTi	J. Less-Common Metals, 39, 347
Fe_2LaSi_2	Cr_2Si_2Th	Solid State Comm., 16, 1005
FeMnP	Co_2P	Mater. Res. Bull., 10, 603

TABLE I

Phase	Structure type	Reference
$FeNbSi_2$	$FeNbSi_2$	C.R. Acad. Sci., Paris, C, 281, 831
Fe_2NdSi_2	Cr_2Si_2Th	Solid State Comm., 16, 1005
Fe_2PrSi_2	Cr_2Si_2Th	Ibid., 16, 1005
FeS_6Sr_2Ta	$BaNiO_3$	Mater. Res. Bull., 10, 57
Fe_2Si_2Sm	Cr_2Si_2Th	Solid State Comm., 16, 1005
$FeSi_2Ti$	$FeNbSi_2$	C.R. Acad. Sci., Paris, C, 281, 831
$Ga_{23}Ge_{23}K_8$	$Ge_{46}K_8$	Z. Naturforsch., 30B, 805
GaGeLi	Mg_2Si	Z. anorg. Chem., 410, 241
$Ga_5Ge_3Li_8$	$Ga_5Ge_3Li_8$	Ibid., 410, 241
Ga_2InPS_3	SZn (sphalerite)	Mater. Res. Bull., 10, 331
Ga_2InPSe_3	SZn (sphalerite)	Ibid., 10, 331
$Ga_{23}K_8Sn_{23}$	$Ge_{46}K_8$	Z. Naturforsch, 30B, 805
GaLiSi	Mg_2Si	Z. anorg. Chem., 410, 241
Ga_2MgTe_4	Ga_2MgTe_4	Z. Naturforsch., 30B, 179
$Ga_{0.67}Mo_2S_4$	Al_2MgO_4 (spinel)	J. Solid State Chem., 13, 298
$GaMo_4Se_8$	$GaMo_4S_8$	C.R. Acad. Sci., Paris, C, 280, 949
Ga_4Ni_3Pd	FeSi	J. Less-Common Metals, 42, 69
$Ga_9Ni_{12}Pd$	$Pd_{13}Tl_9$	Z. Metallk., 66, 542
Ga_4Ni_3Pt	FeSi	J. Less-Commom Metals, 42, 69
$GaNi_2Zr$	$AlCu_2Mn$	Russ. Metallurgy (Metally), 4, 151
Ga_5NiZr_2	$AuCu_3$	Ibid., 4, 151
GaS_2Tl	GaS_2Tl	J. Crystal Growth, 29, 121
$Ga_{0.5}S_4V_2$	Al_2MgO_4 (spinel)	J. Solid State Chem., 13, 298
GaSe	9R	Mater. Res. Bull., 10, 577
GaSe	12R	Ibid., 10, 577
GaSe	15R	Ibid., 10, 577
$Gd_6Ge_{2.5}S_{14}$	$Dy_6Ge_{2.5}S_{14}$	J. Solid State Chem., 13, 65

TABLE I

Phase	Structure type	Reference
$Gd_4Ge_3S_{12}$	$Ge_3La_4S_{12}$	J. Solid State Chem., 13, 65
GdInNi	Fe_2P	J. Less-Common Metals, 39, 185
GdInPd	Fe_2P	Ibid., 39, 185
Gd_2Ni_{17}	Th_2Zn_{17}	Ibid., 41, 165
$Gd_6S_{14}Si_{2.5}$	$Al_{3.3}Ce_6S_{14}$	J. Solid State Chem., 13, 65
Gd_2S_5U	S_5U_3	Mater. Res. Bull., 10, 547
GdS_3Yb	P_4Th_3	Z. Naturforsch., 30B, 896
GdS_3Yb	S_3U_2	Ibid., 30B, 896
Gd_2Se_5Zr	S_5U_3	Mater. Res. Bull., 9, 1333
$GdZn_2$	Cu_2Mg	Fiz. Metal. Metalloved., 38, 102
$Ge_{2.5}Ho_6S_{14}$	$Dy_6Ge_{2.5}S_{14}$	J. Solid State Chem., 13, 65
$Ge_{30}In_{16}K_8$	$Ge_{46}K_8$	Z. Naturforsch., 30B, 805
GeInLi	CaF_2	Z. anorg. Chem., 410, 241
GeMnNi	Co_2P	Inorg. Chem., 14, 1117
$Ge_{2.5}Nd_6S_{14}$	$Dy_6Ge_{2.5}S_{14}$	J. Solid State Chem., 13, 65
$Ge_3Nd_4S_{12}$	$Ge_3La_4S_{12}$	Ibid., 13, 65
GePr	BFe	Izv. Akad. Nauk SSSR, Neorg. Mater., 11, 160
$GePr_3$	$GePr_3$	Ibid., 11, 160
$Ge_{1.6}Pr_{0.4}$ (α)	$GdSi_2$ (α)	Ibid., 11, 160
$Ge_{1.6}Pr_{0.4}$ (β)	Si_2Th (α)	Ibid., 11, 160
Ge_3Pr_4	P_4Th_3	Ibid., 11, 160
Ge_3Pr_5	Mn_5Si_3	Ibid., 11, 160
Ge_4Pr_5	Ge_4Sm_5	Ibid., 11, 160
$Ge_{2.5}Pr_6S_{14}$	$Dy_6Ge_{2.5}S_{14}$	J. Solid State Chem., 13, 65
$Ge_3Pr_4S_{12}$	$Ge_3La_4S_{12}$	Ibid., 13, 65
$Ge_{2.5}S_{14}Sm_6$	$Dy_6Ge_{2.5}S_{14}$	Ibid., 13, 65
$Ge_{2.5}S_{14}Tb_6$	$Dy_6Ge_{2.5}S_{14}$	Ibid., 13, 65

TABLE I

Phase	Structure type	Reference
GeS_3Tl_2	GeS_3Tl_2	J. Appl. Cryst., 8, 391
$Ge_{2.5}S_{14}Y_6$	$Dy_6Ge_{2.5}S_{14}$	J. Solid State Chem., 13, 65
$Ge_3Sm_4S_{12}$	$Ge_3La_4S_{12}$	Ibid., 13, 65
$HfHo_2S_5$	S_5U_3	Mater. Res. Bull., 9, 1333
$HfInPt_2$	ClCs	J. Less-Common Metals, 39, 341
$HfLa_2Se_5$	S_5U_3	Mater. Res. Bull., 9, 1333
$HfLiS_2$	$NaTiS_2$	Ibid., 10, 363
HfPdSn	AgAsMg	J. Less-Common Metals, 39, 341
HfPtSn	AgAsMg	Ibid., 39, 341
HfS_5Sm_2	S_5U_3	Mater. Res. Bull., 9, 1333
HfS_5Y_2	S_5U_3	Ibid., 9, 1333
Ho_2S_3	S_3U_2	Z. Naturforsch., 30B, 889
Ho_2S_5Zr	S_5U_3	Mater. Res. Bull., 9, 1333
$In_5Li_8Si_3$	$In_5Li_8Si_3$	Z. anorg. Chem., 410, 241
InLiSn	CaF_2	Ibid., 410, 241
In_3PS_3	Al_2MgO_4 (spinel)	Mater. Res. Bull., 10, 331
In_3PSe_3	SZn (sphalerite)	Ibid., 10, 331
$InPt_2Ti$	ClCs	J. Less-Common Metals, 39, 341
$InPt_2Zr$	$AlCu_2Mn$	Ibid., 39, 341
$InPt_2Zr$	ClCs	Ibid., 39, 341
LaP_3Si	CeP_3Si (F)	Ibid., 41, 197
LaP_6Si_2	LaP_6Si_2	Ibid., 41, 197
La_3S_4	P_4Th_3	J. Solid State Chem., 15, 203
La_2S_5Si	Ce_2S_5Si	Ibid., 13, 65
LaS_3Ti	$CrLaS_3$	Ibid., 12, 80
La_2S_5U	S_5U_3	Mater. Res. Bull., 10, 547
La_2S_5Zr	S_5U_3	Ibid., 9, 1333

TABLE I

Phase	Structure type	Reference
$LaSe_2$	$AsFe_2$	C.R. Acad. Sci., Paris, C, 280, 1021
La_2Se_5Zr	S_5U_3	Mater. Res. Bull., 9, 1333
LiPt	LiPt	J. Less-Common Metals, 43, 143
Li_2Pt	AlB_2	Ibid., 43, 143
LiS_2Ti	$CrLiS_2$	Mater. Res. Bull., 10, 363
LiS_2Zr	$NaTiS_2$	Ibid., 10, 363
Lu_2S_3	S_3U_2	Z. Naturforsch., 30B, 889
Lu_3S_2 (ε)	S_3Yb_2 (ε)	Russ. J. Inorg. Chem., 20, 973
$Mg_7P_8Zn_5$	La_2O_3 (anti)	Z. Naturforsch., 30B, 132
Mn_2Rh_3Sb	AuCu	Nippon Kinzogu Gakk., 39, 1065
$MnSi_2Ti$	$FeNbSi_2$	C.R. Acad. Sci., Paris, C, 281, 831
Nb_4Se_{18}	Nb_4Se_{18}	Ibid., C, 281, 297
Nb_3Si	PTi_3	J. Less-Common Metals, 43, 105
Nd_2S_5Si	Ce_2S_5Si	J. Solid State Chem., 13, 65
Nd_2S_5U	S_5U_3	Mater. Res. Bull., 10, 547
NiP	AsNi	C.R. Acad. Sci., Paris, C, 281, 777
NiPS	FeS_2 (pyrite)	Inorg. Chem., 14, 2912
$NiS_{0.6}Te_{0.4}$	AsNi	J. Less-Common Metals, 40, 145
Ni_2SbTi	$AlCu_2Mn$	Ibid., 39, 341
Ni_3Si (β_2)	Ni_3Si (β_2)	Z. Metallk., 66, 521
Ni_3Si (β_3)	Ni_3Si (β_3)	Ibid., 66, 521
Os_2Pu	Cu_2Mg	J. Appl. Cryst., 8, 687
Os_2Pu	$MgZn_2$	Ibid., 8, 687
P_2Pr	As_2Nd	J. Less-Common Metals, 41, 197
P_5Pr	P_5Pr	Ibid., 41, 197
P_3PrSi	CeP_3Si (C)	Ibid., 41, 197
P_4Se_4	P_4Se_4	Z. anorg. Chem., 416, 181

TABLE I

Phase	Structure type	Reference
PV_2	Co_2P	Acta Chem. Scand., A29, 641
PoY	ClNa	Izv. Akad. Nauk SSSR, Neorg. Mater., 11, 1230
Pr_3S_4	P_4Th_3	J. Solid State Chem., 15, 203
Pr_2S_5Si	Ce_2S_5Si	Ibid., 13, 65
Pr_2S_5U	S_5U_3	Mater. Res. Bull., 10, 547
PtSnTi	AgAsMg	J. Less-Common Metals, 39, 341
PtSnZr	AgAsMg	Ibid., 39, 341
PtYb	BFe	Ibid., 43, 205
$PtYb_2$	Cl_2Pb	Ibid., 43, 205
Pt_2Yb	Cu_2Mg	Ibid., 43, 205
Pt_2Yb_5	C_2Mn_5	Ibid., 43, 205
Pt_3Yb	$AuCu_3$	Ibid., 43, 205
Pt_3Yb_5	Mn_5Si_3	Ibid., 43, 205
Pt_4Yb_3	Pd_4Pu_3	Ibid., 43, 205
Pt_4Yb_5	Ge_4Sm_5	Ibid., 43, 205
Ru_2Sn_3 (l.t.)	Ge_3Ru_2	Ibid., 40, 139
S_8Sb_5Tl	S_8Sb_5Tl	Miner. (Tschermaks) Petrogr. Mitt., 22, 200
$S_{14}Si_{2.5}Tb_6$	$Al_{3.3}Ce_6S_{14}$	J. Solid State Chem., 13, 65
$S_{14}Si_{2.5}Y_6$	$Al_{3.3}Ce_6S_{14}$	Ibid., 13, 65
S_5Sm_2Zr	S_5U_3	Mater. Res. Bull., 9, 1333
$STi_{0.787}V_{0.213}$	AsNi	J. Solid State Chem., 13, 307
$STi_{0.545}V_{0.455}$	MnP	Ibid., 13, 307
S_3Tm_2 (ε)	S_3Yb_2 (ε)	Russ. J. Inorg. Chem., 20, 973
S_3Y_2	S_3U_2	Z. Naturforsch., 30B, 889
S_5Y_2Zr	S_5U_3	Mater. Res. Bull., 9, 1333
S_4Yb_3	S_4Yb_3	Russ. J. Inorg. Chem., 20, 657

TABLE I

Phase	Structure type	Reference
S_3Yb_2 (ε)	S_3Yb_2 (ε)	Russ. J. Inorg. Chem., 20, 657, 973
S_3Yb_2 (γ)	P_4Th_3	Ibid., 20, 657
S_3Yb_2 (δ)	S_3Yb_2 (δ)	Ibid., 20, 657
S_3Yb_2 (θ)	S_3Yb_2 (θ)	Ibid., 20, 657
Se_5Sm_2Zr	S_5U_3	Mater. Res. Bull., 9, 1333
Se_5Tb_2Zr	S_5U_3	Ibid., 9, 1333
Se_5U_3	S_5U_3	Russ. J. Inorg. Chem., 20, 120
Se_4U_3	Th_3P_4	Ibid., 20, 120
Se_3U_2	S_3Sb_2	Ibid., 20, 120
SnTl	AuCu	Z. Metallk., 66, 504
Te_3U_2	P_4Th_3	Russ. J. Inorg. Chem., 20, 120
Te_5U_3	$Te_{12}Th_7$	Ibid., 20, 120

STRUCTURE REPORTS

SECTION II

INORGANIC COMPOUNDS

Edited by

J. Trotter

(University of British Columbia)

with the assistance of

J. M. Bree

J. C. Speakman

ARRANGEMENT

To find particular inorganic compounds the subject index or formula index should be used. The general arrangement is: elements, boron hydrides, carbonyls, phosphorus-nitrogen and sulphur-nitrogen compounds, halides, cyanides, oxides, double oxides (including titanates, vanadates, niobates, chromates, molybdates, tungstates, manganates , uranates), hydroxides, borates, carbonates, nitrates, phosphates, arsenates, sulphates, perchlorates, iodates, silicates, electron-diffraction studies. Only complete structure analyses are described; compounds for which only lattice parameters are determined have not been reported, and those which have been described only in preliminary communications and for which details will appear at a later date are tabulated.

POTASSIUM GERMYL

RUBIDIUM GERMYL

CAESIUM GERMYL

$KGeH_3$
$RbGeH_3$
$CsGeH_3$

G. THIRASE, E. WEISS, H.J. HENNIG and H. LECHERT, 1975. Z. anorg. Chem., 417, 221-228.

K and Rb compounds
Cubic, [Fm3m], a = 7.245, 7.518 Å, [Z = 4]. Mo radiation, R = 0.053 and 0.082 for 48 and 41 reflexions, respectively. NaCl-type structure.

Cs compound
Orthorhombic, Cmcm, a = 5.168, b = 14.435, c = 5.966 Å, Z = 4. Mo radiation, R = 0.073 for 392 reflexions. TlI-type structure (1), y(Cs) = 0.3777, y(GeH_3) = 0.0993.

1. Strukturbericht, 4, 6.

CYCLOHEXA(CYANOBORANE)

$(BH_2CN)_6$

A.T. McPHAIL and D.L. McFADDEN, 1975. J. Chem. Soc., Dalton, 1784-1786.

Triclinic, P$\bar{1}$, a = 10.04, b = 8.78, c = 4.90 Å, α = 110.2, β = 93.9, γ = 70.5°, D_m = 1.00, Z = 1. Mo radiation, R = 0.055 for 631 reflexions.

The structure contains a chair-like macrocyclic 18-membered ring, disordered over two orientations; B-C,N = 1.56(1), C,N-C,N = 1.14(1), B-H = 1.14(5) Å.

2,2'-BIS[1-THIA-closo-DECABORANE(8)]

$(B_9H_8S)_2$

W.R. PRETZER, T.K. HILTY and R.W. RUDOLPH, 1975. Inorg. Chem., 14, 2459-2462.

Monoclinic, $P2_1/n$, a = 12.184, b = 9.777, c = 6.601 Å, β = 95.72°, D_m = 1.166, Z = 2. Cu radiation, R = 0.044 for 1140 reflexions.

The centrosymmetrical molecule (Fig. 1) contains two bicapped square-antiprism B_9H_8S fragments joined by a B-B bond (1.678(5) Å). Other B-B distances are 1.689-1.940, B-S = 1.918-1.930, and B-H = 1.08-1.12 Å.

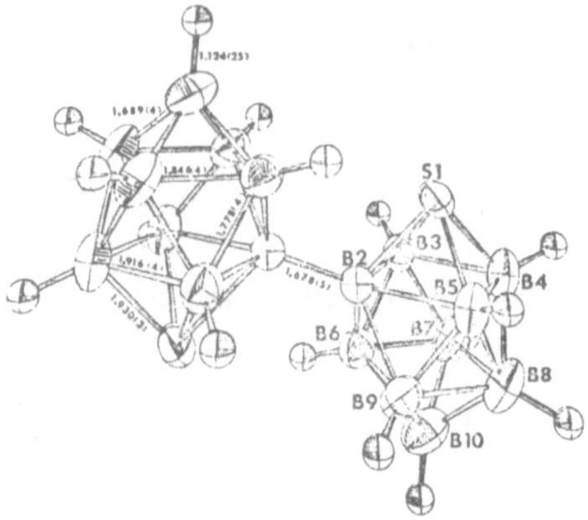

Fig. 1. The $(B_9H_8S)_2$ molecule.

RUBIDIUM DODECAHYDROCLOSODODECABORATE

$Rb_2B_{12}H_{12}$

S.I. USPENSKAJA, K.A. SOLNCEV and N.T. KUZNECOV, 1975. Ž. Strukt. Khim., 16, 482-484 [J. Struct. Chem., 16, 450-451].

Cubic, Fm3, a = 10.85 Å, D_m = 1.63, Z = 4. Cu radiation, R = 0.115 for 232 reflexions. Rb in 8(c); B in 48(h): y = 0.133, z = 0.075; H in 48(h): y = 0.243, z = 0.105.

The structure contains Rb^+ ions and $B_{12}H_{12}^-$ ions with slightly distorted icosahedral symmetry; B-B = 1.65, 1.78, mean B-H = 1.23, mean Rb-H = 3.10 Å.

ALUMINUM BOROHYDRIDE MONOAMMINE

$Al(BH_4)_3.NH_3$

E.B. LOBKOVSKIJ, V.B. POLJAKOVA, S.P. ŠILKIN and K.N. SEMENENKO, 1975. Ž. Strukt. Khim., 16, 77-84 [J. Struct. Chem., 16, 66-72].

Further refinement gives results essentially identical with those in 1.

1. Structure Reports, 40A, 122.

TRICARBONYLMANGANESE TRIDECAHYDROOCTABORATE

$B_8H_{13}Mn(CO)_3$

J.C. CALABRESE, M.B. FISCHER, D.F. GAINES and J.W. LOTT, 1974. J. Amer. Chem. Soc., 96, 6318-6323.

Orthorhombic, Pmcn, a = 11.549, b = 5.506, c = 19.260 Å, D_m = 1.289, Z = 4. Mo radiation, R = 0.033 for 928 reflexions.

The molecule has crystallographic C_s symmetry, and contains a tridentate B_8 ligand bound to Mn by three M-H-B bridge bonds (Fig. 1). Mn-C = 1.815, Mn-B = 2.26, Mn-H = 1.68, 1.81 Å.

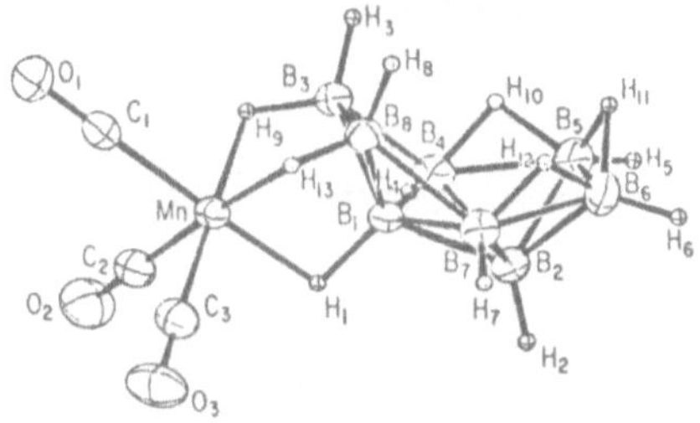

Fig. 1. Structure of $B_8H_{13}Mn(CO)_3$.

CHROMIUM HEXACARBONYL

$Cr(CO)_6$

A. JOST, B. REES and W.B. YELON, 1975. Acta Cryst., B31, 2649-2658.

Orthorhombic, Pnma, a = 11.505, b = 10.916, c = 6.203 Å, at 78°K, Z = 4. Neutron diffraction data, R = 0.044 for 622 reflexions.

The structure is as previously described (1). Cr-C = 1.918(1), C-O = 1.141(1) Å, C-Cr-C = 89.5-90.9, 179.2-179.4, Cr-C-O = 179.1-180.0°.

1. Structure Reports, 32A, 393.

μ-HYDRIDO-NONACARBONYL(NITROSYL)DITUNGSTEN (α and β forms)

$HW_2(CO)_9NO$

J.P. OLSEN, T.F. KOETZLE, S.W. KIRTLEY, M. ANDREWS, D.L. TIPTON and R. BAU, 1974. J. Amer. Chem. Soc., 96, 6621-6627.

α-Form
Triclinic, $P\bar{1}$, a = 12.284, b = 9.621, c = 6.921 Å, α = 112.84, β = 91.31, γ = 97.34°, D_m = 2.89, Z = 2. Mo radiation, R = 0.061 for 1932 reflexions, and neutron diffraction data, R = 0.060 for 2518 reflexions.

β-Form
Monoclinic, C2/c, a = 14.582, b = 6.771, c = 15.627 Å, β = 102.34°, D_m = 2.89, Z = 4. Neutron diffraction data, R = 0.066 for 919 reflexions.

Both forms contain nearly identical $(CO)_5WHW(CO)_4NO$ molecules, with W...W = 3.329, W-H = 1.873 Å, and a bent W-H-W linkage (125.5°). Each W has 4 equatorial CO ligands, with the two sets of ligands staggered, and CO and NO disordered in the axial positions. The axial ligand-metal vector does not point directly at the H atom, but bisects the H-W-W angle; this suggests a closed, three-centre two-electron W-H-W bridge bond with appreciable W-W overlap.

TRICHLOROTINPENTACARBONYLMANGANESE

$Cl_3SnMn(CO)_5$

S. ONAKA, 1975. Bull. Chem. Soc. Japan, 48, 319-323.

Monoclinic, $P2_1/c$, a = 14.10, b = 13.38, c = 13.27 Å, β = 97.39°, D_m = 2.30, Z = 8. Mo radiation, R = 0.10 for 2581 reflexions.

The two molecules in the asymmetric unit have similar geometries. Sn has tetrahedral coordination, and Mn distorted octahedral coordination with the equatorial carbonyl groups displaced towards the Sn atom; two Sn-Cl bonds nearly eclipse two Mn-C bonds. Mn-Sn = 2.59, mean Mn-C = 1.87, mean Sn-Cl = 2.35 Å.

DICHLOROBIS(PENTACARBONYLMANGANESE)TIN

DIBROMOBIS(PENTACARBONYLMANGANESE)TIN

$[Mn(CO)_5]_2SnX_2$ (X = Cl, Br)

H. PREUT, W. WOLFES and H.-J. HAUPT, 1975. Z. anorg. Chem., 412, 121-128.

X = Cl
Monoclinic, $P2_1/n$, a = 14.144, b = 12.223, c = 10.303 Å, β = 100.09°, Z = 4. Mo radiation, R = 0.037 for 2107 reflexions.

X = Br
Monoclinic, $P2_1/c$, a = 15.275, b = 7.650, c = 16.855 Å, β = 110.70°, Z = 4. Mo radiation, R = 0.042 for 1470 reflexions.

Sn has tetrahedral and Mn octahedral coordination; Sn-Mn = 2.64, Sn-Cl = 2.39, Sn-Br = 2.55, Mn-C = 1.85, C-O = 1.13 Å, Mn-Sn-Mn = 126, X-Sn-X = 96 and 98°.

DIHYDRIDOTETRAKIS(PENTACARBONYLMANGANESE)DITIN

$H_2Sn_2[Mn(CO)_5]_4$

K.D. BOS, E.J. BULTEN, J.G. NOLTES and A.L. SPEK, 1975. J. Organometal. Chem., 92, 33-41.

Monoclinic, C2/c, a = 15.71, b = 17.18, c = 12.51 Å, β = 107.3°, D_m = 2.11, Z = 4. Mo radiation, R = 0.20 for 2968 reflexions.

The molecule has C_2 symmetry and contains an Mn_2Sn-$SnMn_2$ skeleton, Sn-Sn = 2.89, Sn-Mn = 2.70(1) Å. Five carbonyl groups complete octahedral coordination at each Mn atom, Mn-C = 1.76-1.91 Å, and the H atoms complete tetrahedral geometry at each Sn atom.

OCTACARBONYL-BIS[μ-(PENTACARBONYLRHENIUM)-INDIUM(III)]-DIRHENIUM

$Re_2(CO)_8[InRe(CO)_5]_2$

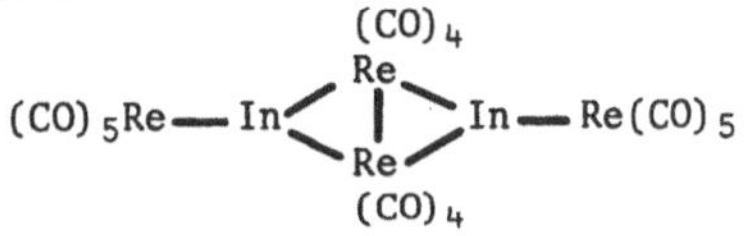

H. PREUT and H.-J. HAUPT, 1975. Chem. Ber., 108, 1447-1453.

Monoclinic, $P2_1/n$, a = 6.788, b = 16.352, c = 12.519 Å, β = 89.23°, Z = 2. Mo radiation, R = 0.048 for 3301 reflexions.

The molecule contains a planar Re_2In_2 ring, with Re-Re = 3.232(1) Å and Re-In-Re = 71.07° indicating a Re-Re bond. The $Re(CO)_5$ groups are trans with respect to the plane of the ring. Mean Re-In = 2.766(1), Re-C = 1.98(2), C-O = 1.14(2) Å.

TRIIRON DODECACARBONYL

$Fe_3(CO)_{12}$

F.A. COTTON and J.M. TROUP, 1974. J. Amer. Chem. Soc., 96, 4155-4159.

Monoclinic, $P2_1/n$, a = 8.359, b = 11.309, c = 8.862 Å, β = 97.00°, Z = 2. Mo radiation, R = 0.046 for 1354 reflexions.

Structure as in 1.

1. Structure Reports, 34A, 160.

TRICHLOROGERMYLTETRACARBONYLCOBALT

$Cl_3GeCo(CO)_4$

G.C. van der BERG, A. OSKAM and K. OLIE, 1974. J. Organometal. Chem., 80, 363-368.

Monoclinic, Cc, a = 26.238, b = 6.623, c = 12.969 Å, β = 106°12', D_m not measured, Z = 8. Cu radiation, R = 0.098 for 1647 reflexions.

Both molecules in the asymmetric unit show slight deviations from C_{3v} symmetry. Co has trigonal bipyramidal coordination with the Ge apical, and Ge coordination is tetrahedral, with the Cl and equatorial CO staggered. Co-Ge = 2.31(1), Ge-Cl = 2.14(1), Co-C = 1.79(4) Å.

TETRACOBALT DECACARBONYL DISULPHIDE

$Co_4(CO)_{10}S_2$

C.H. WEI and L.F. DAHL, 1975. Cryst. Struct. Comm., 4, 583-588.

Monoclinic, $P2_1/n$, a = 10.06, b = 6.81, c = 12.45 Å, β = 97°15', Z = 2. Mo radiation, R = 0.092 for 440 reflexions (films, visual intensities).

The centrosymmetric molecule has close to D_{2h} symmetry, and contains a Co_4S_2 octahedral cluster, with two bridging and eight terminal carbonyl groups (Fig. 1). Co-Co = 2.48(1) (bridged by carbonyl), 2.60(1) (not bridged), Co-S = 2.24-2.27(1), S...S' = 2.74(2), Co-C = 1.86 and 1.93(3) (bridging), 1.64-1.78(4) Å (terminal). The closest intermolecular distance is O...O = 3.12 Å.

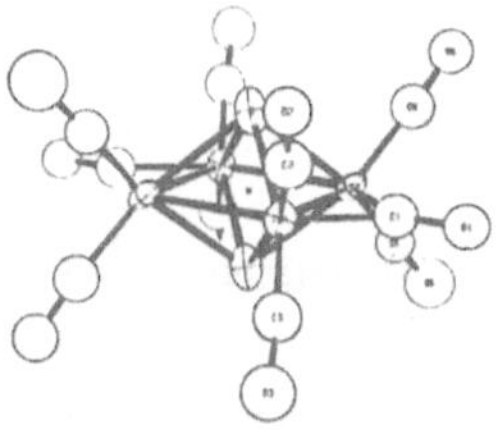

Fig. 1. Tetracobalt decacarbonyl disulphide.

DODECACARBONYLTETRACOBALT

DODECACARBONYLTETRARHODIUM

DODECACARBONYLTETRAIRIDIUM

$Co_4(CO)_{12}$
$Rh_4(CO)_{12}$
$Ir_4(CO)_{12}$

I. C.H. WEI and L.F. DAHL, 1966. J. Amer. Chem. Soc., 88, 1821-1822.

II. C.H. WEI, G.R. WILKES and L.F. DAHL, 1967. Ibid., 89, 4792-4793.

III. C.H. WEI, 1969. Inorg. Chem., 8, 2384-2397.

IV. G.R. WILKES, 1965. Ph.D. Thesis, University of Wisconsin; Diss. Abs., 26, 5029.

Cobalt compound
Orthorhombic, Pccn, a = 8.99, b = 11.70, c = 17.28 Å (in agreement with 1), Z = 4. Mo radiation, R = 0.13 for 529 reflexions (film data, disordered structure).

Rhodium compound
Monoclinic, $P2_1/c$, a = 9.24, b = 12.02, c = 17.74 Å, β = 90°, D_m = 2.58, Z = 4. Mo radiation, R = 0.10 for 962 reflexions (film data, twinned crystal).

The Co and Rh compounds have similar molecular geometries: a tetrahedron of metal atoms, with a basal $M_3(CO)_9$ fragment of three carbonyl-bridged $M(CO)_2$ groups and an $M(CO)_3$ apex. Co-Co = 2.44-2.53(2), Rh-Rh= 2.70-2.80(1) Å. The Ir compound (IV) has a similar metal tetrahedron, but only terminal carbonyl groups.

1. P. CORRADINI, 1959. J. Chem. Phys., 31, 1676; Structure Reports, 23, 284.

TRI-μ-CARBONYL-ENNEACARBONYLDICOBALTDIIRIDIUM (DODECACARBONYLDICOBALTDIIRIDIUM)

$Co_2Ir_2(CO)_{12}$

V.G. ALBANO, G. CIANI and S. MARTINENGO, 1974. J. Organometal. Chem., 78, 265-272.

Monoclinic, $P2_1/c$, a = 9.12, b = 11.62, c = 17.31 Å, β = 90°, D_m = 3.01, Z = 4. Mo radiation, R = 0.051 for 570 reflexions (twinned crystal).

The molecular structure is similar to those of $M_4(CO)_{12}$, M = Co, Rh (1). It contains a tetrahedron of metal atoms with 9 terminal and 3 edge-bridging CO ligands ($Ir_4(CO)_{12}$ has only terminal ligands, 1). Co and Ir are partially disordered, but Ir shows a preference for the position to which only terminal ligands are coordinated. M-M = 2.59-2.64(1) Å.

1. Preceding report.

DI-μ_3-CARBONYL-HEXA-μ-CARBONYL-CARBIDO-UNDECACARBONYL-polyhedro-OCTARHODIUM

$Rh_8(CO)_{19}C$

V.G. ALBANO, M. SANSONI, P. CHINI, S. MARTINENGO and D. STRUMOLO, 1975. J. Chem. Soc., Dalton, 305-309.

Triclinic, $P\bar{1}$, a = 9.18, b = 17.76, c = 10.46 Å, α = 75°57', β = 69°4', γ = 92°22', D_m = 3.01, Z = 2. Mo radiation, R = 0.023 for 3423 reflexions.

The Rh_8 metal atom cluster can be described as a monocapped prism plus one edge-bridging atom (Fig. 1); Rh-Rh = 2.699-2.913(3) Å. The carbide atom occupies the centre of the prism, mean Rh-C = 2.13 Å, and the carbonyl groups are distributed on the cluster surface, two triply-bridging, six edge-bridging, and eleven terminal.

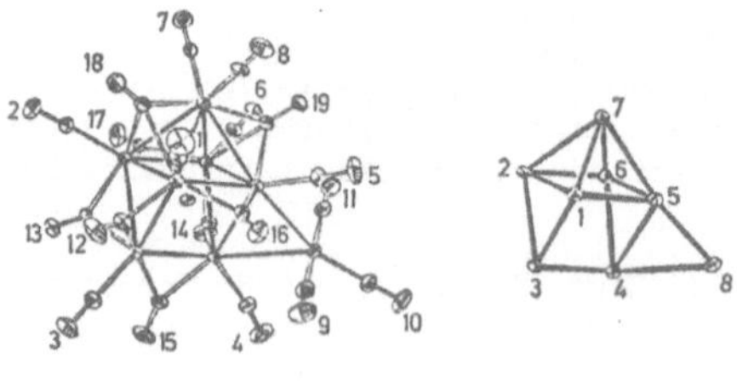

Fig. 1. The molecular structure and metal atom cluster of $Rh_8(CO)_{19}C$.

IRON NITROSYL SULPHIDE TETRAMER

$Fe_4(NO)_4S_4$

R.S. GALL, C.T.-W. CHU and L.F. DAHL, 1974. J. Amer. Chem. Soc., 96, 4019-4023.

Monoclinic, $P2_1/n$, a = 12.350, b = 9.627, c = 10.407 Å, β = 103.66°, Z = 4. R = 0.042 for 1376 reflexions.

The molecule has almost exact T_d symmetry, with a cubane-like Fe_4S_4 core, Fe...Fe = 2.634(1), Fe-S = 2.217(2), S...S = 3.503(2) Å, and terminal nitrosyl ligands, Fe-N = 1.66, N-O = 1.15 Å, Fe-N-O = 177-179°.

trans-DIAZIDOTETRAAMMINECOBALT(III) trans-TETRA-AZIDODIAMMINECOBALTATE(III)

$[Co(NH_3)_4(N_3)_2][Co(NH_3)_2(N_3)_4]$

L.F. DRUDING, F.D. SANCILIO and D.M. LUKASZEWSKI, 1975. Inorg. Chem., 14, 1365-1369.

Triclinic, $P\bar{1}$, a = 7.661, b = 6.928, c = 9.671 Å, α = 87.4, β = 72.7, γ = 116.4°, D_m = 1.87, Z = 1. Cu radiation, R = 0.049 for 583 reflexions.

Cation and anion are each trans-octahedral; Co-N = 1.97-1.98(1), N-N = 1.20 and 1.15(1) Å, Co-N-N = 118-125, N-N-N = 173 (cation), 178° (anion). Ions are joined by N-H...N hydrogen bonds, 3.05-3.48 Å.

POTASSIUM TETRAAZIDOZINCATE

$K_2Zn(N_3)_4$

A.C. BRUNNER and H. KRISCHNER, 1975. Z. Kristallogr., 142, 24-34.

Orthorhombic, Pbca, a = 14.40, b = 11.70, c = 11.29 Å, Z = 8. Cu radiation, R = 0.087 for 1106 reflexions.

The structure (Fig. 1) contains isolated $Zn(N_3)_4{}^{2-}$ tetrahedra, with linear, symmetric azide groups; Zn-N = 1.97-2.01, N-N = 1.18, 1.18 Å, N-Zn-N = 101-117°. K^+ ions have eight-coordination, K-N = 2.77-3.39 Å.

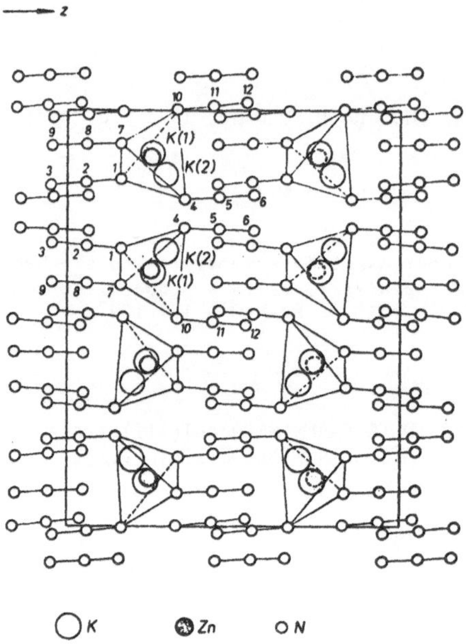

Fig. 1. Structure of $K_2Zn(N_3)_4$.

CAESIUM TETRAAZIDOZINCATE

$Cs_2Zn(N_3)_4$

G.F. PLATZER and H. KRISCHNER, 1975. Z. Kristallogr., 141, 363-374.

Orthorhombic, $Pca2_1$, a = 21.98, b = 6.79, c = 7.45 Å, D_m not given, Z = 4. Cu radiation, R = 0.11 for 891 reflexions (films, photometer intensities).

The structure (Fig. 1) contains isolated tetrahedral $Zn(N_3)_4^{2-}$ ions and Cs^+ ions. Zn-N = 1.99-2.10(6), N-N = 1.17(7) Å, N-Zn-N = 97-128°, N-N-N = linear; Cs-N = 3.00-3.47(9) Å (8- and 9-coordination).

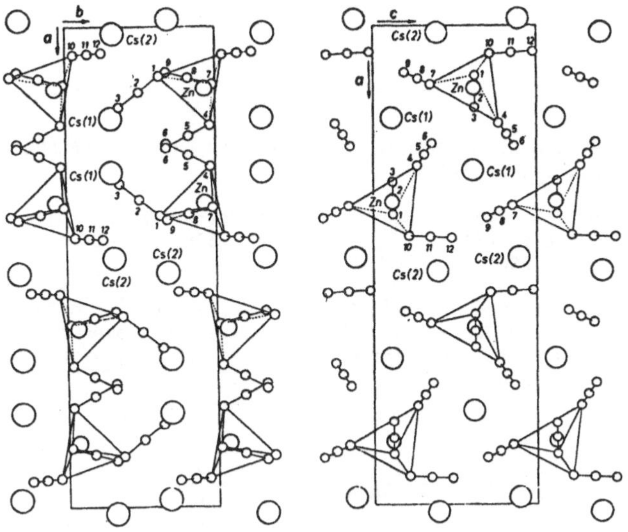

Fig. 1. Structure of caesium tetraazidozincate, viewed along c and b.

BARIUM AMIDE

$Ba(NH_2)_2$

H. JACOBS and C. HADENFELDT, 1975. Z. anorg. Chem., 418, 132-140.

Monoclinic, Cc, a = 8.951, b = 12.67, c = 7.037 Å, β = 123.5°, Z = 8. Mo radiation, R = 0.027 for 1363 reflexions.

Atomic positions

	x	y	z
Ba(1)	0	0.6867	0
Ba(2)	0.2574	0.1078	0.3484
N(1)	0.021	0.494	0.245
N(2)	0.377	0.304	0.603
N(3)	0.114	0.085	0.624
N(4)	0.217	0.262	0.013

The amide ions are relatively close-packed, and Ba ions have 7- and 8-coordination, Ba-N = 2.75-3.24 Å.

HEXAAMMINECALCIUM(0)
(DEUTERATED)

$Ca(ND_3)_6$

R.B. von DREELE, W.S. GLAUNSINGER, A.L. BOWMAN and J.L. YARNELL, 1975.
J. Phys. Chem., 79, 2992-2995.

Cubic, Im3m, a = 9.0137 Å, Z = 2. Neutron powder data. Ca in 2(a); N in 12(e): x = 0.298; 0.25 D in 48(j): y = 0.100, z = 0.328; 0.25 D in 96(ℓ): x = 0.130, y = -0.082, z = 0.286.

The material was obtained by distilling deuterated ammonia onto freshly cut Ca metal. The structure proposed has highly distorted ND_3 molecules (N-D = 0.94 and 1.39 (x 2) Å) arranged in a regular octahedron around Ca, with fourfold rotational disorder for each ND_3.

POTASSIUM TETRAAMIDOMANGANATE(II)

RUBIDIUM TETRAAMIDOZINCATE

$K_2Mn(NH_2)_4$
$Rb_2Zn(NH_2)_4$

M. DREW, L. GUÉMAS, P. CHEVALIER, P. PALVADEAU and J. ROUXEL, 1975.
Rev. Chim. Minér., 12, 419-426.

Monoclinic, $P2_1/c$, a = 7.45, 7.80, b = 7.00, 7.00, c = 13.56, 13.83 Å, β = 105.94, 106.33°, D_m = 1.93, 2.75, Z = 4. Mo radiation, R = 0.08 and 0.07 for 472 and 892 reflexions.

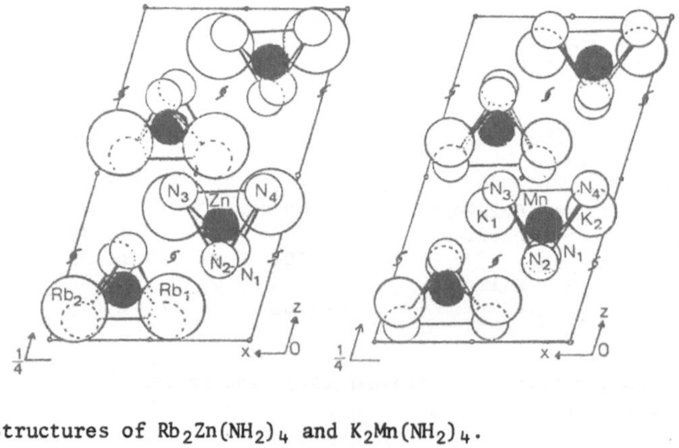

Fig. 1. Structures of $Rb_2Zn(NH_2)_4$ and $K_2Mn(NH_2)_4$.

The crystals are isostructural and contain alkali metal cations and tetrahedral $M(NH_2)_4^{2-}$ anions (Fig. 1); Mn-N = 2.12(4), Zn-N = 2.03(5) Å. The alkali metal cations each have four close neighbours at about 3.0 Å, with others more distant.

SULPHUR NITRIDES

DISULPHUR DINITRIDE

POLY(SULPHUR NITRIDE)

S_2N_2
β-$(SN)_x$

I. M. BOUDEULLE, 1975. Cryst. Struct. Comm., 4, 9-12.

II. C.M. MIKULSKI, P.J. RUSSO, M.S. SARAN, A.G. MacDIARMID, A.F. GARITO and A.J. HEEGER, 1975. J. Amer. Chem. Soc., 97, 6358-6363.

III. A.G. MacDIARMID, C.M. MIKULSKI, P.J. RUSSO, M.S. SARAN, A.F. GARITO and A.J. HEEGER, 1975. Chem. Comm., 476-477.

S_2N_2 (II, III)
Monoclinic, $P2_1/c$, a = 4.485, b = 3.767, c = 8.452 Å, β = 106.43°, at -130°C, Z = 2. R = 0.03 [no details; a complete description is to be published later]. Planar, square four-membered ring, S-N = 1.654 Å.

β-$(SN)_x$
Monoclinic, $P2_1/c$, a = 4.153, 4.12, b = 4.439, 4.43, c = 7.637, 7.64 Å, β = 109.7, 109.5°, in II, I, respectively, Z = 4 [D_x = 2.3, but D_m is given as 1.2 in I]. X-ray (II, III) and electron (I) diffraction patterns of fibres, R = 0.11 and 0.19 [II and III promise a complete description later]. Infinite chain with S-N $\sim$ 1.6 Å.

THIOTRITHIAZYL cis-TETRACHLORODIAQUOINDATE(III)

$[S_4N_3][InCl_4(H_2O)_2]$

M.L. ZIEGLER, H.U. SCHLIMPER, B. NUBER, J. WEISS and G. ERTL, 1975. Z. anorg. Chem., 415, 193-201.

Orthorhombic, Pnam, a = 19.473, b = 6.183, c = 10.814 Å, D_m = 2.353, Z = 4. Mo radiation, R = 0.049 for 1170 reflexions.

The $InCl_4(H_2O)_2^-$ anion is cis-octahedral, In-Cl = 2.42-2.49, In-O = 2.26 Å. The $S_4N_3^+$ ion is a seven-membered ring, but is disordered in the crystal, so that its dimensions have not been determined reliably.

THIODITHIAZYL HEXAFLUOROARSENATE(V)

$[S_3N_2][AsF_6]$

R.J. GILLESPIE, P.R. IRELAND and J.E. VEKRIS, 1975. Canad. J. Chem., 53, 3147-3152.

Monoclinic, $P2_1/c$, a = 8.499, b = 8.298, c = 11.069 Å, β = 94.59°, D_m not measurable, Z = 4. Mo radiation, R = 0.028 for 798 reflexions.

The structure contains planar, cyclic $S_3N_2^+$ and octahedral AsF_6^- ions. In the S-S-N-S-N ring, S-S = 2.147(3), N-SS = 1.60(1), N-SN = 1.56(1) Å, N-S-S = 97, S-N-S = 120, N-S-N = 107°. As-F = 1.66-1.73 Å.

DEUTERIUM FLUORIDE

DF

M.W. JOHNSON, E. SÁNDOR and E. ARZI, 1975. Acta Cryst., B31, 1998-2003.

Orthorhombic, $Bm2_1b$, a = 3.31, 3.33, b = 4.26, 4.27, c = 5.22, 5.27 Å, at 4.2 and 85°K, respectively, Z = 4. Neutron powder data, R = 0.049 and 0.069. Atoms in 4(a): (0yz); y(F) = 1/4, z(F) = 0.126, 0.125; y(D) = 0.444, 0.430, z(D) = 0.036, 0.025, at 4.2 and 85°K.

The F position is similar to that previously determined for HF (1), and the D position corresponds to a parallel zigzag chain model (Fig. 1). F...F = 2.50, F-D = 0.97 Å.

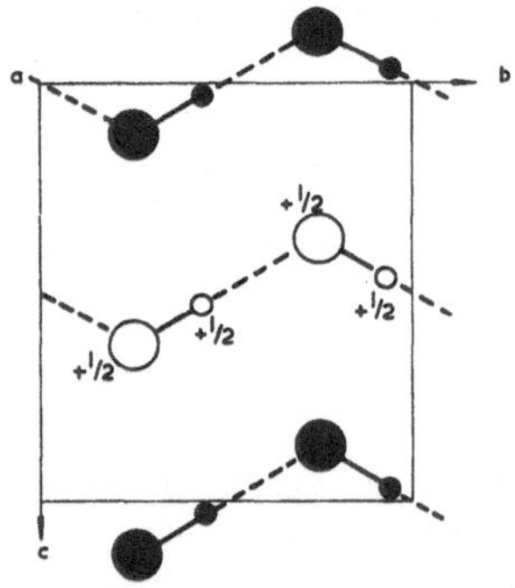

Fig. 1. Structure of deuterium fluoride.

1. Structure Reports, 18, 347; 37A, 175.

MOLYBDENUM HEXAFLUORIDE
(Orthorhombic form)

MoF_6

J.H. LEVY, J.C. TAYLOR and P.W. WILSON, 1975. Acta Cryst., B31, 398-401.

Orthorhombic, Pnma, a = 9.559, b = 8.668, c = 5.015 Å (at 193°K), D_m = 3.27 (at 227°K), Z = 4. Neutron powder data with profile-fitting technique.

Isostructural with UF_6 (1). The structure contains close-packed fluorine layers with Mo in octahedral holes (Fig. 1), Mo-F = 1.77-1.86 Å.

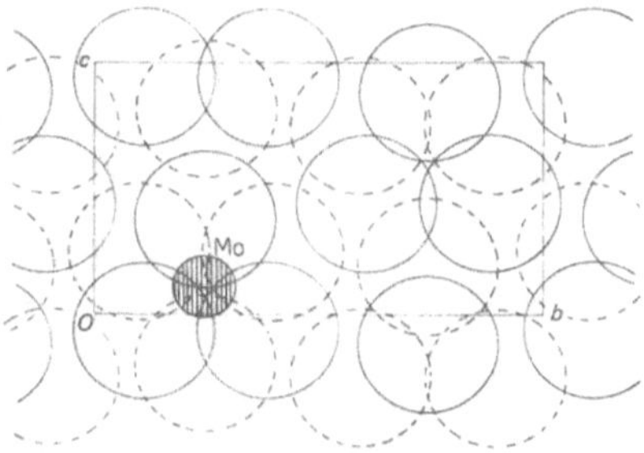

Fig. 1. Packing of fluorine layers in MoF_6.

1. Structure Reports, 14, 33; 39A, 140; this volume, p. 152.

MOLYBDENUM HEXAFLUORIDE
(Cubic form)

MoF_6

J.H. LEVY, P.L. SANGER, J.C. TAYLOR and P.W. WILSON, 1975. Acta Cryst., B31, 1065-1067.

Cubic, [Im3m], a = 6.221 Å (at 266°K), Z = 2. Neutron powder data with profile analysis.

The MoF_6 groups are in rapid rotational disorder, with F atoms distributed over a sphere but with maxima on the fourfold axes. Below 263.4°K the structure is orthorhombic (1), UF_6-type (2). Mo-F = 1.80 Å.

1. Preceding report.
2. Structure Reports, 14, 33; 39A, 140; this volume, p. 152.

TUNGSTEN HEXAFLUORIDE

WF_6

J.H. LEVY, J.C. TAYLOR and P.W. WILSON, 1975. J. Solid State Chem., 15, 360-365.

Orthorhombic, Pnma, a = 9.603, b = 8.713, c = 5.044 Å, at 193°K, Z = 4. Neutron diffraction data, R = 0.075 for 306 profile points.

Atomic positions

	x	y	z
W	0.1247	1/4	0.0999
F(1)	0.0145	1/4	-0.1952
F(2)	0.2452	1/4	0.3717
F(3)	0.0212	0.0967	0.2330
F(4)	0.2357	0.1075	-0.0624

Isostructural with orthorhombic UF_6 (1). W has octahedral coordination, W-F = 1.79-1.83(2) Å.

1. Structure Reports, 14, 33; 39A, 140; this volume, following report.

URANIUM HEXAFLUORIDE

UF_6

J.C. TAYLOR and P.W. WILSON, 1975. J. Solid State Chem., 14, 378-382.

Orthorhombic, Pnma, a = 9.843, b = 8.920, c = 5.173 Å, at 193°K, Z = 4. Neutron powder data with profile fitting.

Structure as previously determined at room temperature (1). U-F = 1.86-1.99 at 293°K, 1.95-2.03(2) Å at 193°K.

1. Structure Reports, 14, 33; 39A, 140.

LITHIUM HYDRAZINIUM TETRAFLUOROBERYLLATE

$LiN_2H_5BeF_4$

M.R. ANDERSON and I.D. BROWN, 1975. Acta Cryst., B31, 1500-1501.

Orthorhombic, $Pna2_1$, a = 9.811, b = 8.880, c = 5.139 Å, Z = 4 (1). Neutron diffraction data, R_w = 0.044 for 470 reflexions.

The structure is as determined by X-ray methods (1), except that the N-H bond lengths are about 0.1 Å longer, as expected, and an additional weak interaction, N-H...F = 3.12 Å, is included as a hydrogen bond.

1. Structure Reports, 39A, 141.

AMMONIUM COBALT(II) TETRAFLUOROBERYLLATE HEXAHYDRATE

$(NH_4)_2Co(BeF_4)_2.6H_2O$

J. VICAT, D. TRAN QUI, A. FILHOL, E. ROUDAUT, M. THOMAS and S. ALÉONARD, 1975. Acta Cryst., B31, 1895-1903.

Monoclinic, $P2_1/a$, a = 9.269, b = 12.541, c = 6.136 Å, β = 106.80°, D_m not given, Z = 4. Neutron diffraction data, R = 0.051 for 2086 reflexions.

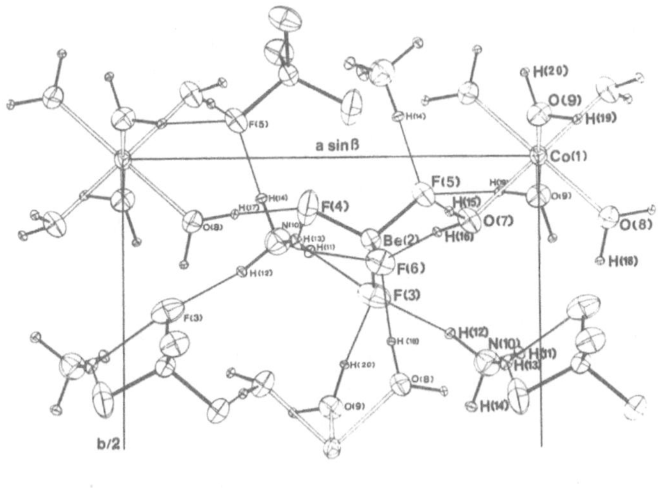

Fig. 1. Hydrogen bonding in ammonium cobalt(II) tetrafluoroberyllate hexahydrate.

Tutton's salt structure (1); Co-O = 2.058-2.117(1), Be-F = 1.528-1.550(2) Å. The ions are joined by a system of O-H...F and N-H...F hydrogen bonds (Fig. 1).

1. Strukturbericht, 2, 93, 436; Structure Reports, 27, 619; 29, 354.

TETRAAMMINECOPPER(II) TETRAFLUOROBERYLLATE MONOHYDRATE

$Cu(NH_3)_4BeF_4.H_2O$

J.-C. TEDENAC, É. PHILIPPOT and M. MAURIN, 1975. Bull. Soc. Fr. Minér. Crist., 98, 36-42.

Orthorhombic, Pnma, a = 23.942, b = 7.030, c = 20.979 Å, D_m = 1.75, Z = 16. Cu radiation, R = 0.094 for 927 reflexions (films, visual intensities).

The structure consists of Cu coordination polyhedra and BeF_4^{2-} tetrahedra, joined by O-H...F and N-H...F hydrogen bonds. The Cu coordination polyhedra are square pyramids with 4 NH_3 basal and 1 H_2O axial, Cu-N = 1.95-2.11, Cu-O = 2.35-2.50(2) Å.

CALCIUM TETRAFLUOROBORATE

$Ca(BF_4)_2$

T.H. JORDAN, B. DICKENS, L.W. SCHROEDER and W.E. BROWN, 1975. Acta Cryst., B31, 669-672.

Orthorhombic, Pbca, a = 9.2792, b = 8.9103, c = 13.3719 Å, D_m = 2.560, Z = 8. Mo radiation, R = 0.024 for 2896 reflexions.

The structure contains columns of tetrahedral BF_4^- ions and columns of alternating Ca^{2+} and BF_4^- ions. The columns are linked by Ca...F bonds. Ca has square-antiprismatic coordination, Ca-F = 2.330-2.401(2) Å. B-F = 1.398-1.408(3) Å, F-B-F = 108.4-110.5°.

CRYOLITE

Na_3AlF_6

F.C. HAWTHORNE and R.B. FERGUSON, 1975. Canad. Miner., 13, 377-382.

Monoclinic, $P2_1/n$, a = 5.4024, b = 5.5959, c = 7.7564 Å, β = 90.278°, Z = 2. Mo radiation, R = 0.033 for 726 reflexions.

Structure as previously described (1). Al-F = 1.805-1.813(1), Na-F = 2.227-2.816(2) Å.

1. Strukturbericht, 6, 29, 120.

LITHIUM MANGANESE(II) HEXAFLUOROGALLATE

$LiMnGaF_6$

W. VIEBAHN, 1975. Z. anorg. Chem., 413, 77-84.

Trigonal, P312, a = 8.638, c = 4.738 Å, D_m = 3.94, Z = 3. Mo radiation, R = 0.066 for 947 reflexions (film data).

The structure (Fig. 1) is an ordered variant of the Na_2SiF_6 structure-type (1) (Ga in 1(a) and 2(d), Mn in 3(e), Li in 3(f)). Ga-F = 1.89, Mn-F = 2.06-2.19, Li-F = 1.88-2.23(2) Å.

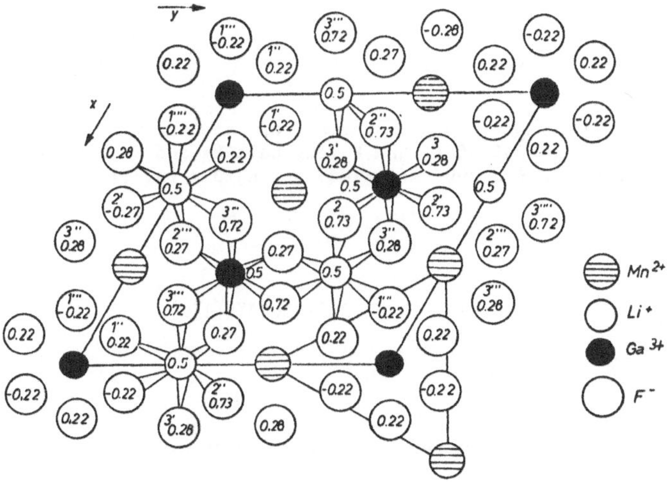

Fig. 1. Structure of $LiMnGaF_6$.

1. Structure Reports, 19, 325; 29, 264.

POTASSIUM TETRAFLUOROTHALLATE(III)

$KTlF_4$

C. HEBECKER, 1975. Z. Naturforsch., 30B, 305-312.

Trigonal, $P3_1$, a = 8.025, c = 10.16 Å, D_m = 5.65, Z = 6. Mo radiation, R = 0.11 for 491 reflexions.

Atomic positions

Atoms in 3(a)

	x	y	z
K(1)	0.479	0.305	0.011
K(2)	0.321	0.477	0.702
Tl(1)	0.0005	0.3604	0.016
Tl(2)	0.1758	0.1761	0.350
F(1)	0.05	0.32	0.25
F(2)	0.47	0.34	0.25
F(3)	0.35	0.08	0.47
F(4)	0.35	0.46	0.46
F(5)	0.28	0.47	-0.04
F(6)	-0.19	0.51	0.43
F(7)	0.20	0.01	0.18
F(8)	-0.07	0.04	0.09

The structure is a superstructure variant of the CaF_2 type. Each Tl has 7-coordination (Tl-F = 2.02-2.49 Å) and K ions have 8- and 9-coordination (K-F = 2.45-3.12 Å). Other ABF_4 compounds (A = Na, Ag, K; B = In, Tl, Ln) are isostructural with $KTlF_4$.

RUBIDIUM TETRAFLUOROTHALLATE(III)

$RbTlF_4$

C. HEBECKER, 1975. Z. anorg. Chem., 412, 37-46.

Orthorhombic, $Pb2_1a$, a = 8.252, b = 8.359, c = 6.244 Å, D_m = 5.77, Z = 4. Mo radiation, R = 0.097 for 374 reflexions (film data).

The structure (Fig. 1) contains layers of corner-sharing TlF_6 octahedra, Tl-F(terminal) = 2.01 and 2.05, Tl-F(shared) = 2.00, 2.12, 2.38, and 2.46 Å. The layers are joined by Rb ions, Rb-F = 2.68-3.45 Å (10-coordination). $CsTlF_4$ and $Tl(I)Tl(III)F_4$ (TlF_2) are isostructural.

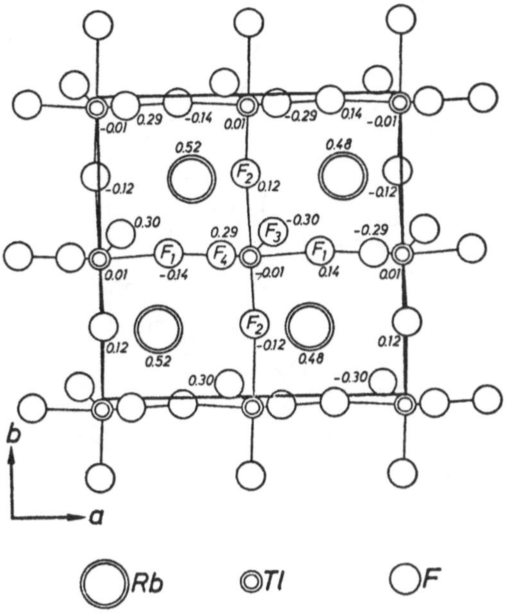

Fig. 1. Structure of $RbTlF_4$.

SODIUM DECAFLUOROTRIANTIMONATE(III)

$NaSb_3F_{10}$

R. FOURCADE, G. MASCHERPA and E. PHILIPPOT, 1975. Acta Cryst., B31, 2322-2326.

Hexagonal, $P6_3$, a = 8.285, c = 7.600 Å, D_m = 4.23, Z = 2. Mo radiation, R = 0.049 for 589 reflexions.

The structure (Fig. 1) contains infinite $[Sb_3F_{10}^-]_n$ sheets perpendicular to c. These contain SbF_5E octahedra (E = lone pair) linked by two asymmetric equatorial F bridges (Sb-F = 1.98 and 2.60(1) Å) and one symmetric, planar, triply-bridging F atom (Sb-F = 2.38(1) Å). Na ions are between the layers and have distorted octahedral coordination, Na-F = 2.32 and 2.36(1) Å.

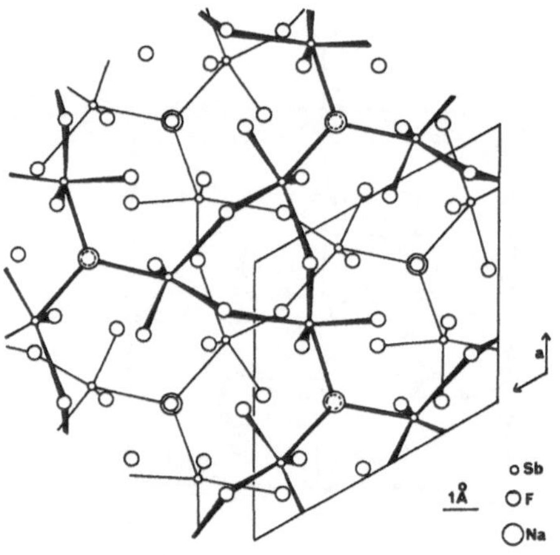

Fig. 1. Structure of $NaSb_3F_{10}$.

POTASSIUM CHLOROTRIFLUOROANTIMONATE(III)

$KSbF_3Cl$

B. DUCOURANT, R. FOURCADE, É. PHILIPPOT and G. MASCHERPA, 1975. Rev. Chim. Minér., 12, 485-492.

Orthorhombic, Pbca, a = 8.117, b = 14.755, c = 8.012 Å, D_m = 3.50, Z = 8. Mo radiation, R = 0.042 for 615 reflexions.

Atomic positions

	x	y	z
Sb	0.0048	0.3246	0.1271
K	0.1955	0.0581	0.1817
Cl	0.3730	0.3562	0.0137
F(1)	0.4980	0.1514	0.1129
F(2)	0.7713	0.3530	0.1093
F(3)	0.0360	0.4530	0.1600

The structure contains distorted SbF_3Cl_3E monocapped octahedra (E = lone pair), with the stereochemically active lone pair capping a Cl_3 face; these units are linked through three chlorine bridges to form a layer structure, with K^+ ions between the layers. Sb-F = 1.94-1.96(1), Sb-Cl = 3.09-3.16 Å, F-Sb-F = 82-88, Cl-Sb-Cl = 110-125°; K-F = 2.68-2.87, K-Cl = 3.28-3.32 Å.

POTASSIUM TRIDECAFLUOROTETRAANTIMONATE(III)

KSb_4F_{13}

B. DUCOURANT, R. FOURCADE, É. PHILIPPOT and G. MASCHERPA, 1975. Rev. Chim. Minér., 12, 553-562.

Tetragonal, I$\bar{4}$, a = 9.636, c = 6.365 Å, D_m = 4.32, Z = 2. Mo radiation, R = 0.079.

Atomic positions

			x	y	z
Sb	in	8(g)	0.2971	-0.0203	0.0046
K		2(b)	1/2	1/2	0
F(1)		2(d)	1/2	0	1/4
F(2)		8(g)	0.622	0.086	0.603
F(3)		8(g)	0.929	0.150	0.163
F(4)		8(g)	0.596	0.268	0.270

The structure (Fig. 1) contains tetrahedral $F(SbF_3)_4$ units, which contain SbF_3 groups (Sb-F = 1.93-1.98(1) Å) bridged by F atoms (Sb...F = 2.71 Å) and with a central F^- ion (Sb...F 2.51 Å); these units are linked by a further Sb...F contact (2.72 Å). Sb therefore has monocapped octahedral MA_3B_3E geometry (E = lone pair). The K^+ ion has eight-coordination (distorted cube, K-F = 2.67, 2.96 Å).

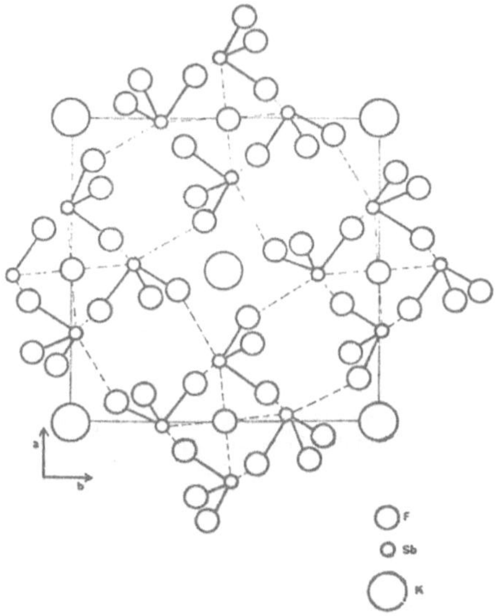

Fig. 1. Structure of KSb_4F_{13}.

TETRAFLUOROIODINE(V) μ-FLUORO-BIS[PENTAFLUOROANTIMONATE(V)]

$IF_4.Sb_2F_{11}$

A.J. EDWARDS and P. TAYLOR, 1975. J. Chem. Soc., Dalton, 2174-2177.

Monoclinic, $P2_1/c$, a = 8.52, b = 14.82, c = 9.98 Å, β = 112.7°, D_m not measured, Z = 4. Mo radiation, R = 0.089 for 1299 reflexions (films, photometer intensities).

The structure (Fig. 1) contains IF_4^+ and $Sb_2F_{11}^-$ ions, with four additional interionic I...F contacts leading to a complex three-dimensional arrangement. The IF_4^+ ion is a trigonal bipyramid with the lone pair occupying one equatorial site; I-F = 1.77 (equatorial), 1.84 (axial), I...F = 2.51-2.94(3) Å. The $Sb_2F_{11}^-$ ion contains two SbF_6 octahedra sharing a corner; Sb-F = 1.83-1.90 (terminal), 2.02(3) Å (bridging).

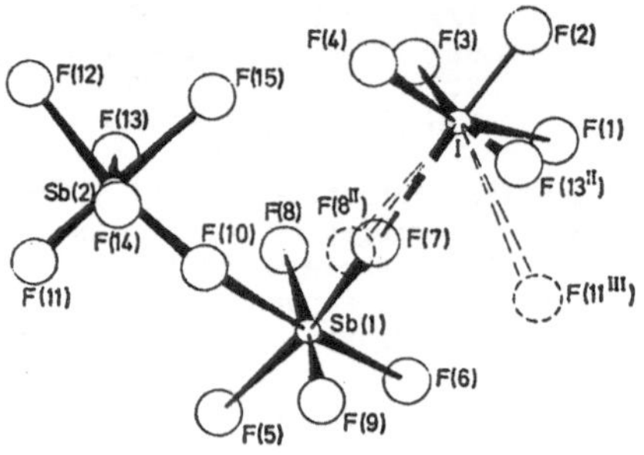

Fig. 1. Structure of $IF_4^+Sb_2F_{11}^-$.

LEAD(II) HEXAFLUOROZIRCONATE(IV)

$PbZrF_6$

J.-P. LAVAL, D. MERCURIO-LAVAUD and B. GAUDREAU, 1974. Rev. Chim. Minér., 11, 742-750.

Orthorhombic, Cmma, a = 7.549, b = 11.123, c = 5.310 Å, D_m = 5.90, Z = 4. Cu radiation, R = 0.091 for 77 reflexions (film data).

Atomic positions

		x	y	z
Zr	in 4(a)	1/4	0	0
Pb	4(g)	0	1/4	0.488
F(1)	8(m)	0	0.427	0.145
F(2)	16(o)	0.275	0.365	0.244

Isostructural with $RbPaF_6$ (1). The structure contains ZrF_8 dodecahedra which share edges to form chains along a; Zr-F = 2.00 and 2.19 Å. Pb has ten F neighbours at 2.56-2.74 Å, four on one side and six on the other, so that the lone-pair is possibly sterically active.

1. Structure Reports, 33A, 195.

HEXAFLUOROVANADATES(III)

CAESIUM THALLIUM(I) HEXAFLUOROVANADATE(III)

CAESIUM POTASSIUM HEXAFLUOROVANADATE(III)

RUBIDIUM POTASSIUM HEXAFLUOROVANADATE(III)

RUBIDIUM SODIUM HEXAFLUOROVANADATE(III)

CAESIUM SODIUM HEXAFLUOROVANADATE(III)

SODIUM HEXAFLUOROVANADATE(III)

Cs_2TlVF_6
Cs_2KVF_6
Rb_2KVF_6
Rb_2NaVF_6
Cs_2NaVF_6
Na_3VF_6

E. ALTER and R. HOPPE, 1975. Z. anorg. Chem., 412, 110-120.

A_2BVF_6, cubic, Fm3m, Z = 4, powder data. Elpasolite structure (1), V-F = 1.94-1.95 Å.

A	B	a(Å)	D_m	x(F)
Cs	Tl	9.234	5.38	0.210
Cs	K	9.047	4.20	0.215
Rb	K	8.855	3.59	0.220
Rb	Na	8.468	3.92	0.230

Cs_2NaVF_6, rhombohedral, [$R\bar{3}m$], a = 6.24, c = 30.48 Å, Z = 6, powder data. Cs_2NaCrF_6-type structure (2).

Na_3VF_6, monoclinic, $P2_1/n$, a = 5.513, b = 5.721, c = 7.963 Å, β = 90.47°, D_m = 3.09, Z = 2, powder data. Cryolite-structure (3).

1. Strukturbericht, 2, 498; Structure Reports, 40A, 137.
2. Structure Reports, 38A, 349; 40A, 140.
3. Strukturbericht, 6, 29, 120; this volume, p. 154.

HEXAFLUOROMOLYBDATES(III)

Cs_2KMoF_6
Cs_2TlMoF_6
Rb_2KMoF_6
Rb_2NaMoF_6
Tl_2KMoF_6
Tl_2NaMoF_6
K_2NaMoF_6

R. HOPPE and K. LEHR, 1975. Z. anorg. Chem., 416, 240-250.

Cubic, Fm3m, a = 9.21, 9.39, 8.91, 8.63, 8.98, 8.65, 8.50 Å, D_m = 4.32, 5.32, 3.92, 4.13, 6.00, 6.45, 3.20, Z = 4. Powder data. Elpasolite (1) structures, x(F) = 0.222, 0.218, 0.227, 0.235, 0.227, 0.235, 0.240.

1. Strukturbericht, 2, 498; Structure Reports, 40A, 137.

POTASSIUM TRIFLUOROMANGANATE(II)

$KMnF_3$

M. HIDAKA, 1975. J. Phys. Soc. Japan, 39, 180-186.

Tetragonal, P4/mbm, a = 5.894, c = 8.348 Å, at 50°K, Z = 4. Mo radiation, R = 0.17 for 122 superlattice reflexions (films, densitometer intensities).

Atomic positions

		x	y	z
K	in 4(f)	1/2	0	0.246
Mn	4(a)	0	0	0
F(1)	4(e)	0	0	0.250
F(2)	4(h)	0.273	0.773	1/2
F(3)	4(g)	0.250	0.750	0

Slightly-distorted perovskite structure.

HEXAFLUORIDES

RUBIDIUM SODIUM HEXAFLUOROFERRATE(III)

RUBIDIUM POTASSIUM HEXAFLUOROFERRATE(III)

CAESIUM SODIUM HEXAFLUOROFERRATE(III)

CAESIUM SODIUM HEXAFLUOROCHROMATE(III)

CAESIUM LITHIUM HEXAFLUOROGALLATE(III)

Rb_2NaFeF_6
Rb_2KFeF_6
Cs_2NaFeF_6
Cs_2NaCrF_6
Cs_2LiGaF_6

R. HAEGELE, W. VERSCHAREN and D. BABEL, 1975. Z. Naturforsch., 30B, 462-464.

Rb_2NaFeF_6, Rb_2KFeF_6
Cubic, Fm3m, a = 8.462, 8.869 Å, Z = 4. Elpasolite structure (1), x(F) = 0.2283, 0.2156.

Cs_2NaFeF_6, Cs_2NaCrF_6
Rhombohedral, R$\bar{3}$m, a = 6.267, 6.243, c = 30.48, 30.33 Å, Z = 6. Cs(1), Cs(2), Na in 6(c): z = 0.1278 (0.1280); 0.2813 (0.2812); 0.4024 (0.4023); Fe/Cr(1) in 3(a); Fe/Cr(2) in 3(b); F(1) in 18(h): x = 0.1412 (0.1413), z = 0.4620 (0.4620); F(2) in 18(h): x = 0.1868 (0.1883), z = 0.6308 (0.6310).

Cs_2LiGaF_6
Trigonal, P$\bar{3}$m1, a = 6.249, c = 5.086 Å, Z = 1. Cs in 2(d): z = 0.2699; Li in 1(b); Ga in 1(a); F in 6(i): x = 0.1388, z = 0.7618.

The materials have elpasolite-type structures. Film and diffractometer data.

1. Strukturbericht, 2, 498; 40A, 137.

CALCIUM PENTAFLUOROFERRATE(III)

$CaFeF_5$

R. von der MÜHLL and J. RAVEZ, 1974. Rev. Chim. Minér., 11, 652-663.

Monoclinic, $P2_1/c$, a = 5.500, b = 10.050, c = 7.584 Å, β = 110.49°, D_m = 3.22, Z = 4. Mo radiation, R = 0.056 for 561 reflexions.

The structure contains zigzag chains along c of FeF_6 octahedra sharing opposite corners; Fe-F = 1.82-2.04(1) Å. Ca has 7 F neighbours at 2.00-2.68 Å.

HEXAFLUORORHODATES(III)

A_2BRhF_6

V. WILHELM and R. HOPPE, 1975. Z. anorg. Chem., 414, 91-96.

Cubic, Fm3m, Z = 4, powder data. Elpasolite structure (1), Rh-F = 1.99 Å.

A	B	a(Å)	D_m	x(F)
Cs	K	9.049	4.59	0.220
Rb	K	8.876	4.01	0.224
Tl	Na	8.526	7.00	0.233
Rb	Na	8.492	4.42	0.234
K	Na	8.362	3.57	0.238

1. Strukturbericht, 2, 498; Structure Reports, 40A, 137.

HEXAFLUOROPLATINATES(IV)

$APtF_6$

V. WILHELM and R. HOPPE, 1975. Z. anorg. Chem., 414, 130-136.

$PbPtF_6$, rhombohedral, [R$\bar{3}$m], a = 7.23, c = 7.07 Å, D_m = 7.98, [Z = 3], powder data. $BaGeF_6$-type structure (1), F at (0.128, 0.128, 0.144), Pt-F = 1.90, Pb-F = 2.57, 2.98 Å.

$APtF_6$, A = Ca, Zn, Cd, Hg, Mn, Co, Ni, rhombohedral, [R$\bar{3}$], a = 5.25, 4.98, 5.12, 5.13, 5.09, 5.00, 4.94, c = 14.78, 13.83, 14.62, 14.81, 14.23, 13.82, 13.69 Å, D_m = 4.86, 6.31, 6.28, 7.55, 5.64, 6.09, 6.35, [Z = 3], powder data. $LiSbF_6$-type structure (2), F at (0.315 to 0.325, 0.014 to 0.048, -0.070 to -0.085), Pt-F = 1.88-1.92 Å.

1. Structure Reports, 9, 196.
2. Ibid., 27, 460.

ERBIUM NEODYMIUM FLUORIDE

$(Er_{0.3}Nd_{0.7})F_3$

K. OKAMURA and S. YAJIMA, 1974. Bull. Chem. Soc. Japan, 47, 1531-1532.

Orthorhombic, Pnma, a = 6.660, b = 7.053, c = 4.393 Å, Z = 4. Cu radiation, R = 0.10 for powder data.

Isostructural with YF_3 and ErF_3 (1).

1. Structure Reports, 17, 328; 21, 209.

THALLIUM DICHLORIDE

(THALLIUM(I) TETRACHLOROTHALLATE(III))

$TlCl_2$

G. THIELE and W. RINK, 1975. Z. anorg. Chem., 414, 231-235.

Tetragonal, $I4_1/a$, a = 6.965, c = 15.528 Å, D_m = 4.64, Z = 8. Mo radiation, R = 0.033 for 1160 reflexions. Tl(1) in 4(a); Tl(2) in 4(b): Cl in 16(f): (0.2452, 0.1550, 0.0856) [origin at $\bar{4}$].

Scheelite-type ($CaWO_4$) structure (1), so that the compound is $Tl(I)^+Tl(III)Cl_4^-$. Tl(III)-Cl = 2.42(1); Tl(I)-Cl = 3.27 and 3.29 Å (8-coordination).

1. Strukturbericht, 1, 347, 385; Structure Reports, 9, 171; 29, 331.

THALLIUM CHLORIDE BROMIDE

$TlCl_{2-x}Br_x$ (x = 0.83)

G. THIELE and W. RINK, 1975. Z. anorg. Chem., 414, 47-55.

Orthorhombic, $P2_12_12_1$, a = 6.823, b = 11.867 [also given as 11.891], c = 9.344 [also given as 9.506] Å, D_m = 5.45, Z = 8. Mo radiation, R = 0.038 for 733 reflexions.

Atomic positions

	x	y	z
Tl(III)	0.2680	0.0007	0.4867
Tl(I)	0.6905	0.1720	0.1998
X(1)	0.0421	0.8240	0.0319
X(2)	0.1825	0.1802	0.1285
X(3)	0.0743	0.4897	0.0963
X(4)	0.4541	0.9302	0.2654

The material is a thallium(I) tetrahalogenothallate(III). The Tl(III)X_4^- group is a distorted tetrahedron (Tl-X = 2.47-2.57(1) Å, X-Tl-X = 95-126°), with an X atom from a neighbouring group completing a distorted trigonal bipyramid (Fig. 1a), and forming chains along c; Tl...X = 3.12 Å, X-Tl...X = 169°. Tl(I) has nine-coordination (Fig. 1b), Tl-$\bar{X}$ = 3.24-2.59 Å.

Phases with x < 0.28 have a $TlCl_2$-type ($CaWO_4$-type) structure (1); those with x > 1.83 have a $TlBr_2$-type structure (2).

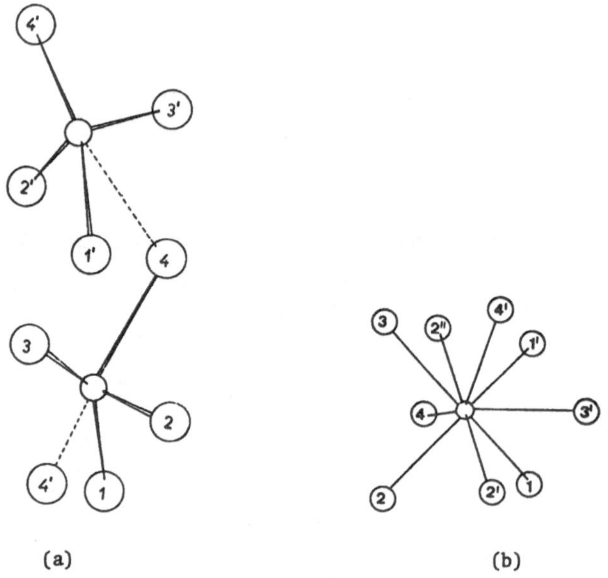

Fig. 1. Coordination of (a) Tl(III) and (b) Tl(I) in $TlCl_{1.17}Br_{0.83}$.

1. Preceding report.
2. Structure Reports, 28, 99.

NITROGEN TRICHLORIDE

NCl_3

H. HARTL, J. SCHÖNER, J. JANDER and H. SCHULZ, 1975. Z. anorg. Chem., 413, 61-71.

Orthorhombic, Pnma, a = 7.48, b = 9.35, c = 16.48 Å (at -125°C), D_m = 2.05, Z = 12. Cu radiation, R = 0.091 for 898 reflexions (film data).

The structure (Fig. 1) contains trigonal-pyramidal molecules, N-Cl = 1.75(1) Å, Cl-N-Cl = 107°, with weak intermolecular interactions, Cl...N = 3.19-3.68, Cl...Cl = 3.37-3.55 Å.

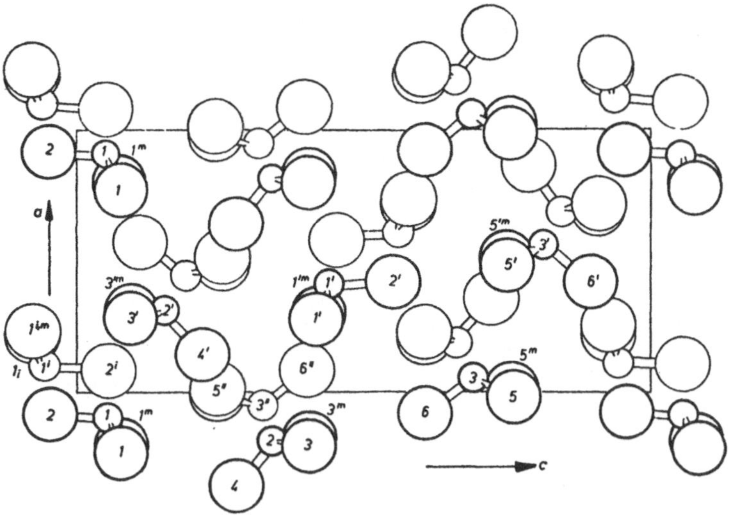

Fig. 1. Structure of nitrogen trichloride.

VANADIUM(III) BROMIDE HEXAHYDRATE
(trans-DIBROMOTETRAAQUOVANADIUM(III) BROMIDE DIHYDRATE)

VANADIUM(III) CHLORIDE HEXAHYDRATE
(trans-DICHLOROTETRAAQUOVANADIUM(III) CHLORIDE DIHYDRATE)

$VBr_3.6H_2O$ $[VBr_2(H_2O)_4]Br.2H_2O$
$VCl_3.6H_2O$ $[VCl_2(H_2O)_4]Cl.2H_2O$

W.F. DONOVAN and P.W. SMITH, 1975. J. Chem. Soc., Dalton, 894-896.

Bromide
Monoclinic, $P2_1/c$, a = 6.408, b = 6.550, c = 12.300 Å, β = 96.15°, D_m = 2.48, Z = 2. Cu radiation, R = 0.11 for 372 reflexions (film data).

Chloride
Monoclinic, $P2_1/c$, a = 6.430, b = 6.439, c = 11.901 Å, β = 98.8°, D_m = 1.65, Z = 2. Cu radiation, R = 0.12 for 126 reflexions (film data).

The compounds are isostructural and contain trans-octahedral $[VX_2(H_2O)_4]^+$ cations (Fig. 1); V-Br = 2.539, V-Cl = 2.361, V-OH_2 = 1.96-2.02 Å. The cations, halide anions, and water of crystallization are linked by hydrogen bonds, O-H...O and O-H...Br^-.

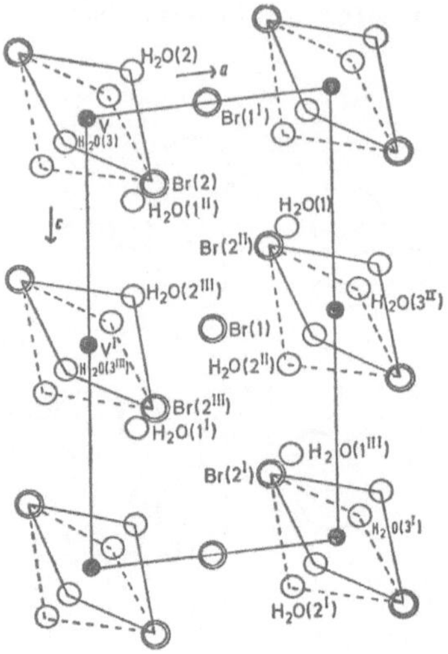

Fig. 1. Structure of $[VBr_2(H_2O)_4]Br.2H_2O$.

IRON(II) CHLORIDE

$FeCl_2$ (high pressure)

C. VETTIER and W.B. YELON, 1975. J. Phys. Chem. Solids, 36, 401-405.

1.04 kbar
Rhombohedral, $R\bar{3}m$, a = 3.598, c = 17.481 Å, Z = 3. Neutron powder data. Fe in 3(a); Cl in 6(c): z = 0.2550.

6.40 kbar
Trigonal, $P\bar{3}m1$, a = 3.585, c = 5.735 Å, Z = 1. Neutron powder data. Fe in 1(a); Cl in 2(d): z = 0.2393.

At 1.04 kbar the structure is as previously described for $FeCl_2$ (1); at 6.40 kbar the material has the $FeBr_2$-type structure (2), the transition occurring at 5.7 kbar.

1. Strukturbericht, 1, 743, 773; 2, 245; Structure Reports, 23, 296.
2. Strukturbericht, 2, 246.

μ-AMIDO-μ-NITRITO-BIS[TETRAAMMINECOBALT(III)] TETRACHLORIDE DIHYDRATE

$[Co_2(NH_3)_8(NH_2)(NO_2)]Cl_4.2H_2O$

C.Y. HUANG, J.A. WEIL and B.E. ROBERTSON, 1975. Acta Cryst., B31, 914-916.

Orthorhombic, $Ccm2_1$, a = 11.221, b = 17.529, c = 9.729 Å, D_m = 1.722, Z = 4. Mo radiation, R = 0.038 for 1068 reflexions.

The cation contains two octahedrally coordinated Co atoms, with bridging NH_2 and ON(O) groups; it lies across the crystallographic mirror plane, so that the O-N bridge is disordered. Co-NH_3 = 1.933-1.998, Co-NH_2 = 1.918, Co-(N,O) = 1.933(4) Å. The structure is held together by an extensive system of hydrogen bonds. A monoclinic form has been described previously (1).

1. Structure Reports, 35A, 409.

CADMIUM CHLORIDE HYDRATE

$CdCl_2.2\cdot 5H_2O$

H. LELIGNY and J.C. MONIER, 1975. Acta Cryst., B31, 728-732.

Monoclinic, $P2_1/n$, a = 9.21, b = 11.88, c = 10.08 Å, β = 93°30', D_m = 2.84, Z = 8. Mo radiation, R = 0.053 for 932 reflexions (films, photometer intensities).

The structure contains a framework of $CdOCl_5$ and CdO_2Cl_4 octahedra sharing edges and corners (Fig. 1), and strengthened by hydrogen bonds. Cd-Cl = 2.54-2.71(1), Cd-O = 2.30-2.35(2), O-H...O = 2.72-3.13 Å.

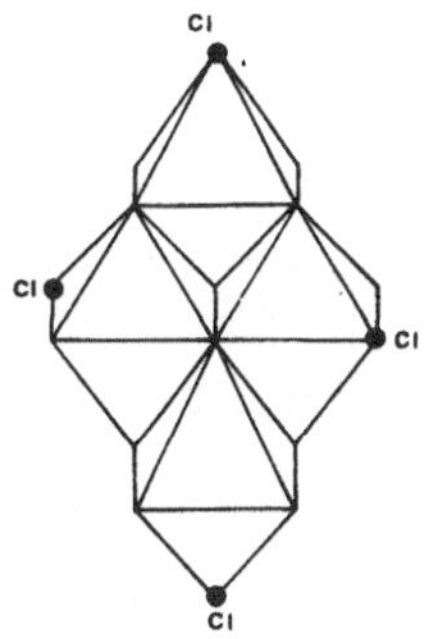

Fig. 1. A group of octahedra in cadmium chloride hydrate; atoms marked with a black circle are shared with neighbouring octahedral groups.

EUROPIUM DICHLORIDE DIHYDRATE

$EuCl_2.2H_2O$

A. HAASE and G. BRAUER, 1975. Acta Cryst., B31, 290-292.

Monoclinic, C2/c, a = 11.661, b = 6.404, c = 6.694 Å, β = 105.37°, D_m = 3.58, Z = 4. Mo radiation, R = 0.038 for 504 reflexions.

Isostructural with $SrCl_2.2H_2O$ (1). Eu has 8-coordination, Eu-4 O = 2.66, 2.82, Eu-4 Cl = 2.84, 2.98 Å.

1. Structure Reports, 9, 161.

PLUTONIUM(III) CHLORIDE

CURIUM(III) BROMIDE

CALIFORNIUM(III) BROMIDE

$PuCl_3$
$CmBr_3$
$CfBr_3$

J.H. BURNS, J.R. PETERSON and J.N. STEVENSON, 1975. J. Inorg. Nucl. Chem., 37, 743-749.

$PuCl_3$
Hexagonal, $P6_3/m$, a = 7.394, c = 4.243 Å, Z = 2. Mo radiation, R = 0.046 for 1112 reflexions. UCl_3-type structure, as previously reported (1).

$CmBr_3$
Orthorhombic, Cmcm, a = 4.041, b = 12.70, c = 9.135 Å, Z = 4. Mo radiation, R = 0.075 for 334 reflexions. $PuBr_3$-type structure (2), as previously suggested (3).

$CfBr_3$
Monoclinic, C2/m, a = 7.214, b = 12.423, c = 6.825 Å, β = 110.7°, Z = 4. Mo radiation, R = 0.095 for 273 reflexions. $AlCl_3$-type structure (4).

Pu-Cl = 2.886-2.919(1), Cm-Br = 2.865-3.137(4), Cf-Br = 2.795-2.828(11) Å.

1. Structure Reports, 11, 278.
2. Ibid., 11, 282.
3. L.B. ASPREY, T.K. KEENAN and F.H. KRUSE, 1965. Inorg. Chem., 4, 985.
4. Structure Reports, 11, 273.

POTASSIUM MAGNESIUM CHLORIDE

CAESIUM MAGNESIUM CHLORIDE

K_2MgCl_4
Cs_2MgCl_4

C.S. GIBBONS, V.C. REINSBOROUGH and W.A. WHITLA, 1975. Canad. J. Chem., 53, 114-118.

K_2MgCl_4
Tetragonal, I4/mmm, a = 4.94, c = 15.58 Å, D_m not given, Z = 2. Mo radiation, R = 0.045 for 180 reflexions.

Cs_2MgCl_4
Orthorhombic, Pnma, a = 9.777, b = 7.514, c = 13.234 Å, D_m not given, Z = 4. Mo radiation, R = 0.074 for 1084 reflexions.

The potassium salt has the K_2NiF_4-type structure (1), and the caesium salt is isostructural with β-K_2SO_4 (2) and Cs_2CuCl_4 (3).

1. Structure Reports, 17, 332; 19, 323.
2. Strukturbericht, 2, 86, 423; Structure Reports, 22, 447.
3. Structure Reports, 16, 188.

RUBIDIUM MAGNESIUM CHLORIDE

$RbMgCl_3$

H.J. SEIFERT and H. FINK, 1975. Rev. Chim. Minér., 12, 466-475.

Hexagonal, $P6_3/mmc$, a = 7.090, c = 11.844 Å, D_m = 2.77, Z = 4. Film data, R = 0.064 for 128 reflexions. Rb in 2(a) and 2(d); Mg in 4(f): z = 0.1171; Cl in 6(g) and 6(h): x = 0.1767.

Hexagonal variant of the perovskite structure. Pairs of face-sharing octahedra are connected by corners; Mg-Cl = 2.48, Rb-Cl = 3.54-3.67 Å.

CAESIUM TETRACHLOROGALLATE

CAESIUM TETRABROMOGALLATE

$CsGaCl_4$
$CsGaBr_4$

R.C. GEARHART, J.D. BECK and R.H. WOOD, 1975. Inorg. Chem., 14, 2413-2416.

Orthorhombic, Pnma, a = 11.66, 12.15, b = 7.16, 7.48, c = 9.41, 9.88 Å, Z = 4. Mo radiation, R = 0.13 for 399 reflexions and 0.21 for 131 reflexions (films, densitometer intensities).

$BaSO_4$-type structure (1).

1. Strukturbericht, 1, 343.

LITHIUM TETRACHLOROMANGANATE(II)

Li_2MnCl_4

C.J.J. van LOON and J. de JONG, 1975. Acta Cryst., B31, 2549-2550.

Cubic, Fd3m, a = 10.503 Å, D_m not given, Z = 8. Neutron powder data. 8 Li + 8 Mn in 16(d), 8 Li in 8(a), 32 Cl in 32(e): x = 0.2564 (origin at centre).

Inverse spinel structure.

SODIUM MANGANESE(II) CHLORIDES

Na_6MnCl_8
$Na_2Mn_3Cl_8$

C.J.J. van LOON and D.J.W. IJDO, 1975. Acta Cryst., B31, 770-773.

Na_6MnCl_8
Cubic, Fm3m, a = 11.2274 Å, Z = 4. X-ray and neutron powder data. Na in 24(d); Mn in 4(a); Cl(1) in 8(c); Cl(2) in 24(e): x = 0.2287.

$Na_2Mn_3Cl_8$
Rhombohedral, $R\bar{3}m$, a = 7.4563, c = 19.591 Å, Z = 3. X-ray and neutron powder data. Na in 6(c): z = 0.157; Mn in 9(e); Cl(1) in 6(c): z = 0.4072; Cl(2) in 18(h): x = 0.4987, z = 0.4052.

Na_6MCl_8 compounds (M = Mg, Mn, Fe, Cd) have the Mg_6MnO_8 structure (1). The structure contains cubic close-packed anions, with cations ordered in octahedral sites; for the Mn compound, Mn-Cl = 2.568, Na-Cl = 2.807 and 2.817(2) Å.

$Na_2M_3Cl_8$ (M = Mg, Mn, Fe, Cd) represents a new structure type, which can be considered as an equal contribution of cubic close-packed and simple hexagonal anion stacking. Between identical layers Na ions occupy sites with trigonal prismatic coordination. M ions occupy three-quarters of the octahedral sites between different anion layers. For the Mn compound, Mn-Cl = 2.565 and 2.594, Na-Cl = 2.738 and 2.957(6) Å.

1. Structure Reports, 18, 524.

HEXACHLOROTECHNETIC ACID NONAHYDRATE

$H_2TcCl_6.9H_2O$

P.A. KOZ'MIN and G.N. NOVICKAJA, 1975. Koordin. Khim., 1, 473-476.

Triclinic, P1, a = 8.30, b = 7.35, c = 8.04 Å, α = 95, β = 98, γ = 117.5°, Z = 1. R = 0.19 for 1300 reflexions.

The structure contains octahedral $TcCl_6^{2-}$ ions, Tc-Cl = 2.22-2.51(3) Å, and H_2O and H_3O^+ species [only 6 O were located].

POTASSIUM OCTACHLORODITECHNETATE HYDRATE

$K_3Tc_2Cl_8.nH_2O$

I. F.A. COTTON and L.W. SHIVE, 1975. Inorg. Chem., 14, 2032-2035.

Trigonal, $P3_121$, a = 12.838, c = 8.187 Å, D_m = 2.79, Z = 3. Mo radiation, R = 0.051 for 1402 reflexions.

Isostructural with $(NH_4)_3Tc_2Cl_8.2H_2O$ (1), but with one of the water molecules having an occupancy of only 0.8. The $[Cl_4Tc\text{-}TcCl_4]^{3-}$ ion has approximately D_{4h} symmetry, Tc having square-pyramidal coordination, Tc-Tc = 2.117(2), Tc-Cl = 2.36 Å, Tc-Tc-Cl = 104-106°.

$K_8[Tc_2Cl_8]_3(H_3O).3H_2O$

II. P.A. KOZ'MIN and G.N. NOVICKAJA, 1975. Koordin. Khim., 1, 248-251.

Trigonal, $P3_121$, a = 12.80, c = 8.30 Å, D_m = 2.79, Z = 1. R = 0.16 for 450 reflexions (film data).

This compound and the Cs analogue were previously described as $M_3Tc_2Cl_8.2H_2O$ (2). The present results give a similar crystal structure, but with only 2 K distributed in position 3(a) and partial occupancy of water positions.

1. Structure Reports, 35A, 426.
2. Ibid., 39A, 176.

POTASSIUM TRICHLOROFERRATE(II)

POTASSIUM TRIBROMOFERRATE(II)

$KFeCl_3$
$KFeBr_3$

I. E. GUREWITZ, J. MAKOVSKY and H. SHAKED, 1974. Phys. Rev., B, 9, 1071-1076.

II. M. AMIT, A. HOROWITZ and J. MAKOVSKY, 1974. Israel J. Chem., 12, 827-830.

Orthorhombic, Pnma, a = 8.712, 9.220, b = 3.845, 4.026, c = 14.15, 14.899 Å, D_m not given, Z = 4. X-ray and neutron powder data.

Isostructural with NH_4CdCl_3 (1) and $KCdCl_3$ (2).

1. Strukturbericht, 6, 13, 79; 7, 19, 115.
2. Structure Reports, 11, 434.

CAESIUM PENTABROMOFERRATE(II)

CAESIUM PENTABROMOMANGANATE(II)

THALLIUM(I) PENTACHLOROFERRATE(II)

THALLIUM(I) PENTACHLOROCOBALTATE(II)

AMMONIUM PENTACHLOROFERRATE(II)

AMMONIUM PENTACHLOROCOBALTATE(II)

Cs_3FeBr_5
Cs_3MnBr_5
Tl_3FeCl_5
Tl_3CoCl_5
$(NH_4)_3FeCl_5$
$(NH_4)_3CoCl_5$

M. AMIT, A. HOROWITZ (ZODKEVITZ), E. RON and J. MAKOVSKY, 1973. Israel J. Chem., 11, 749-763.

Cs_3FeBr_5, Cs_3MnBr_5, Tl_3FeCl_5, Tl_3CoCl_5
Tetragonal, I4/mcm, a = 9.600, 9.596, 8.400, 8.431, c = 15.35, 15.57, 14.74, 14.50 Å, D_m not given, 3.23, 5.21, 5.27 [poor agreement with calculated values], Z = 4 , powder data. Cs_3CoCl_5-type structure (1).

$(NH_4)_3FeCl_5$, $(NH_4)_3CoCl_5$
Orthorhombic, Pnma, a = 8.687, 8.714, b = 10.006, 9.914, c = 12.687, 12.669 Å, D_m = 1.70, 1.77, Z = 4 , powder data. $(NH_4)_2ZnCl_5$-type structure (2).

1. Strukturbericht, 3, 134, 498; Structure Reports, 29, 277.
2. Structure Reports, 9, 202.

POTASSIUM HEXACHLORORHODATE(III) MONOHYDRATE

$K_3RhCl_6.H_2O$

P. MURRAY-RUST and J. MURRAY-RUST, 1975. Acta Cryst., B31, 1037-1040.

Orthorhombic, Pbcn, a = 12.40, b = 15.65, c = 12.05 Å, D_m not given, Z = 8. Mo radiation, R = 0.046 for 2042 reflexions.

The results are in agreement with those of an independent study (1).

1. Structure Reports, 38A, 215.

POTASSIUM PENTACHLORONITROSYLIRIDATE MONOHYDRATE

POTASSIUM PENTABROMONITROSYLIRIDATE MONOHYDRATE

$KIrCl_5NO.H_2O$
$KIrBr_5NO.H_2O$

F. BOTTOMLEY, 1975. J. Chem. Soc., Dalton, 2538-2541.

Orthorhombic, Pnma, a = 22.416, 23.272, b = 6.935, 7.261, c = 6.069, 6.302 Å, D_m not measured, Z = 4. Mo radiation, R = 0.049 and 0.096 for 1672 and 1291 reflexions, respectively.

The structures contain well-separated K^+ and IrX_5NO^- ions and water molecules. The anions have crystallographic C_s symmetry, and show small deviations from C_{4v} symmetry; Ir-Cl = 2.286 (trans to NO), 2.339(3) (cis to NO), Ir-Br = 2.419, 2.480(4), Ir-N = 1.76(1), 1.71(3) Å, Ir-N-O = 174, 170°.

MILLON'S SALT

BECTON'S SALT

$Cu(NH_3)_4PtCl_4$
$Pt(NH_3)_4CuCl_4$

B. MOROSIN, P. FALLON and J.S. VALENTINE, 1975. Acta Cryst., B31, 2220-2223.

Millon's salt, $Cu(NH_3)_4PtCl_4$
Tetragonal, P4/mnc, a = 9.036, c = 6.441 Å, D_m not given, Z = 2. Mo radiation, R = 0.029 for 273 reflexions.

Becton's salt, $Pt(NH_3)_4CuCl_4$
Monoclinic, $P2_1/c$, a = 7.687, b = 7.941, c = 8.057 Å, β = 91.61°, D_m not given, Z = 2. Mo radiation, R = 0.038 for 852 reflexions.

Atomic positions

Millon's salt, P4/mnc

	x	y	z
Pt	0	0	0
Cu	1/2	1/2	0
Cl	0.0572	0.2477	0
N	0.559	0.713	0

Becton's salt, $P2_1/c$

	x	y	z
Pt	1/2	0	1/2
Cu	0	0	0
Cl(1)	-0.0103	-0.0356	0.2792
Cl(2)	0.2068	0.2073	0.0368
N(1)	0.519	-0.032	0.251
N(2)	0.317	0.184	0.459

Millon's salt is isostructural with Magnus's green salt, $Pt(NH_3)_4PtCl_4$ (1). The structure (Fig. 1) contains stacks along c of alternating square-planar $Cu(NH_3)_4^{2+}$ and $PtCl_4^{2-}$ ions; Cu-N = 2.00(1), Pt-Cl = 2.298(4) Å. N-H...Cl hydrogen bonds link the ions within and between stacks (Fig. 1).

Becton's salt (Fig. 2) contains square-planar $Pt(NH_3)_4^{2+}$ ions and infinite Cu-Cl layers in which Cu has distorted octahedral coordination, with four Cu-Cl = 2.271 and 2.302 Å, and two Cu-Cl = 3.257(4) Å.

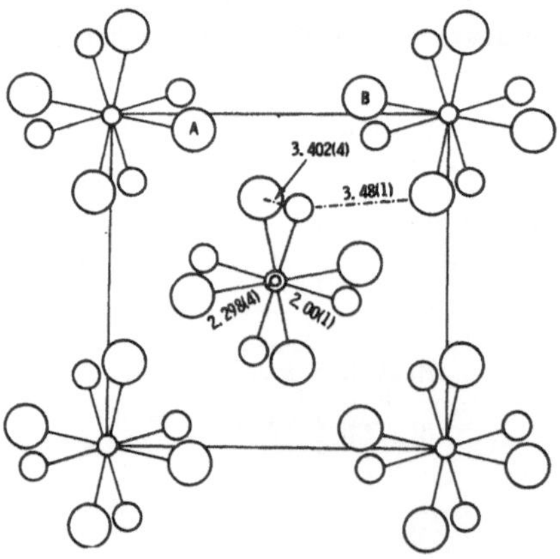

Fig. 1. Structure of Millon's salt, $Cu(NH_3)_4PtCl_4$.

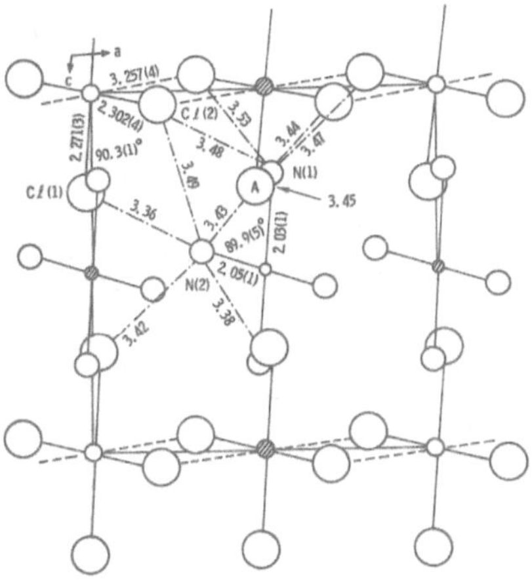

Fig. 2. Structure of Becton's salt, $Pt(NH_3)_4CuCl_4$.

1. Structure Reports, 21, 419.

HEXAAMMINECOBALT(III) CHLOROCUPRATE(I,II)

$[Co(NH_3)_6][CuCl_{4.76}]$

P. MURRAY-RUST, 1975. Acta Cryst., B31, 978-981.

Cubic, Fd3c, a = 21.81 Å, D_m = 2.00, Z = 32. Mo radiation, R = 0.08 for 171 reflexions (film data).

The structure is similar to that of $[Cr(NH_3)_6][CuCl_5]$ (1), with replacement of $Cu(II)Cl_5^{3-}$ trigonal bipyramids by disordered $Cu(I)Cl_4^{3-}$ tetrahedra. The Cu(I) end-member is described in 2.

1. Structure Reports, 33A, 207.
2. Ibid., 39A, 178.

AMMONIUM TRISTETRACHLOROAURATE(III) µ-CHLORO-BISDICHLOROARGENTATE(I)

$(NH_4)_6(AuCl_4)_3Ag_2Cl_5$

J.C. BOWLES and D. HALL, 1975. Acta Cryst., B31, 2149-2150.

Orthorhombic, Immm, a = 20.86, b = 11.20, c = 6.61 Å, D_m = 3.18, Z = 2. Cu radiation, R = 0.11 for 625 reflexions (films, visual intensities).

Two of the three $AuCl_4^-$ ions stack alternately with the chlorine-bridged $Ag_2Cl_5^{3-}$ ion along b (Fig. 1) to form a double-stranded analogue of the chain structure in $Cs_2(AuCl_4)(AgCl_2)$ (1). The third $AuCl_4^-$ ion interleaves the $Ag_2Cl_5^{3-}$ ions along c. Au-Cl = 2.18-2.29, Au...Cl = 2.21, 2.31, Ag-Cl = 2.46 (terminal), 2.69 Å (bridging), Cl(terminal)-Ag-Cl(terminal) = 153°. N...Cl contacts range upwards from 3.26 Å, and weak hydrogen bonding may be involved.

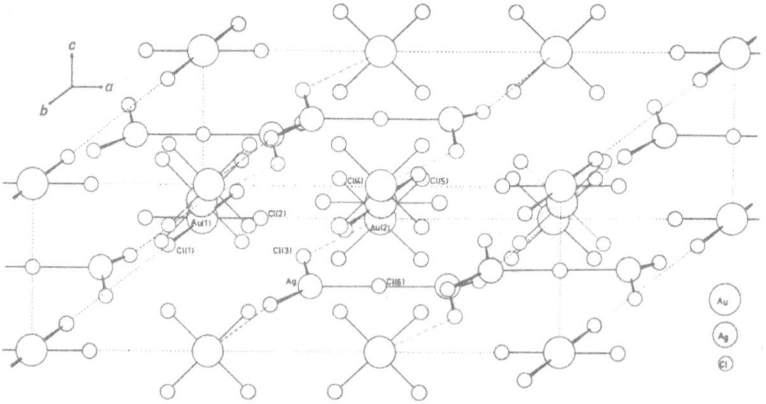

Fig. 1. Structure of $(NH_4)_6(AuCl_4)_3Ag_2Cl_5$; ammonium ions are omitted for clarity.

1. Strukturbericht, 6, 34, 126.

CAESIUM CADMIUM CHLORIDE

$CsCdCl_3$

J.R. CHANG, G.L. McPHERSON and J.L. ATWOOD, 1975. Inorg. Chem., 14, 3079-3085.

Hexagonal, $P6_3/mmc$, a = 7.403, c = 18.406 Å, D_m = 3.96, Z = 6. Mo radiation, R = 0.044 for 354 reflexions.

Atomic positions

		x	y	z
Cs(1)	in 2(b)	0	0	1/4
Cs(2)	4(f)	1/3	2/3	0.0890
Cd(1)	2(a)	0	0	0
Cd(2)	4(f)	1/3	2/3	0.8405
Cl(1)	6(h)	0.5069	1.0138	1/4
Cl(2)	12(k)	0.8349	1.6698	0.0816

The structure is as previously described (1). Cd(1)-Cl = 2.598, Cd-Cl(2) = 2.588 and 2.639 Å.

1. Structure Reports, 29, 280.

HEXAAMMINECHROMIUM(III) PENTACHLOROMERCURATE(II)

$[Cr(NH_3)_6][HgCl_5]$

W. CLEGG, D.A. GREENHALGH and B.P. STRAUGHAN, 1975. J. Chem. Soc., Dalton, 2591-2593.

Cubic, Fd3c, a = 22.653 Å, D_m = 2.44, Z = 32. Mo radiation, R = 0.045 for 249 reflexions.

Atomic positions

	x	y	z
Hg	1/4	1/4	1/4
Cl(ax)	0.1858	0.1858	0.1858
Cl(eq)	1/4	0.1676	0.3324
Cr	0	0	0
N	0.0715	0.0498	-0.0288

The structure contains octahedral $[Cr(NH_3)_6]^{3+}$ cations, Cr-N = 2.079(8) Å, and trigonal bipyramidal $[HgCl_5]^{3-}$ anions, Hg-Cl = 2.518 (axial), 2.640(4) Å (equatorial).

ZIRCONIUM(III) BROMIDE

$ZrBr_3$

I. E.M. LARSEN, J.W. MOYER, F. GIL-ARNAO and M.J. CAMP, 1974. Inorg. Chem., 13, 574-581.

II. J. KLEPPINGER, J.C. CALABRESE and E.M. LARSEN, 1975. Ibid., 14, 3128-3130.

Hexagonal, $P6_3/mcm$, a = 6.728, c = 6.299 Å, Z = 2. Mo radiation, R = 0.083 and 0.023 for 190 and 92 reflexions. Zr in 2(b); Br in 6(g): x = 0.3212.

The structure is as previously described (1). Zr-Br = 2.674(1) Å.

1. Structure Reports, 29, 249.

MANGANESE(II) BROMIDE TETRAHYDRATE

$MnBr_2.4H_2O$

K. SUDARSANAN, 1975. Acta Cryst., B31, 2720-2721.

Monoclinic, $P2_1/n$, a = 11.668, b = 9.824, c = 6.316 Å, β = 99.43°, D_m not given, Z = 4. Mo radiation, R = 0.028 for 605 reflexions.

Isostructural with the chloride (1). The Mn coordination octahedron is slightly more distorted in the bromide; Mn-Br = 2.80, 2.91, Mn-O = 2.22-2.48(1) Å, angles at Mn = 74-101, 171-179°.

1. Structure Reports, 29, 273; 37A, 195.

GOLD(III) BROMIDE

Au_2Br_6

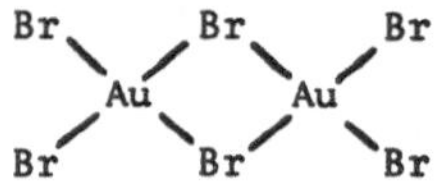

K.-P. LÖRCHER and J. STRÄHLE, 1975. Z. Naturforsch, 30B, 662-664.

Monoclinic, $P2_1/c$, a = 6.831, b = 20.41, c = 8.105 Å, β = 119.74°, Z = 4. Mo radiation, R = 0.113 for 1169 reflexions.

The structure contains well-separated, nearly-planar Au_2Br_6 molecules, in which Au has square-planar coordination, Au-Br = 2.38-2.41 (terminal), 2.46-2.47 Å (bridging).

URANIUM(III) BROMIDE

UBr_3

J.H. LEVY, J.C. TAYLOR and P.W. WILSON, 1975. J. Less-Common Metals, 39, 265-270.

Hexagonal, $P6_3/m$, a = 7.942, c = 4.441 Å, D_m not given, Z = 2. Neutron powder data with profile analysis; U in 2(d); Br in 6(h): x = 0.2996, y = 0.3859.

UCl_3-type structure (1). U has 9-coordination (tricapped trigonal prism), U-Br = 3.062 (x 6), 3.145 (x 3) Å.

1. Structure Reports, 11, 277.

AMMONIUM HEPTABROMODIALUMINATE

$NH_4Al_2Br_7$

E. RYTTER, B.E.D. RYTTER, H.A. ØYE and J. KROGH-MOE, 1975. Acta Cryst., B31, 2177-2181.

Orthorhombic, $Pna2_1$, a = 12.560, b = 9.501, c = 11.423 Å, D_m not given, Z = 4. Mo radiation, R = 0.13 for 1231 reflexions.

The structure (Fig. 1) contains NH_4^+ and $Al_2Br_7^-$ ions, the latter consisting of two tetrahedra sharing a corner; Al-Br = 2.23-2.29 (terminal), 2.41 and 2.45(2) Å (bridging), Al-Br-Al = 108°. The ammonium ion has seven bromine neighbours at 3.30-3.75 Å.

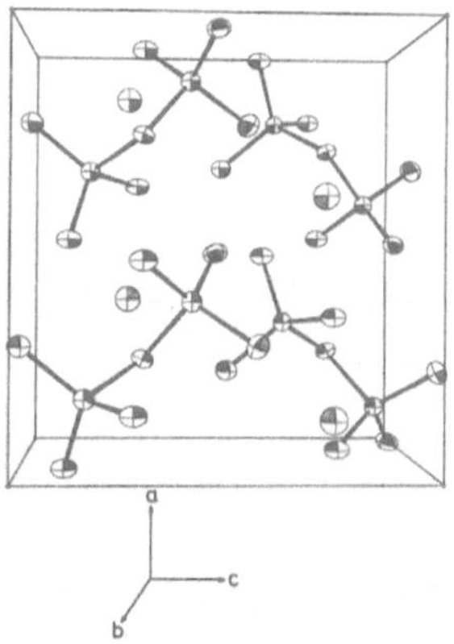

Fig. 1. Structure of $NH_4Al_2Br_7$.

CAESIUM TRIBROMOSTANNATE(II)

$CsSnBr_3$

J.D. DONALDSON, J. SILVER, S. HADJIMINOLIS and S.D. ROSS, 1975. J. Chem. Soc., Dalton, 1500-1506.

Cubic, Pm3m, a = 5.795 Å, D_m = 4.2, Z = 1. Mo radiation, R = 0.11 for 97 reflexions (film data). Cs in 1(a), Sn in 1(b), Br in 3(c).

Ideal perovskite structure (1).

1. Strukturbericht, 1, 300.

POTASSIUM TETRABROMOPALLADATE(II)

K_2PdBr_4

D.S. MARTIN, J.L. BONTE, R.M. RUSH and R.A. JACOBSON, 1975. Acta Cryst., B31, 2538-2539.

Tetragonal, P4/mmm, a = 7.409, c = 4.309 Å, D_m not given, Z = 1. Mo radiation, R = 0.040 for 910 reflexions. Pd in 1(a); K in 2(e); Br in 4(j): x = 0.23323.

K_2PtCl_4-type structure (1). Pd-Br = 2.444(3), K...Br = 3.396(3) Å.

1. Strukturbericht, 1, 358, 424; Structure Reports, 38A, 216.

LEAD IODIDE

PbI_2

I. M. CHAND and G.C. TRIGUNAYAT, 1975. Acta Cryst., B31, 1222-1223.

II. Idem, 1976. Ibid., B32, 1619.

Three polytypes are described, $12R_2$, $18R_1$, and $18R_2$, with structures $(13)_3$, $(1311)_3$, and $[(21)_2]_3$, respectively, in Ždanov notation.

III. T. MINAGAWA, 1975. Acta Cryst., A31, 823-824.

Trigonal, P$\bar{3}$m1, a = 4.557, c = 6.979 Å, Z = 1. Cu radiation, film data, R = 0.11 for 53 reflexions. Pb in 1(a); I in 2(d): z = 0.268.

The material is the 2H polytype (1), and transforms to the $12R[13]_3$ form at higher temperatures.

1. Strukturbericht, 1, 163, 191; Structure Reports, 9, 146; 23, 319.

TETRAAMMINECOPPER(II) TETRAIODIDE

$Cu(NH_3)_4I_4$

E. DUBLER and L. LINOWSKY, 1975. Helv. Chim. Acta, 58, 2604-2609.

Monoclinic, C2/m, a = 14.172, b = 8.926, c = 6.558 Å, β = 128.65°, D_m not given, Z = 2. Mo radiation, R = 0.031 for 830 reflexions.

Atomic positions

	x	y	z
Cu	0	1/2	0
I(1)	0.0872	0	0.4628
I(2)	0.2951	0	0.3739
N	0.4072	0.1594	0.0251

The structure contains Cu atoms with tetragonally-distorted octahedral coordination, Cu-4N = 2.013 (x 4), Cu-2I = 3.224 (x 2) Å. The coordinated I atoms are the terminal atoms of linear, centrosymmetric I_4^{2-} ions, I-I = 2.802 (central), 3.342(1) Å (terminal) (corresponding to bond orders of 0.80 and 0.43, respectively); the I_4^{2-} ions thus bridge Cu atoms to form infinite $[Cu(NH_3)_4I_4]$ chains.

SILVER IODIDE

AgI

I. Q. JOHNSON and R.N. SCHOCK, 1975. Acta Cryst., B31, 1482-1483.

II. Idem, 1976. Ibid., B32, 1304.

Hexagonal, $P6_3mc$, a = 4.598, c = 15.029 Å, D_m not given, Z = 4. Mo radiation, R = 0.024 for 52 reflexions (0.053 if the 110 reflexion is included). Ag(1) in 2(a): z = 1/4; Ag(2) in 2(b): z = 0; I(1) in 2(a): z = 0.4384; I(2) in 2(b): z = z(I(1))-1/4.

The material is a 4H polytype, and has essentially the wurtzite-type structure apart from stacking differences. Ag-I = 2.811 (x 3), 2.832 Å.

URANIUM(III) IODIDE

UI_3

J.H. LEVY, J.C. TAYLOR and P.W. WILSON, 1975. Acta Cryst., B31, 880-882.

Orthorhombic, Ccmm, a = 14.011, b = 4.328, c = 10.005 Å, D_m not given, Z = 4. Neutron powder data with profile analysis.

$PuBr_3$-type structure, as previously proposed (1). U has eight-coordination (trigonal prism, with two rectangular faces capped), U-F = 3.244 (x 4), 3.165 (x 2), 3.456(11) Å (x 2).

1. Structure Reports, 11, 282.

CHROMIUM(II) IRON(II) IODIDE

β-$(Cr,Fe)I_2$

L. GUEN, NGUYEN HUY DUNG, R. EHOLIE and J. FLAHAUT, 1975. Ann. Chim., 10, 11-16.

Monoclinic, B2/m, a = 7.388, b = 7.410, c = 3.965 Å, γ = 114°, Z = 2. Mo radiation, R = 0.15 for 89 reflexions (film data). (Cr,Fe) in 2(a); I in 4(i): (0.735, 0.239, 0).

Isostructural with CrI_2 (1).

1. Structure Reports, 27, 432.

DIPOTASSIUM SILVER TRIIODIDE

K_2AgI_3

I. M.M. THACKERAY and J. COETZER, 1975. Acta Cryst., B31, 2339-2340.

II. C.B. SHOEMAKER, 1976. Ibid., B32, 1619.

Orthorhombic, Pnma, a = 10.01, b = 4.78, c = 19.32 Å, D_m not given, Z = 4. Mo radiation, R = 0.062 for 649 reflexions.

Atomic positions

Atoms in 4(c)

	x	y	z
I(1)	0.4895	1/4	0.4013
I(2)	0.8807	3/4	0.4294
I(3)	0.6884	3/4	0.2226
Ag	0.6280	3/4	0.3646
K(1)	0.9223	1/4	0.2880
K(2)	0.7526	1/4	0.5403

The structure is essentially as previously determined (1). The I^- ions form three distinct geometrical units (Fig. 1), tetrahedra, square pyramids, and square prisms. The tetrahedra are face-shared and one out of every three contains a Ag^+ ion. Square prisms accommodate the K^+ ions. Ag-I = 2.81-2.85(1), K-I = 3.46-3.72(1) Å.

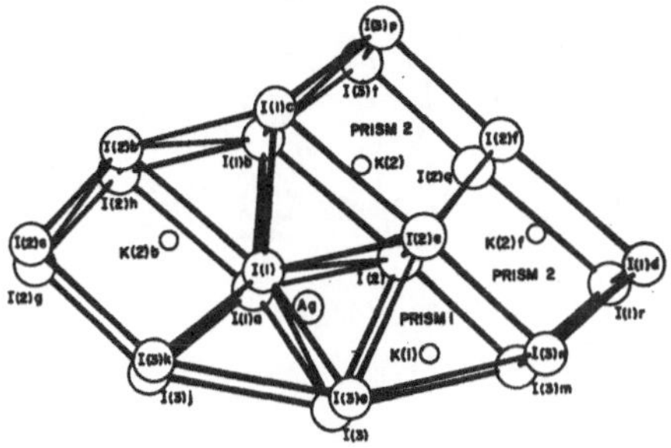

Fig. 1. Structure of K_2AgI_3.

1. Structure Reports, 15, 169.

CADMIUM ARSENIC IODIDE

$Cd_4As_2I_3$

J. GALLAY, G. ALLAIS and A. DESCHANVRES, 1975. Acta Cryst., B31, 2274-2276.

Cubic, Pa3, a = 12.993 Å, D_m = 5.93, Z = 8. Mo radiation, R = 0.08 for 608 reflexions.

Atomic positions

			x	y	z
I	in	24(d)	0.1862	0.4369	0.2567
Cd(2)		24(d)	0.0341	0.0048	0.2583
Cd(1)		8(c)	0.2176		
As(1)		8(c)	0.1033		
As(2)		8(c)	0.4467		

The structure (Fig. 1) is characterized by As(2)-As(2) bonds, 2.40(1) Å. Cd(2) has octahedral coordination to 4 I (3.00-3.74 Å) and 2 As (2.55, 2.57 Å) atoms; Cd(1) has 3 I at 2.92, 3 I at 3.89, and 1 As at 2.57 Å.

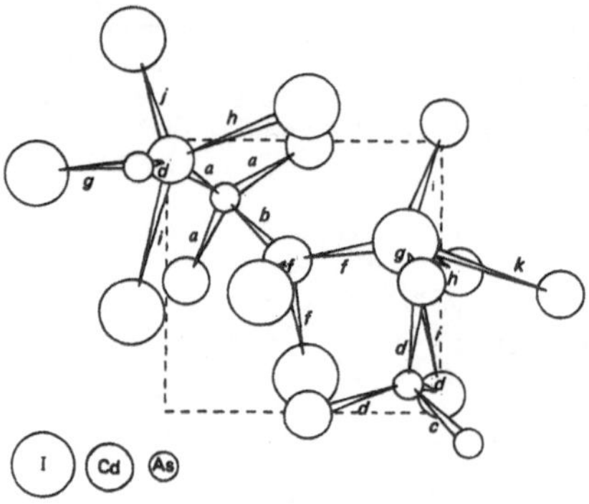

Fig. 1. Structure of $Cd_4As_2I_3$.

CAESIUM OCTAIODOTRIMERCURATE(II)

$Cs_2Hg_3I_8$

P.M. FEDEROV and V.I. PAKHOMOV, 1975. Koordin. Khim., 1, 670-674.

Monoclinic, Cm, a = 7.43, b = 21.70, c = 7.68 Å, β = 108°, D_m = 5.14, Z = 2.
Mo radiation, R = 0.15 for 670 reflexions (film data).

Atomic positions

			x	y	z
Cs	in	4(b)	0.247	0.1258	0.561
Hg(1)		2(a)	0	0	0
Hg(2)		4(b)	0.889	0.1950	0.915
I(1)		2(a)	0.336	0	0.233
I(2)		2(a)	0.882	0	0.631
I(3)		4(b)	0.286	0.1931	0.057
I(4)		4(b)	0.727	0.1952	0.556
I(5)		4(b)	0.285	0.4014	0.111

The structure contains HgI_4 tetrahedra which share corners to form a polymeric $Hg_3I_8^{2-}$ layer parallel to (001); Hg-I = 2.65-2.95(2) Å, I-Hg-I = 93-129°. Cs has eight iodine neighbours at 3.88-4.23 Å.

SILVER MERCURY(II) IODIDE

α-Ag_2HgI_4

J.S. KASPER and K.W. BROWALL, 1975. J. Solid State Chem., 13, 49-56.

Cubic, [F$\bar{4}$3m], a = 6.35 Å at 66°C, Z = 1. Mo radiation, R = 0.037 for 28 reflexions.

Structure as in 1, with high thermal parameters (2) interpreted in terms of anharmonic vibrations.

1. Strukturbericht, 3, 7, 235.
2. S. HOSHINO, 1955. J. Phys. Soc. Japan, 10, 197.

BISMUTH OXIDE FLUORIDE CHLORIDE

$Bi_6O_7FCl_3$

F. HOPFGARTEN, 1975. Acta Cryst., B31, 1087-1092.

Orthorhombic, Pnma, a = 20.105, b = 3.892, c = 15.432 Å, D_m = 8.27, Z = 4. Mo radiation, R = 0.055 for 1214 reflexions.

The Bi atoms are of two types. One type has four O, or O and F, at the basal corners of a square pyramid, with the lone-pair of electrons at the apex; the other type has 2 F + 3 O at five corners of an octahedron, with the lone-pair at the sixth corner. If Cl atoms are included, most of the Bi atom coordinations can be considered as square antiprisms (8-coordination). Bi-O or F = 2.19-2.51, Bi-Cl = 3.14-3.25 Å. The coordination polyhedra share edges and corners to form infinite zigzag nets parallel to b (Fig. 1), between which are trigonal prism columns of chloride ions.

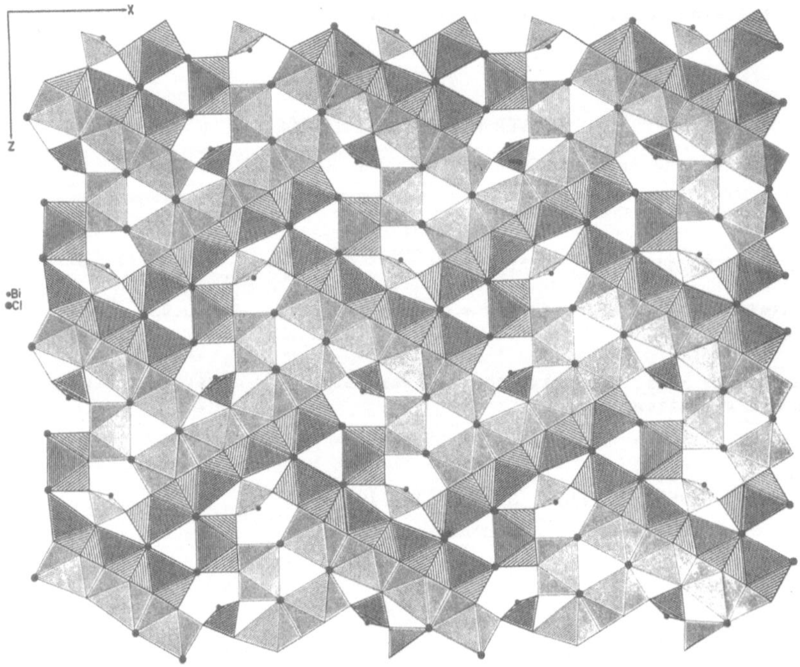

Fig. 1. Structure of $Bi_6O_7FCl_3$.

IODINE OXIDE TRIFLUORIDE

IOF_3

A.J. EDWARDS and P. TAYLOR, 1974. J. Fluor. Chem., 4, 173-179.

Orthorhombic, $P2_12_12_1$, a = 5.70, b = 5.62, c = 10.49 Å, D_m not given, Z = 4. Mo radiation, R = 0.060 for 306 reflexions (film data).

The structure is as previously described (1) except that the equatorial O and F atoms are reversed in the present description (with the comment that the X-ray evidence alone is not conclusive). The structure contains IOF_3 molecules, which may be considered as trigonal bipyramids with O and a non-bonding electron-pair in two of the equatorial positions; I-O = 1.71(4), I-F = 1.84-1.91(3) Å. The molecules are joined by weak bridges, I...O = 2.62, I...F = 2.81 and 3.11 Å.

1. J.W. VIERS and H.W. BAIRD, 1967. Chem. Comm., 1093.

XENON DIFLUORIDE - TUNGSTEN OXYFLUORIDE

$XeF_2.WOF_4$

P.A. TUCKER, P.A. TAYLOR, J.H. HOLLOWAY and D.R. RUSSELL, 1975. Acta Cryst., B31, 906-908.

Monoclinic, $P2_1/c$, a = 5.44, b = 9.97, c = 12.17 Å, β = 92.0°, D_m not given, Z = 4. Mo radiation, R = 0.081 for 817 reflexions.

The structure contains fluorine-bridged molecules (Fig. 1) in which W has octahedral and Xe linear coordination; W-O = 1.65, W-F = 1.79 (terminal), 2.18 (bridging), Xe-F = 1.89 (terminal), 2.04(3) Å (bridging), W-F-Xe = 147°.

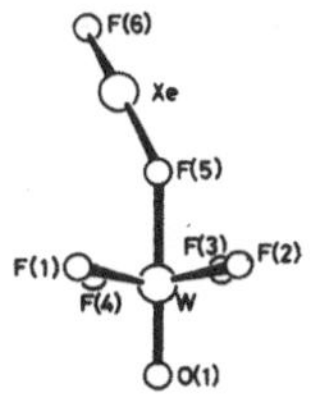

Fig. 1. Structure of the $XeF_2.WOF_4$ adduct.

YTTRIUM OXYFLUORIDE

$Y_7O_6F_9$

D.J.M. BEVAN and A.W. MANN, 1975. Acta Cryst., B31, 1406-1411.

Orthorhombic, Abm2, a = 5.420, b = 38.58, c = 5.527 Å, D_m = 5.09, Z = 4. Cu radiation, R = 0.084 for 349 reflexions.

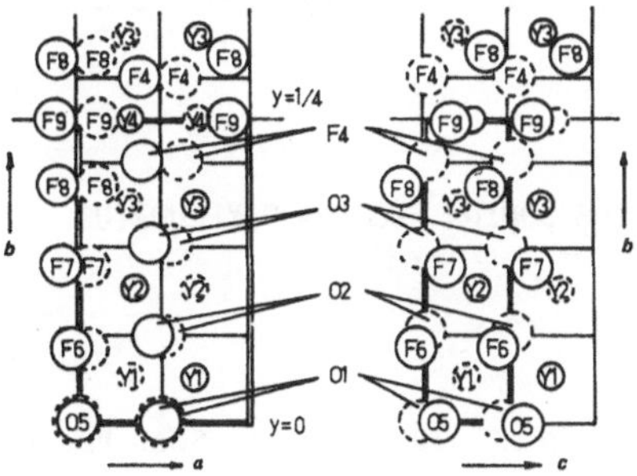

Fig. 1. Projections of the $Y_7O_6F_9$ structure.

The compound is an anion-excess, fluorite-related phase. The structure (Fig. 1) contains two regions of almost undistorted fluorite and YF_3 structure types. Y ions have 7- and 8-coordination, Y-(O,F) = 2.13-2.68(6) Å.

ZIRCONIUM OXYFLUORIDE

$ZrO_{0.67}F_{2.67}$

P. JOUBERT and B. GAUDREAU, 1975. Rev. Chim. Minér., 12, 289-302.

Cubic, Pm3m, a = 3.997 Å, Z = 1. Mo radiation, R = 0.053 for 266 reflexions. 1 Zr in 6(e): x = 0.051; 3(O/F) (partial occupancy) in 3(d); 12(h): x = 0.096 and 0.304; and 24(ℓ): y = 0.376, z = 0.292.

ReO_3-type structure (1).

1. Strukturbericht, 2, 31, 299.

HYDRAZINIUM OXOHEPTAFLUORONIOBATE(V)

(HYDRAZINIUM OXOPENTAFLUORONIOBATE(V) FLUORIDE)

$(N_2H_6)_2NbOF_7$ $N_2H_6NbOF_5.N_2H_6F_2$

V.I. PAKHOMOV, T.A. KAIDALOVA, E.G. IL'IN and Ju. A. BUSLAEV, 1975. Koordin. Khim., 1, 37-42 [Coordin. Chem., 1, 28-32].

Monoclinic, $P2_1$, a = 14.62, b = 10.82, c = 5.56 Å, γ = 111.7°, D_m = 2.34, Z = 4. R = 0.16 for 1250 reflexions.

The material is a double salt and contains $N_2H_6^{2+}$ and F^- ions, and distorted octahedral $NbOF_5^{2-}$ ions, one of which is disordered; Nb-O = 1.7, 1.8, Nb-F = 1.88-2.05 Å.

BARIUM SODIUM NIOBIUM OXYFLUORIDE

$BaNa_2Nb_5O_{14}F$

R. von der MÜHLL and J. RAVEZ, 1975. Bull. Soc. Fr. Minér, Crist., 98, 118-120.

Tetragonal, P4/mbm, a = 12.369, c = 3.928 Å, D_m = 4.88, Z = 2. Mo radiation, R = 0.067 for 743 reflexions.

The structure is similar to that of $K_{0.60}WO_3$ (1), with disorder of Ba,Na in 4(g), and disorder of one of the oxygen atoms.

1. Structure Reports, 12, 292.

SODIUM TANTALUM OXYFLUORIDE

β-$Na_2Ta_2O_5F_2$

M. VLASSE, J.-P. CHAMINADE, J.-C. MASSIES and M. POUCHARD, 1975. J. Solid State Chem., 12, 102-109.

Monoclinic, C2/m, a = 12.855, b = 7.349, c = 12.833 Å, β = 108.97°, D_m = 6.02, Z = 8. Mo radiation, R = 0.072 for 1351 reflexions.

The structure consists of two interpenetrating sub-lattices: one of overall formula $Ta_{16}X_{52}$, composed of TaX_6 octahedra, and the second, $Na_{14}X_4$, composed of Na_4X tetrahedra. The arrangement can be described as a succession of weberite and pyrochlore slabs along a. Ta-O,F = 1.93-2.04(1), Na-O,F = 2.14-3.05 Å.

AMMONIUM TRIOXODIFLUOROMOLYBDATE

$(NH_4)_2MoO_3F_2$

R. MATTES, G. MÜLLER and H.J. BECHER, 1975. Z. anorg. Chem., 416, 256-262.

Orthorhombic, $P2_12_12_1$, a = 6.361, b = 7.488, c = 11.103 Å, D_m = 2.73, Z = 4. Mo radiation, R = 0.066 for 596 reflexions.

The structure contains infinite chains consisting of planar cis-MoO_2F_2 units (Mo-O = 1.84, Mo-F = 1.94 Å) linked by alternating short (1.88 Å) and long (2.01(1) Å) Mo-O bonds; Mo thus has distorted octahedral coordination. The chains are joined by hydrogen bonding via the NH_4^+ ions, N-H...F = 2.84, N-H...O = 2.86 Å.

AMMONIUM RHENIUM OXYFLUORIDE HYDRATE

$NH_4ReO_{1.5}F_3.H_2O$

E. PINTCHOVSKI, S. SOLED, R.G. LAWLER and A. WOLD, 1975. Inorg. Chem., 14, 1390-1394.

Cubic, Fm3m, a = 16.563 Å, D_m = 3.57, Z = 32. Mo radiation, R = 0.034 for 170 reflexions.

The structure (Fig. 1) contains isolated $Re_8O_{12}F_{24}$ units, ammonium ions, and body-centred rhombic dodecahedra of water molecules, the water sites having only 50% occupancy. Re has distorted octahedral coordination; Re-O = 1.83, Re-F = 2.02 Å.

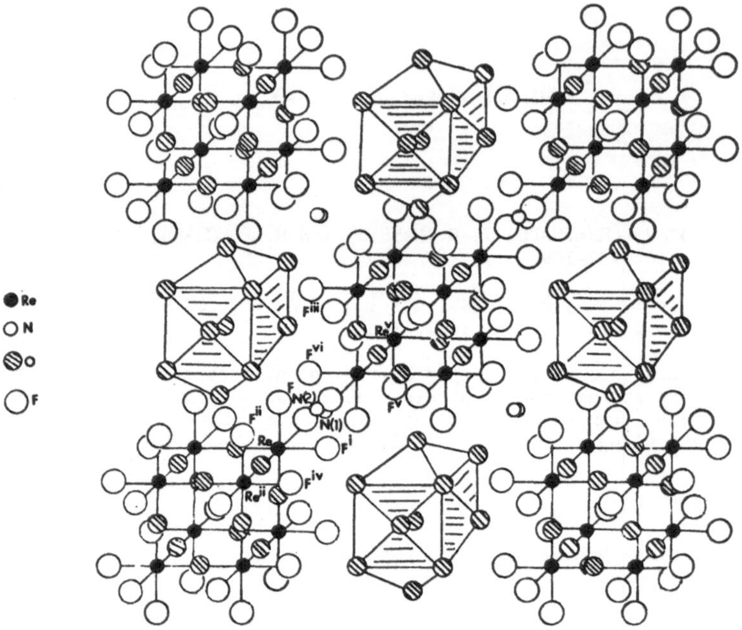

Fig. 1. Structure of $NH_4ReO_{1.5}F_3.H_2O$.

PYROCHLORES

$Hg_2M_2F_6S$ (M = Ni, Cu, Mn)
$Hg_2M_2F_6O$ (M = Ni, Zn)

D. BERNARD, J. PANNETIER and J. LUCAS, 1975. J. Solid State Chem., 14, 328-334.

Cubic, Fd3m, a ∿ 10.3 Å, Z = 8. Powder data. Pyrochlore structures (1), x = 0.312-0.319.

1. Structure Reports, 33A, 190.

URANIUM OXIDE TETRAFLUORIDE

UOF_4

R.T. PAINE, R.R. RYAN and L.B. ASPREY, 1975. Inorg. Chem., 14, 1113-1117.

Rhombohedral, R3m, a = 13.22, c = 5.72 Å, D_m not measurable, Z = 9. Mo radiation, R = 0.042 for 300 reflexions.

Atomic positions

		x	y	z
U	in 9(b)	0.86183	0.13817	0
X(1)	9(b)	0.466	0.534	0.141
X(2)	9(b)	0.589	0.411	0.535
F(2)	18(c)	0.442	0.755	0.911
F(3)	9(b)	0.249	0.751	0.400

X = O or F

U has pentagonal bipyramidal coordination, with O and one F axial (each at 1.78 Å, so probably disordered) and one terminal and four bridging F atoms equatorial (U-F = 1.98 and 2.25-2.29 Å). A tetragonal form is described in 1.

1. Structure Reports, 40A, 161.

LAURIONITE

Pb(OH)Cl

C.C. VENETOPOULOS and P.J. RENTZEPERIS, 1975. Z. Kristallogr., 141, 246-259.

Orthorhombic, Pcmn, a = 9.699, b = 4.020, c = 7.111 Å, Z = 4. Mo radiation, R = 0.057 for 686 reflexions.

Atomic positions

	x	y	z
Pb	0.0877	1/4	-0.2026
Cl	-0.1798	1/4	-0.4436
O	0.0422	1/4	0.1223
H	0.110	1/4	0.223

The structure is roughly as previously described (1), but with a rather large shift in oxygen position. Pb has strongly distorted square-antiprismatic coordination (5 Cl at 3.11-3.44 Å, 3 OH at 2.35-2.44 Å).

1. Strukturbericht, 7, 125; Structure Reports, 8, 131.

VANADIUM OXYCHLORIDE

VOCl

A. HAASE and G. BRAUER, 1975. Acta Cryst., B31, 2521-2522.

Orthorhombic, Pmmn, a = 3.780, b = 3.300, c = 7.91 Å, D_m not given, Z = 2. Mo radiation, R = 0.091 for 491 reflexions. V in 2(b): z = 0.1148; Cl and O in 2(a): z = 0.3279 and -0.0458.

Isostructural with FeOCl (1). V has distorted octahedral coordination, V-O = 1.97, 2.08, V-Cl = 2.36 Å.

1. Strukturbericht, 3, 67, 376; Structure Reports, 35A, 181.

CHROMIUM OXIDE CHLORIDE

CrOCl

A.N. CHRISTENSEN, T. JOHANSSON and S. QUÉZEL, 1974. Acta Chem. Scand., A28, 1171-1174.

Orthorhombic, Pmmn, a = 3.863, b = 3.182, c = 7.694 Å, Z = 2. Neutron powder data, R = 0.066 for 13 reflexions.

Structure as in 1.

1. Structure Reports, 27, 469.

POTASSIUM μ-OXO-DECACHLORODITUNGSTATE(IV)

$K_4W_2OCl_{10}$

T. GLOWIAK, M. SABAT and B. JEŻOWSKA-TRZEBIATOWSKA, 1975. Acta Cryst., B31, 1783-1784.

Tetragonal, I4/mmm, a = 7.132, c = 17.648 Å, D_m not given, Z = 2. Mo radiation, R = 0.038 for 505 reflexions.

The structure is similar to that of $K_4Re_2OCl_{10}.H_2O$ (1), except for the missing water molecule. It contains K^+ ions and $Cl_5W-O-WCl_5^{4-}$ ions, W-Cl = 2.408(5), W-O = 1.871(1) Å, W-O-W = 180°.

1. Structure Reports, 27, 670.

PARATACAMITE

$Cu_2(OH)_3Cl$

M.E. FLEET, 1975. Acta Cryst., B31, 183-187.

Rhombohedral, R$\bar{3}$, a = 9.168 Å, α = 96.263° (hexagonal cell has a = 13.654, c = 14.041 Å), D_m not given, Z = 8. Mo radiation, R = 0.051 for 1027 reflexions.

Of the four independent Cu atoms, two have (4 O + 2 Cl) coordination (Cu-O = 1.93-2.07, Cu-Cl = 2.75-2.82 Å), one has (2 O + 4 O) coordination (Cu-O = 1.93 and 2.20 Å), and one has 6 O at 2.12 Å (Fig. 1). The structure contains O-H...Cl hydrogen bonds. There is a $Co_2(OH)_3Cl$-type (1) sub-structure.

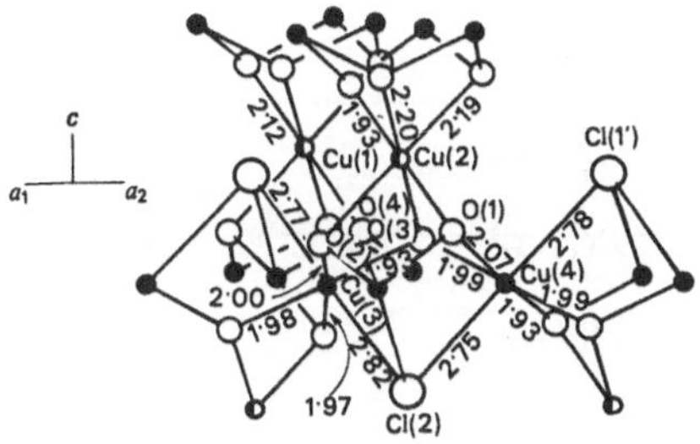

Fig. 1. Part of the structure of paratacamite.

1. Structure Reports, 17, 366.

STRONTIUM COPPER OXYCHLORIDE

$Sr_2CuO_2Cl_2$

B. GRANDE and H. MÜLLER-BUSCHBAUM, 1975. Z. anorg. Chem., 417, 68-74.

Tetragonal, I4/mmm, a = 3.975, c = 15.618 Å, Z = 2. R = 0.093 for 278 reflexions. Cu in 2(a); Sr and Cl in 4(e): z = 0.392 and 0.183; O in 4(c).

Isostructural with K_2NiF_4 (1). Cu has tetragonally-distorted octahedral coordination, Cu-O = 1.99 (x 4), Cu-Cl = 2.86 (x 2) Å, and Sr has four oxygen and five chlorine neighbours, Sr-O = 2.60, Sr-Cl = 3.05, 3.27 Å.

1. Structure Reports, 17, 332; 19, 323.

PROTACTINIUM OXYTRIBROMIDE

$PaOBr_3$

D. BROWN, T.J. PETCHER and A.J. SMITH, 1975. Acta Cryst., B31, 1382-1385.

Monoclinic, C2/m, a = 16.911, b = 3.871, c = 9.334 Å, β = 113.67°, D_m not measurable, Z = 4. Mo radiation, R = 0.066 for 266 reflexions (film data).

Pa has sevenfold approximately pentagonal bipyramidal coordination, with bridging O and Br giving chains along b (Fig. 1). Pa-Br = 2.57-2.91(2), Pa-O = 2.14-2.27(11) Å.

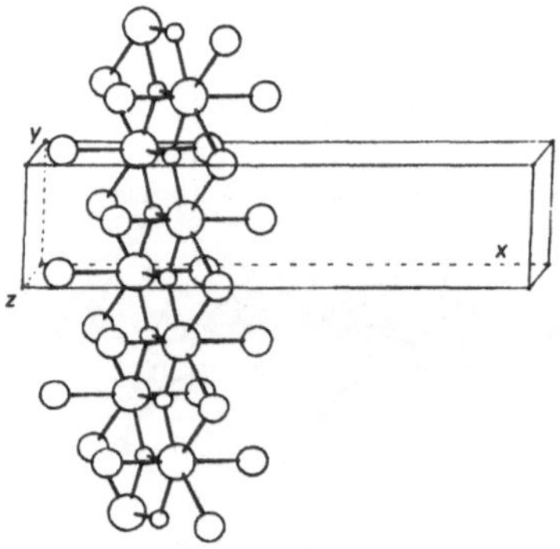

Fig. 1. Structure of $PaOBr_3$.

ANTIMONY(III) OXIDE IODIDE

α-Sb_5O_7I

V. KRÄMER, 1975. Acta Cryst., B31, 234-237.

Monoclinic, $P2_1/c$, a = 6.772, b = 12.726, c = 13.392 Å, β = 120.1°, D_m = 5.55, Z = 4. Mo radiation, R = 0.037 for 1401 reflexions.

The structure contains pseudohexagonal $Sb_2[Sb_3O_7]^+$ sheets (Fig. 1), connected by I which is at the centre of a slightly distorted cubo-octahedron of Sb atoms. Each Sb has trigonal pyramidal coordination, Sb-O = 1.94-2.09 Å.

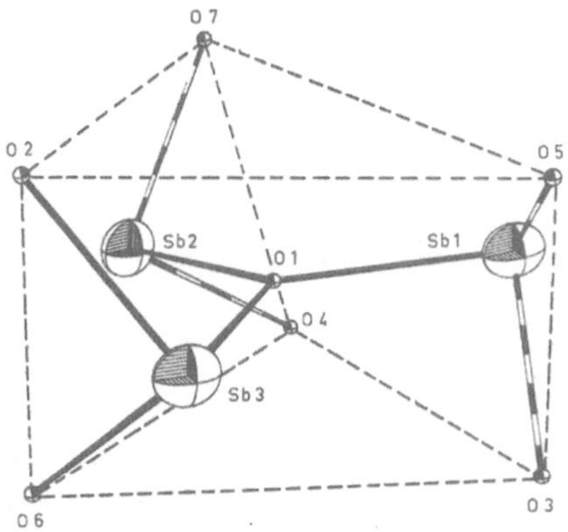

Fig. 1. The $[Sb_3O_7]^{5-}$ unit in α-Sb_5O_7I.

BISMUTH COPPER SULPHIDE BROMIDE

$Bi_2Cu_3S_4Br$

K. MARIOLACOS and V. KUPČÍK, 1975. Acta Cryst., B31, 1762-1763.

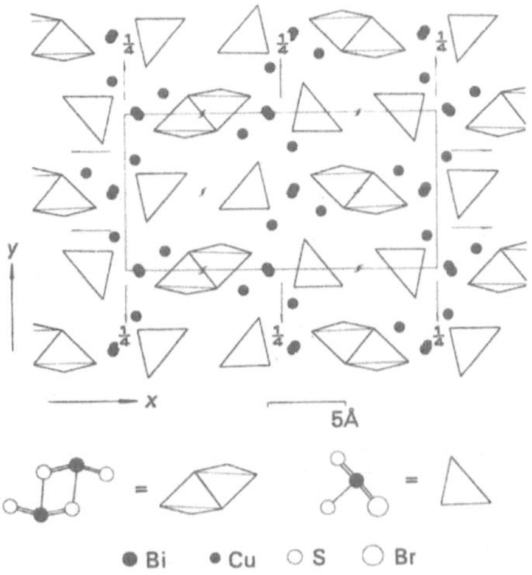

Fig. 1. Structure of $Bi_2Cu_3S_4Br$.

Orthorhombic, $P2_12_12_1$, a = 20.929, b = 10.440, c = 4.015 Å, D_m not given, Z = 4. Mo radiation, R = 0.079 for 1854 reflexions.

Isostructural with $Bi_2Cu_3S_4Cl$ (1). The structure (Fig. 1) contains $(Bi_2S_4{}^{2-})_\infty$ and $(BiS_2Br^{2-})_\infty$ chains along c, connected through Br atoms.

1. Structure Reports, 40A, 167.

YTTRIUM FLUOROSELENIDE

YSeF (4M-polytype)

NGUYEN-HUY-DUNG, C. DAGRON and P. LARUELLE, 1975. Acta Cryst., B31, 519-521.

Monoclinic, $P2_1/m$, a = 9.962, b = 13.001, c = 4.106 Å, γ = 104.92°, D_m = 4.80, Z = 8. Mo radiation, R = 0.096 for 232 reflexions (films, densitometer intensities).

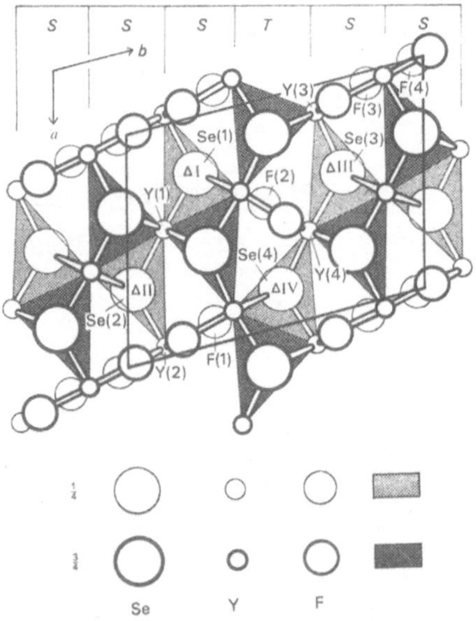

Fig. 1. Structure of the YSeF 4M polytype.

The structure contains layers (designated S and T) in the order ...SSST... (Fig. 1); the orthorhombic modification has a two-layer structure ..ST.. (1). The Y coordination polyhedra are similar to those of Er in the ErSeF 6O polytype (2).

1. Structure Reports, 39A, 199.
2. Following report.

ERBIUM FLUOROSELENIDE

ErSeF (6O-polytype)

NGUYEN-HUY-DUNG, C. DAGRON and P. LARUELLE, 1975. Acta Cryst., B31, 514-518.

Orthorhombic, Pnam, a = 9.902, b = 18.700, c = 4.095 Å, D_m = 6.94, Z = 12. Mo radiation, R = 0.085 for 200 reflexions (films, densitometer intensities).

The structure contains two types of layer (Fig. 1). Er(1) is coordinated to six Se at 2.81-2.84 Å, Er(2) to 6 F at 2.11-2.56 Å and 2 Se at 2.79 and 2.85 Å, and Er(3) to 3 F at 2.21-2.30 Å and 4 Se at 2.80-2.93 Å; the Er(3) coordination is similar to that in the two-layer YSeF polytype (1).

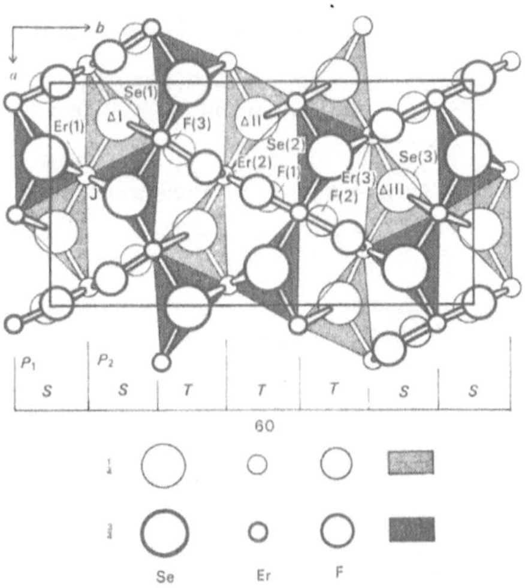

Fig. 1. Structure of the ErSeF 6O polytype.

1. Structure Reports, 39A, 199.

POTASSIUM CYANIDE

KCN (high-pressure)

D.L. DECKER, R.A. BEYERLEIN, G. ROULT and T.G. WORLTON, 1974. Phys. Rev., B, 10, 3584-3593.

KCN-III (74°C and 22 kbar)
Cubic, Pm3m, a = 3.808 Å. Neutron diffraction powder data give results in agreement with previous X-ray study (1). CN^- ions are arranged randomly along <111>.

KCN-IV (23°C and 25 kbar)
Monoclinic, (cf. 2), Cm, a = 5.531, b = 5.209, c = 3.743, β = 94.4°, Z = 2. Neutron diffraction powder data; K at (0, 0, 0), C at (0.426, 0, 0.365), N at (0.583, 0, 0.544).

1. P.W. RICHTER and C.W.F.T. PISTORIUS, 1972. Acta Cryst., B28, 3105.
2. C.W.F.T. PISTORIUS, 1971. J. Phys. Chem. Solids, 32, 2761.

POTASSIUM HEXACYANOVANADATE(II)

$K_4V(CN)_6$

S. JAGNER, 1975. Acta Chem. Scand., A29, 255-264.

Monoclinic, $P2_1/a$, a = 8.5770, b = 21.5271, c = 7.5270 Å, β = 106.55°, D_m = 1.83, Z = 4 (for the ordered MDO_2 form). Mo radiation, R = 0.053 for 757 reflexions (116 of these from diffuse streaks).

The structure is disordered, and the idealized structure is shown in Fig. 1.

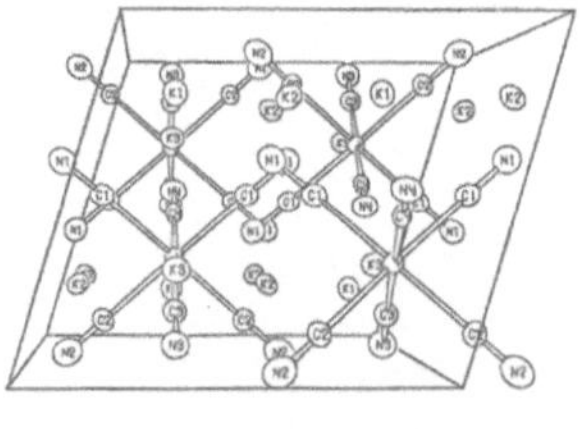

Fig. 1. View of the idealized (MDO_2) structure of $K_4V(CN)_6$.

PENTACHLORO(HYDROGEN CYANIDE)NIOBIUM(V)

$NbCl_5(NCH)$

C. CHAVANT, G. CONSTANT, Y. JEANNIN and R. MORANCHO, 1975. Acta Cryst., B31, 1823-1827.

Orthorhombic, Pnma, a = 12.075, b = 10.450, c = 6.551 Å, D_m = 2.28, Z = 4. Mo radiation, R = 0.06 for 761 reflexions.

The structure contains isolated octahedral molecules (Fig. 1), Nb-Cl = 2.24, 2.31, Nb-N = 2.31, N-C = 1.09 Å. Possible hydrogen bonding has not been definitely established.

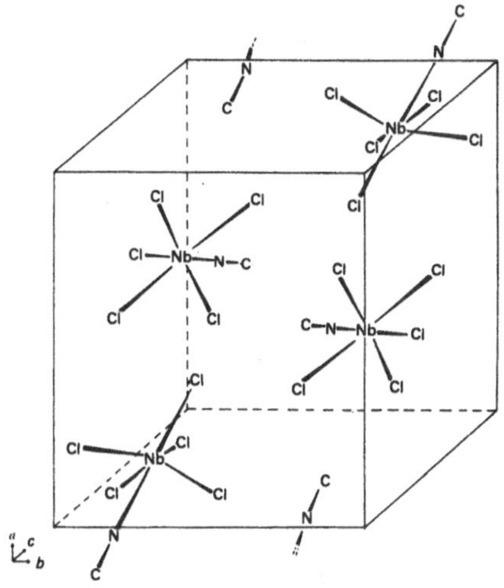

Fig. 1. Structure of $NbCl_5(NCH)$.

HEPTAPOTASSIUM μ-OXO-BIS[PENTACYANOMANGANATE(III)] CYANIDE

$K_7[(CN)_5MnOMn(CN)_5]CN$

R.F. ZIOLO, R.H. STANFORD, G.R. ROSSMAN and H.B. GRAY, 1974. J. Amer. Chem. Soc., 96, 7910-7915.

Orthorhombic, Ibam, a = 12.397, b = 12.772, c = 14.618 Å, D_m = 1.98, Z = 4. Mo radiation, R = 0.091 for 426 reflexions.

The $[Mn_2O(CN)_{10}]^{6-}$ anion has C_{2h} crystallographic symmetry, and consists of two eclipsed Mn coordination octahedra joined by a linear Mn-O-Mn bridge; Mn-O = 1.72, Mn-C = 2.06 (axial), 1.99 Å (equatorial). The crystal contains layers of these anions, and K^+ ions with 8- (square-antiprism) and 6-coordinations (octahedron and trigonal prism); the free cyanide group is disordered in a void in the crystal lattice.

HEXA(DEUTERIUM CYANIDE)IRON(II) BIS(TETRACHLOROFERRATE(III))

$[Fe(NCD)_6](FeCl_4)_2$

J.C. DARAN, Y. JEANNIN, H. FUESS and W. YELON, 1975. Acta Cryst., B31, 1838-1841.

Trigonal, $P\bar{3}$, a = 10.29, c = 6.283 Å, D_m = 1.80, Z = 1. Neutron diffraction data, R = 0.05 for 584 reflexions.

The structure (Fig. 1) contains octahedral cations and tetrahedral anions, joined by C-D...Cl bonds. Fe-N = 2.14, N-C = 1.13, C-D = 1.04, Fe-Cl = 2.19, C...Cl = 3.43, D...Cl = 2.63 Å, C-D...Cl = 137°

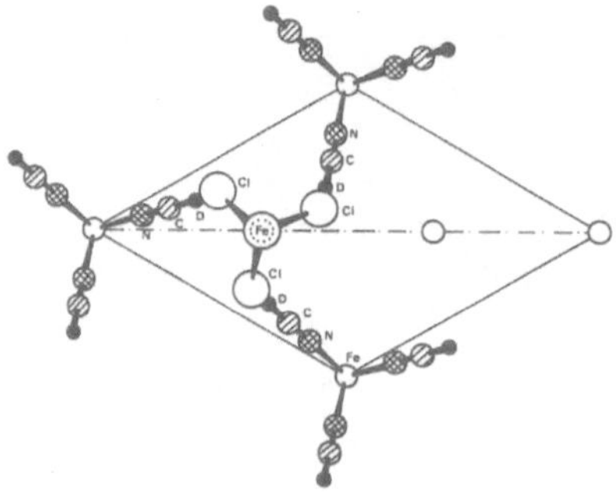

Fig. 1. Structure of $[Fe(NCD)_6](FeCl_4)_2$.

NICKEL SODIUM HEXACYANOFERRATE(III)

$(Na_{0.28}Ni_{1.86})Fe(CN)_6$

M.V. ZIL'BERMAN, V.G. KUZNECOV and V.V. VOL'KHIN, 1974. Ž. Neorg. Khim., 19, 1838-1841 [Russ. J. Inorg. Chem., 19, 1002-1004].

Cubic, Fm3m, a = 10.07 Å, Z = 2.6. Fe and Co radiations, powder data. 2.6 Fe in 4(a); 4 Ni(1) in 4(b); 2.6 (Na+Ni(2)) in 8(c); 15.6 C in 24(e): x = 0.193; 15.6 N in 24(e): x = 0.307.

The material has a structure similar to that of other hexacyanoferrates(III), with a framework of $Fe(CN)_6$ and $Ni(NC)_6$ octahedra with bridging CN groups and additional (Na,Ni) ions. There are vacancies at the $Fe(CN)_6$ sites, as in $K_{0.33}Cu_{1.83}Fe(CN)_6$ (1).

1. M.V. ZIL'BERMAN and V.V. VOL'KHIN, 1971. Ž. Strukt. Khim., 12, 644.

fac-TRICYANOTRIAMMINECOBALT(III)

$Co(NH_3)_3(CN)_3$

Y. YAMAMOTO, 1974. Nippon Kagaku Kaishi, 259-262.

Orthorhombic, $P2_12_12_1$, a = 9.95, b = 11.05, c = 6.58 Å, D_m = 1.74, Z = 4. Fe radiation, R = 0.11 for 473 reflexions (film data).

Co has fac-octahedral coordination, Co-NH_3 = 2.01, Co-C = 1.86, C-N = 1.18 Å.

POTASSIUM DECACYANO-μ-SUPEROXO-DICOBALTATE(III) MONOHYDRATE

$K_5[(CN)_5CoO_2Co(CN)_5].H_2O$

F.R. FRONCZEK, W.P. SCHAEFER and R.E. MARSH, 1975. Inorg. Chem., 14, 611-617.

Triclinic, $P\bar{1}$, a = 11.707, b = 19.423, c = 7.664 Å, α = 93.92, β = 110.36, γ = 94.71°, D_m = 1.91, Z = 3. Mo radiation, R = 0.065 for 3399 reflexions.

The structure contains two independent anions, one on a centre of symmetry, with eclipsed cyanide ligands, and the other non-centrosymmetric, with staggered cyanide ligands. Both types of anion have a bridging superoxo ligand in a staggered arrangement. Co-O = 1.94, Co-N = 1.83-1.90, O-O = 1.26 Å. There is substitutional disorder involving one K ion and one water molecule.

μ-THIOCYANATO-[PENTAAMMINECOBALT(III)]-PENTACYANOCOBALT(III) MONOHYDRATE

$(NH_3)_5CoNCSCo(CN)_5.H_2O$

F.R. FRONCZEK and W.P. SCHAEFER, 1975. Inorg. Chem., 14, 2066-2070.

Orthorhombic, Pbca, a = 14.166, b = 14.549, c = 15.187 Å, D_m = 1.77, Z = 8. Mo radiation, R = 0.050 for 2453 reflexions.

The binuclear molecule contains two octahedrally-coordinated Co atoms bridged by the thiocyanate ligand (Fig. 1). An intramolecular N(11)-H...N(3) contact of 3.03 Å is probably a hydrogen bond. The molecules are joined by N-H...N hydrogen bonds, a possible N-H...S bond (3.39 Å), and hydrogen bonds involving the water molecule, whose H atoms appear to be statistically disordered.

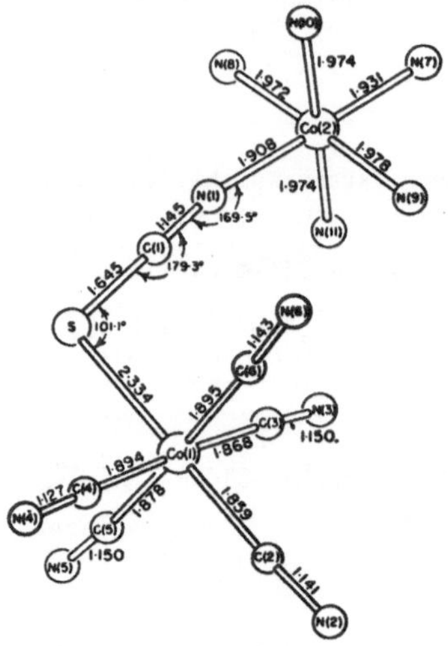

Fig. 1. Structure of $(NH_3)_5CoNCSCo(CN)_5$.

SODIUM TETRACYANOPALLADATE(II) TRIHYDRATE

$Na_2Pd(CN)_4.3H_2O$

J. LEDENT, 1974. Bull. Soc. R. Sci. Liège, 43, 172-189.

Refinement of the structure (1), R = 0.094 for 3051 reflexions.

The structure contains columns of square planar $Pd(CN)_4^{2-}$ ions, Pd-C = 1.98-2.04(1), C-N = 1.14-1.20(2) Å, Pd-C-N = 174-179°. Na ions have octahedral coordination, the octahedra sharing edges to form layers; Na-O = 2.43-2.59, Na-N = 2.48-2.66 Å.

1. Structure Reports, 32A, 235; 39A, 362.

POTASSIUM TETRACYANOPLATINATE(II) CHLORIDE TRIHYDRATE

POTASSIUM TETRACYANOPLATINATE(II) BROMIDE TRIHYDRATE

$K_2Pt(CN)_4Cl_{0.3}\cdot 3H_2O$
$K_2Pt(CN)_4Br_{0.3}\cdot 3H_2O$

I. H.J. DEISEROTH and H. SCHULZ, 1974. Phys. Rev. Letters, 33, 963-965.

II. J.M. WILLIAMS, J.L. PETERSEN, H.M. GERDES and S.W. PETERSON, 1974. Ibid., 33, 1079-1081.

III. A. FREUND, S. ROTH and R. RANVAUD, 1974. J. Appl. Cryst., 7, 631-632.

IV. W. DIETERICH, 1974. Z. Phys., 270, 239-243.

V. G. HEGER, B. RENKER, H.J. DEISEROTH, H. SCHULZ and G. SCHEIBER, 1975. Mater. Res. Bull., 10, 217-224.

VI. H.J. DEISEROTH and H. SCHULZ, 1975. Ibid., 10, 225-228.

VII. J.M. WILLIAMS, M. IWATA, F.K. ROSS, J.L. PETERSEN and S.W. PETERSON, 1975. Ibid., 10, 411-416.

VIII. A.H. REIS, S.W. PETERSON, N. ENRIGHT and J.M. WILLIAMS, 1975. Ibid., 10, 921-926.

IX. C. PETERS and C.F. EAGEN, 1975. Phys. Res. Letters, 34, 1132-1134.

V. J.M. WILLIAMS, F.K. ROSS, M. IWATA, J.L. PETERSEN, S.W. PETERSON, S.C. LIN and K. KEEFER, 1975. Solid State Comm., 17, 45-48.

The general features of the structures are essentially as previously described in space group P4/mmm (1). A model in P4mm accounts for the superstructure reflexions; in this model atoms are displaced from their previous fixed z coordinates, but the Pt...Pt separation is still c/2. Further work is proposed.

1. Structure Reports, 33A, 250.

ZINC NITRATE BIS(MERCURY(II) CYANIDE) HEPTAHYDRATE

(TETRAAQUOBIS[DICYANOMERCURY(II)]ZINC NITRATE TRIHYDRATE)

$[Zn(H_2O)_4(Hg(CN)_2)_2](NO_3)_2.3H_2O$ $Zn(NO_3)_2.2Hg(CN)_2.7H_2O$

L.F. POWER, J.A. KING and F.H. MOORE, 1975. J. Chem. Soc., Dalton, 2072-2075.

Monoclinic, C2/c, a = 17.587, b = 6.659, c = 16.144 Å, β = 94.84°, D_m = 2.89, Z = 4. Neutron diffraction data, R = 0.051 for 1093 reflexions.

The structure is as determined by X-ray methods (1). This neutron diffraction study confirms that the CN groups are carbon-bonded to Hg, and also confirms six of the seven hydrogen bonds proposed in 1. Zn is octahedrally coordinated by four water molecules and two trans, nearly-linear NC-Hg-CN groups, Zn-O = 2.106(3), Zn-N = 2.125(1), Hg-C = 2.030(2) Å.

1. Structure Reports, 37A, 218.

POTASSIUM TRIS(ISOCYANATO)CADMATE

$KCd(NCO)_3$

G. THIELE and P. HILFRICH, 1975. Z. Naturforsch., 30B, 19-21.

Orthorhombic, Pnma, a = 9.993, b = 3.671, c = 18.354 Å [in abstract, slightly different values in text], D_m = 2.65, Z = 4. Mo radiation, R = 0.035 for 1306 reflexions.

The structure contains N-bridged double chains along c, similar to the chains in $KCdCl_3$ (1); Cd thus has octahedral coordination, Cd-N = 2.47 (bridging), 2.33 Å (terminal), Cd-N-C = 118-127, N-C-O = 175-178°. The chains are linked by the K^+ ions.

1. Structure Reports, 11, 434.

SODIUM THIOCYANATE

NaSCN

P.H. van ROOYEN and J.C.A. BOEYENS, 1974. Acta Cryst., B31, 2933-2934.

Orthorhombic, Pnma, a = 13.38, b = 4.09, c = 5.66 Å, D_m not given, Z = 4. Mo radiation, R = 0.075 for 431 reflexions.

Atomic positions

Atoms in 4(c)

	x	y	z
Na	0.3919	1/4	0.9337
S	0.1856	1/4	1.3337
C	0.1127	1/4	0.8758
N	0.0627	1/4	0.7066

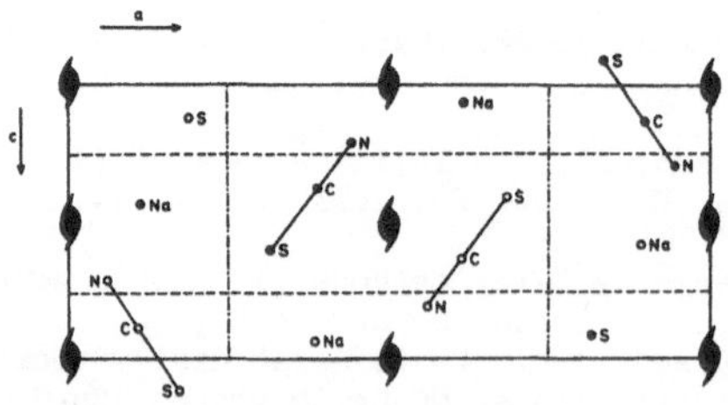

Fig. 1. Structure of sodium thiocyanate.

The structure (Fig. 1) is as previously proposed (1). It contains linear SCN^- ions, and Na^+ ions coordinated octahedrally to 3 S and 3 N in a fac arrangement. S-C = 1.663(9), C-N = 1.168(18), Na-S = 2.92, 2.94, Na-N = 2.42, 2.63 Å.

1. Structure Reports, 33A, 253.

LEAD(II) THIOCYANATE

$Pb(NCS)_2$

J.A.A. MOKUOLU and J.C. SPEAKMAN, 1975. Acta Cryst., B31, 172-176.

Monoclinic, C2/c, a = 9.661, b = 6.544, c = 8.253 Å, β = 92.37°, D_m = 4.082, Z = 4. Mo radiation, R = 0.047 for 974 reflexions.

Pb is coordinated to 4 S at 3.00 and 3.14 Å, and 4 N at 2.69 and 2.78 Å (Fig. 1). The NCS group is linear, N-C = 1.17, C-S = 1.64(1) Å.

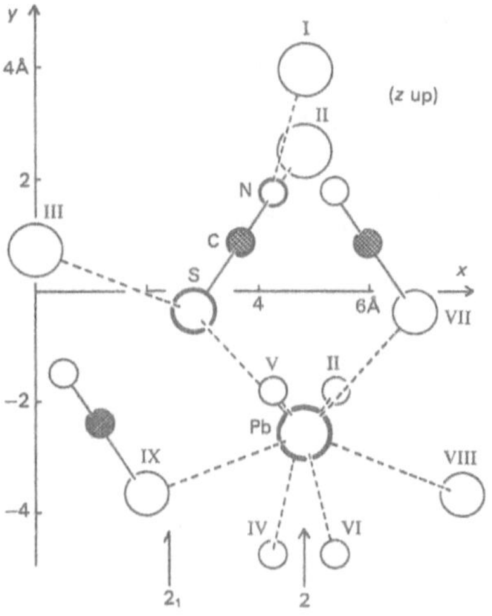

Fig. 1. Structure of lead(II) thiocyanate.

POTASSIUM SELENIUM-TRISELENOCYANATE HEMIHYDRATE

$KSe(SeCN)_3.0{\cdot}5H_2O$

S. HAUGE, 1975. Acta Chem. Scand., A29, 771-777.

Triclinic, P$\bar{1}$, a = 9.170, b = 13.377, c = 9.057 Å, α = 106.22, β = 100.64, γ = 99.07°, D_m = 2.89, Z = 4. Mo radiation, R = 0.042 for 3150 reflexions.

The structure (Fig. 1) includes a (dimeric) di-μ-selenocyanato-bis{diselenocyanatoselenate(II)} anion, some details of which are given in Fig. 2; the eight Se atoms are nearly coplanar, and all six SeCN units are on one side of the plane. Pairs of such layers are connected via rather close Se...Se contacts, with distances ranging upwards from 3.34 Å. SeCN are linear, or nearly so, with Se-C averaging 1.839 Å and C-N 1.13 Å. The K ion and the H_2O molecule lie between pairs of these double layers.

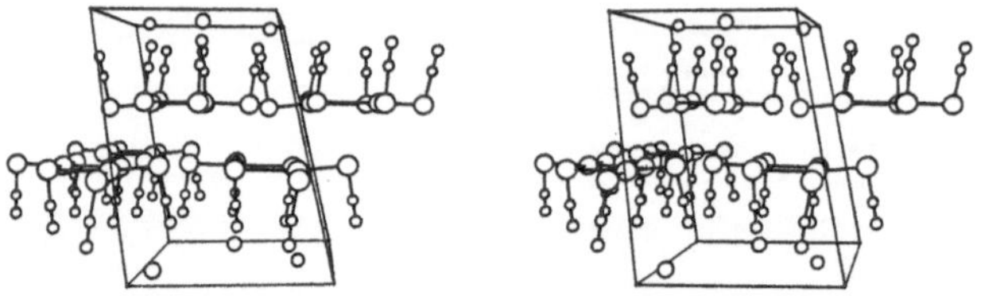

Fig. 1. Stereoscopic view, along the b-axis, of the structure of $KSe(SeCN)_3.0{\cdot}5H_2O$.

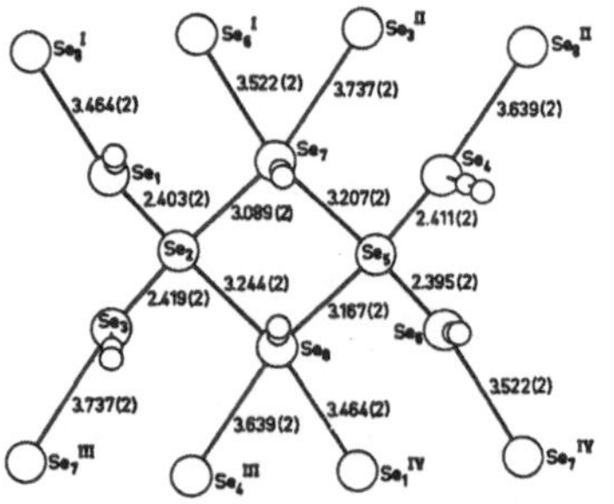

Fig. 2. Some interatomic distances within the $Se_8C_6N_6$ anion, and some close Se...Se contacts to other anions in the same layer.

RUBIDIUM SELENIUM-TRISELENOCYANATE HEMIHYDRATE

$RbSe(SeCN)_3.0{\cdot}5H_2O$

S. HAUGE, 1975. Acta Chem. Scand., A29, 843-848.

Triclinic, P$\bar{1}$, a = 9.180, b = 13.227, c = 9.144 Å, α = 83.89, β = 100.02, γ = 74.97°, D_m = 3.12, Z = 4. Mo radiation, R = 0.043 for 2664 reflexions.

This material is isostructural, though not isomorphous, with the potassium salt (1). The short Se...Se contacts (from 3.37 Å upwards) between anion layers are similar.

1. Preceding report.

POTASSIUM SELENIUM-TRITHIOCYANATE HEMIHYDRATE

$KSe(SCN)_3 \cdot 0 \cdot 5H_2O$

S. HAUGE and R.A. HENRIKSEN, 1975. Acta Chem. Scand., A29, 778-782.

Triclinic, $P\bar{1}$, a = 8.775, b = 15.067, c = 8.956 Å, α = 119.64, β = 101.09, γ = 99.62°, D_m = 2.08, Z = 4. Mo radiation, R = 0.046 for 2945 reflexions.

The material is isostructural with the triselenocyanates described in the two preceding reports, but the contacts between the anion layers are weaker, the shortest Se...S and S...S distances being 3.542 and 3.486 Å, respectively.

CAESIUM TRISELENOCYANATE

$Cs(SeCN)_3$

S. HAUGE, 1975. Acta Chem. Scand., A29, 163-169.

Monoclinic, C2/c, a = 7.969, b = 21.156, c = 5.593 Å, β = 98.84°, D_m = 3.18, Z = 4. Cu radiation, R = 0.062 for about 700 reflexions.

The structure contains discrete $(SeCN)_3^-$ anions, with their central Se-C-N units lying on twofold axes; some dimensions are shown in Fig. 1. Six N atoms constitute the environment of the Cs^+ ion, with Cs...N in the range 3.20-3.29 Å.

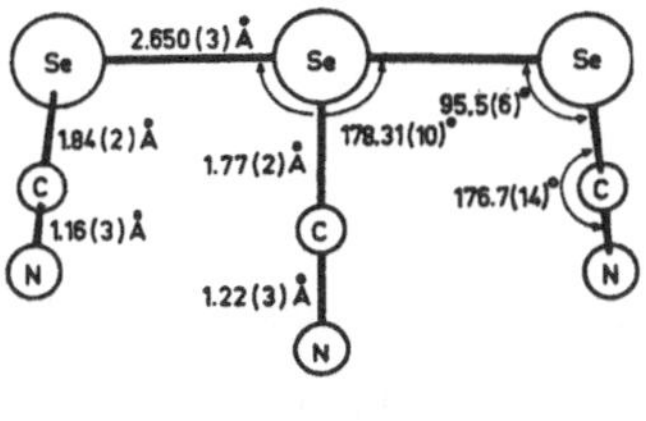

Fig. 1. Some dimensions of the $(SeCN)_3^-$ anion in caesium triselenocyanate.

POTASSIUM TETRAKIS(ISOTHIOCYANATO)COBALTATE(II) TRIHYDRATE

$K_2Co(NCS)_4 \cdot 3H_2O$

M.G.B. DREW and A. HAMID BIN OTHMAN, 1975. Acta Cryst., B31, 613-614.

Orthorhombic, $P2_12_12$, a = 11.116, b = 12.981, c = 5.354 Å, D_m = 1.80, Z = 2. Mo radiation, R = 0.054 for 886 reflexions.

The structure contains tetrahedral $Co(NCS)_4^{2-}$ anions, Co-N = 1.96, N-C = 1.15, C-S = 1.64(1) Å, N-Co-N = 103-114, Co-N-C = 166 (not 111 as in 1), N-C-S = 179°. The closest contact involving the K^+ ion is K...N = 3.14 Å.

1. Structure Reports, 13, 291.

CADMIUM MERCURY(II) THIOCYANATE

$CdHg(SCN)_4$

P.M. FEDEROV, L.S. ANDREJANOVA and V.I. PAKHOMOV, 1975. Koordin. Khim., 1, 252-254.

Tetragonal, [I$\bar{4}$], a = 11.49, c = 4.25 Å, D_m = 3.3, Z = 2. Mo radiation, R = 0.137 for film data.

Atomic positions

		x	y	z
Hg	in 2(a)	0	0	0
Cd	2(d)	1/2	0	1/4
S	8(g)	0.130	0.142	0.311
N	8(g)	0.058	0.352	0.069
C	8(g)	0.071	0.262	0.223

[The structure is essentially as previously described (1).]

1. Structure Reports, 33A, 253.

POTASSIUM NEODYMIUM THIOCYANATE TETRAHYDRATE

POTASSIUM EUROPIUM(III) THIOCYANATE HEXAHYDRATE

$K_4Nd(NCS)_7.4H_2O$
$K_4Eu(NCS)_7.6H_2O$

P.I. LAZAREV, L.A. ASLANOV, V.M. IONOV and M.A. PORAJ-KOŠIC, 1975. Koordin. Khim., 1, 710-715.

Monoclinic, B2/b, a = 25.25, 25.41, b = 20.62, 20.32, c = 6.86, 6.52 Å, γ = 122.8, 122.0°, Z = 4.

The crystals are nearly isostructural, with the additional water molecules in the hexahydrate coordinated to K^+ ions. The Ln ions have 8-coordination (Archimedean antiprism) to four thiocyanate N atoms and four water molecules. The other thiocyanate ions are not coordinated to Ln, and one of them is disordered about a centre of symmetry. The formulae are therefore $K_4[Ln(H_2O)_4(NCS)_4]$-$(NCS)_3.nH_2O$. K ions have 8- and 9-coordination to O, N, and S atoms. Nd-N = 2.45, 2.52, Nd-O = 2.56, 2.65; Eu-N = 2.43, 2.47, Eu-O = 2.47, 2.55(4) Å.

CAESIUM OCTATHIOCYANATOURANATE(IV)

$Cs_4U(NCS)_8$

G. BOMBIERI, P.T. MOSELEY and D. BROWN, 1975. J. Chem. Soc., Dalton, 1520-1522.

Tetragonal, P4/n, a = 11.958, c = 11.170 Å, D_m not given, Z = 2. Mo radiation, R = 0.06 for 845 reflexions.

U has square-antiprismatic coordination to eight N atoms (Fig. 1), U-N = 2.38 and 2.46(3) Å; compare the Et_4N salt, where the coordination is cubic (1). Cs has short contacts to S, N, and C atoms.

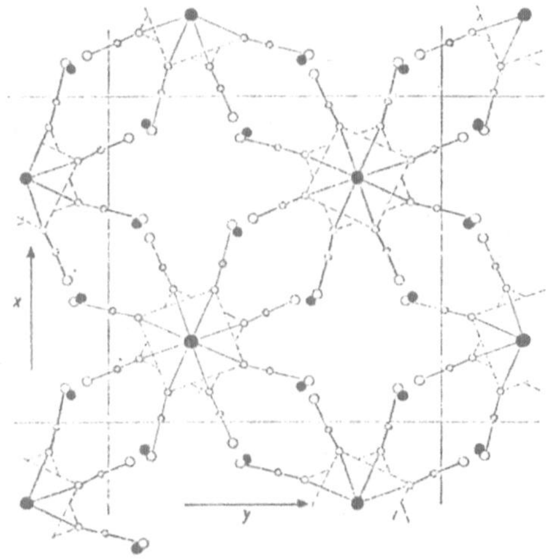

Fig. 1. Structure of $Cs_4U(NCS)_8$.

1. Structure Reports, 37B, 559.

DEUTERIUM PEROXIDE

D_2O_2

E. PRINCE, S.F. TREVINO, C.S. CHOI and M.K. FARR, 1975. J. Chem. Phys., 63, 2620-2624.

Tetragonal, $P4_12_12$, a = 4.035, c = 7.97 Å, at -15°C, Z = 4. Neutron diffraction data, R = 0.030 for 106 reflexions.

The structure is as determined for H_2O_2 by X-ray methods (1). O-O = 1.49, O-D = 0.99 Å (corrected for thermal libration).

1. Structure Reports, 15, 171; 30A, 303.

RED LEAD

(MINIUM)

Pb_3O_4

I. J.-R. GAVARRI and D. WEIGEL, 1975. J. Solid State Chem., 13, 252-257.

Tetragonal, $P4_2/mbc$, a = 8.811, c = 6.563 Å, Z = 4. Neutron powder data, R = 0.06 for 29 reflexions.

Atomic positions

	x	y	z
4 Pb in 4(d)	0	1/2	1/4
8 Pb 8(h)	0.140	0.163	0
8 O 8(h)	0.096	0.637	0
8 O 8(g)	0.671	0.171	1/4

The structure is as previously described (1), but with some shifts in oxygen parameters, so that now Pb(IV)-O = 2.13 (x 2), 2.20 (x 4), Pb(II)-O = 2.22 (x 2), 2.34, 2.73 Å.

Pb_3O_4 (low-temperature)

II. J.R. GAVARRI, G. CALVARIN and D. WEIGEL, 1975. J. Solid State Chem., 14, 91-98.

Orthorhombic, Pbam, a = 9.124, b = 8.467, c = 6.567 Å, at 5°K, Z = 4. X-ray and neutron powder data.

Atomic positions

			x	y	z
Pb(IV)	in	4(f)	0	1/2	0.241
Pb(II)(1)		4(h)	0.1571	0.1527	1/2
Pb(II)(2)		4(g)	0.1624	0.8655	0
O(1)		8(i)	0.6642	0.1655	0.2517
O(2)		4(g)	0.1263	0.6013	0
O(3)		4(h)	0.0951	0.6353	1/2

The structure (Fig. 1) is a distortion of the tetragonal room-temperature form (1 and I). Pb(IV)-O = 2.05-2.23, Pb(II)-O = 2.26-2.91 Å.

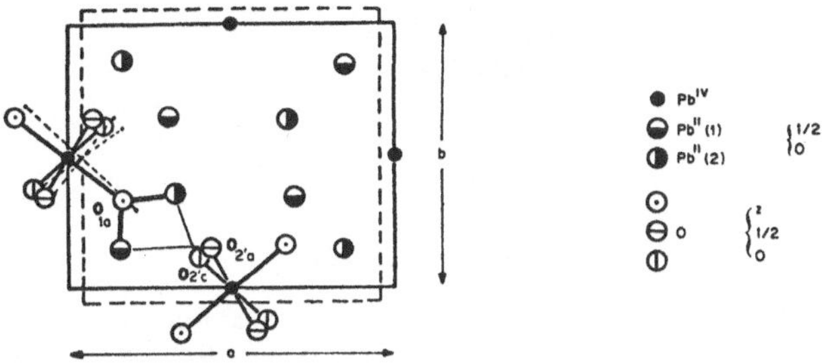

Fig. 1. Structure of Pb_3O_4 at 5°K, showing the relationship to the tetragonal room-temperature cell.

1. Structure Reports, 11, 239; 30A, 313.

CLAUDETITE

(MONOCLINIC FORM)

As_2O_3

F. PERTLIK, 1975. Mh. Chem., 106, 755-762.

Monoclinic, $P2_1/n$, a = 7.99, b = 4.65, c = 9.12 Å, β = 78.3°, D_m not given, Z = 4. Mo radiation, R = 0.07 for 568 reflexions.

Atomic positions

Atoms in 4(e): ±(x,y,z; ½+x, ½-y, ½+z)

	x	y	z
As(1)	0.6163	0.8311	0.3013
As(2)	0.1841	0.2920	0.3717
O(1)	0.677	0.459	0.291
O(2)	0.238	0.140	0.184
O(3)	0.966	0.349	0.367

Each As atom has trigonal pyramidal coordination, As-O = 1.77-1.82(2) Å, O-As-O = 93-99°. The pyramids share corners to form sheets parallel to (001).

SENARMONTITE

Sb_2O_3

C. SVENSSON, 1975. Acta Cryst., B31, 2016-2018.

Cubic, Fd3m, a = 11.1519 Å, D_m not given, Z = 16. Mo radiation, R = 0.021 for 507 reflexions. Sb in 32(e): x = 0.88527; O in 48(f): x = 0.1863.

Isostructural with cubic As_2O_3 (1), as previously determined (2); Sb-O = 1.977(1) Å.

1. Strukturbericht, 1, 246, 264; 2, 315; Structure Reports, 9, 164.
2. Strukturbericht, 1, 245, 264; Structure Reports, 9, 164.

CERVANTITE

α-Sb_2O_4

P.S. GOPALAKRISHNAN and H. MANOHAR, 1975. Cryst. Struct. Comm., 4, 203-206.

Orthorhombic, $Pna2_1$, a = 5.436, b = 4.810, c = 11.76 Å, Z = 4. Mo radiation, R = 0.107 for 550 reflexions (films, visual intensities).

$Sb^V(2)$ and $Sb^{III}(1)$ atoms lie in alternate planes parallel to (001), with distorted octahedral and one-sided four coordination (plus a lone electron-pair), respectively (Fig. 1). Sb(V)-O = 1.93-2.02, Sb(III)-O = 2.01-2.26(4) Å.

Atomic positions

	x	y	z
Sb(1)	0.0214	-0.0346	0
Sb(2)	0.3728	-0.0006	0.2527
O(1)	0.3247	0.1608	0.4070
O(2)	0.3546	-0.1621	0.0940
O(3)	0.1660	-0.3013	0.3043
O(4)	0.0744	0.1880	0.1907

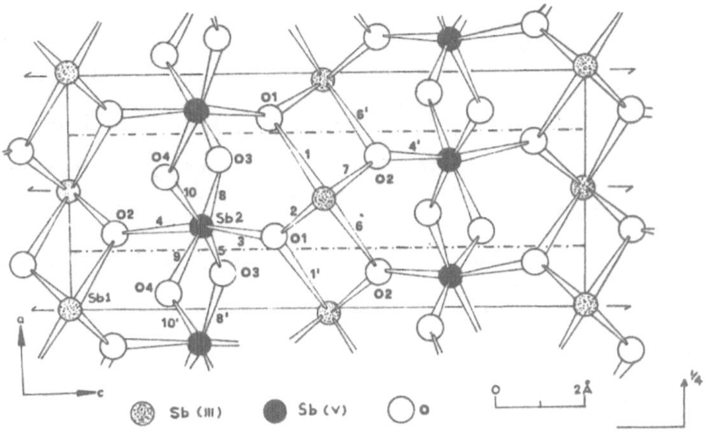

Fig. 1. Structure of cervantite.

TELLURIUM OXIDE

Te_4O_9

O. LINDQVIST, W. MARK and J. MORET, 1975. Acta Cryst., B31, 1255-1259.

Rhombohedral, R$\bar{3}$, a = 9.320, c = 14.486 Å, D_m = 5.9, Z = 6. Mo radiation, R = 0.042 (unusual extinctions indicate disorder).

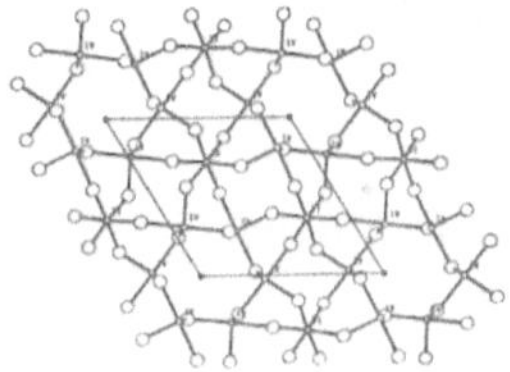

Fig. 1. Structure of Te_4O_9.

The structure contains layers (Fig. 1) of corner-sharing Te(VI)O_6 octahedra, Te-O = 1.90, 1.95 Å, and Te(IV)O_4 polyhedra (trigonal bipyramids with lone-pairs equatorial), Te-O = 1.88, 1.90 (equatorial), 2.02, 2.14 Å (axial). The layers are stacked in the c direction, with inversion of some layers resulting in disorder.

YTTRIUM OXIDE

Y_2O_3

I. B.H. O'CONNOR and T.M. VALENTINE, 1969. Acta Cryst., B25, 2140-2144.

II. M. BONNET, A. DELAPALME and H. FUESS, 1975. Ibid., A31, 264-265.

Cubic, Ia3, a = 10.604 Å (1), Z = 16. Neutron diffraction; I, single-crystal data, R = 0.12 for 179 reflexions; II, powder data. x(Y) = -0.0327, O at (0.3907, 0.1520, 0.3804).

The structure is as previously determined by X-ray methods (1).

1. Structure Reports, 30A, 309.

VANADIUM(III) OXIDE

V_2O_3
α-$(Cr_{0.01}V_{0.99})_2O_3$
β-$(Cr_{0.01}V_{0.99})_2O_3$

W.R. ROBINSON, 1975. Acta Cryst., B31, 1153-1160.

Rhombohedral, R$\bar{3}$c, a $\sim$ 5, c $\sim$ 14 Å, for the three materials at 23-600°C, Z = 6. Mo radiation, R = 0.020-0.039 for 10 sets of data.

All the samples are isostructural with α-alumina (1), V-O distances increase slightly with increasing temperature, except for one distance in β-$(Cr_{0.01}V_{0.99})_2$-O_3 which shows a small decrease.

1. Strukturbericht, 1, 240.

VANADIUM DIOXIDE

VO_2

D.B. McWHAN, M. MAREZIO, J.P. REMEIKA and P.D. DERNIER, 1974. Phys. Rev., B, 10, 490-495.

Tetragonal, $P4_2/mnm$, a = 4.555, 4.556, c = 2.851, 2.860 Å, at 360, 470°K, Z = 2. Mo radiation, R = 0.017 and 0.016 for 69 reflexions at 360 and 470°K. V in 2(a), O in 4(f): x = 0.3001, 0.3003.

Rutile structure (1), as previously described (2). The V...V distance along c (c-axis translation) is anomalously short, accounting for the metallic properties.

1. Strukturbericht, 1, 155, 204; Structure Reports, 17, 387; 19, 361; 20, 263; 37A, 222.
2. Strukturbericht, 1, 158, 211; Structure Reports, 20, 264; 26, 352.

NIOBIUM(V) OXIDE

T-Nb_2O_5

K. KATO and S. TAMURA, 1975. Acta Cryst., B31, 673-677.

Orthorhombic, Pbam, a = 6.175, b = 29.175, c = 3.930 Å, D_m not given, Z = 8.4. Mo radiation, R = 0.054 for 725 reflexions.

The unit cell contains $Nb_{16.8}O_{42}$. 16 Nb atoms lie in a sheet parallel to (001) and are surrounded by seven or six oxygen atoms (pentagonal bipyramids and distorted octahedra). These polyhedra are joined by edge- and corner-sharing within the (001) sheet (Fig. 1), and by corner-sharing along c. The remaining Nb atoms occupy interstitial sites between the sheets at random, and are surrounded by nine oxygen atoms.

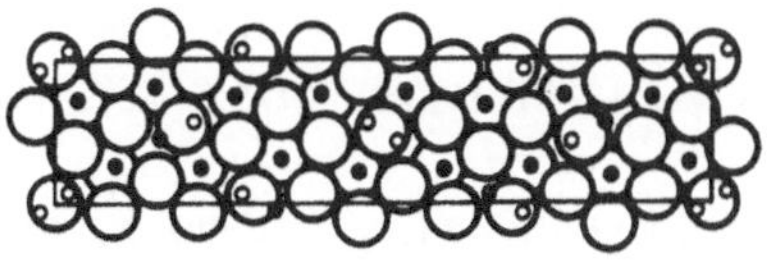

Fig. 1. Structure of T-Nb_2O_5 viewed along c; small open circles are the interstitial Nb atoms.

MOLYBDENUM TANTALUM OXIDE

$(Mo,Ta)_5O_{14}$

N. YAMAZOE and L. KIHLBORG, 1975. Acta Cryst., B31, 1666-1672.

Orthorhombic, a = 45.75, b = 22.87, c = 8.005 Å, Z = 32. There is a tetragonal subcell, a' = a/2 = b, c' = c/2, and the structure is given in terms of an orthorhombic subcell, $Pb2_1a$, a'' = a, b'' = b, c'' = c/2. Mo and Cu radiations, R = 0.040.

The present sample has composition $(Mo_{0.93}Ta_{0.07})_5O_{14}$. The structure is similar to that previously reported for the substructure of Mo_5O_{14} (Fig. 1, 1); the superstructure results from displacements of the atoms along c. Ta substitutes exclusively in the pentagonal bipyramids.

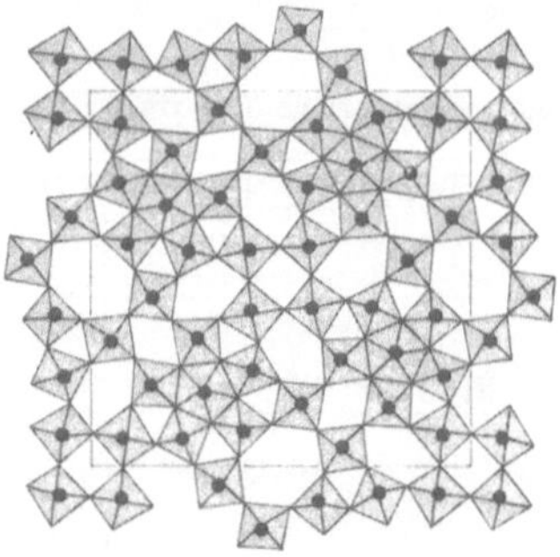

Fig. 1. Structure of Mo_5O_{14}; the tetragonal subcell is outlined.

1. Structure Reports, 29, 308.

MOLYBDENUM VANADIUM OXIDE

$(Mo_{0.93}V_{0.07})_{17}O_{47}$

N. YAMOZOE, T. EKSTRÖM and L. KIHLBORG, 1975. Acta Chem. Scand., A29, 404-408.

Orthorhombic, Pba2, a = 21.531, b = 19.534, c = 4.001 Å, [Z = 2]. Mo radiation, R = 0.064 for 986 reflexions.

The structure differs only in minor details from that of $Mo_{17}O_{47}$ (1), notably an increase (about 0.17 Å) in the mean displacement of the Mo/V atoms from the centres of the oxygen octahedra or pentagonal bipyramids. The V is preferentially substituted in two of the discrete cation sites, Mo(4) and Mo(9).

1. Structure Reports, 28, 127.

MOLYBDENUM(VI) OXIDE MONOHYDRATE

(MOLYBDIC ACID)

α-$MoO_3.H_2O$

H.R. OSWALD, J.R. GÜNTER and E. DUBLER, 1975. J. Solid State Chem., 13, 330-338.

Triclinic, $P\bar{1}$, a = 7.388, b = 3.700, c = 6.673 Å, α = 107.8, β = 113.6, γ = 91.2°, D_m not given, Z = 2. Mo radiation, R = 0.088 for 204 reflexions. Independent study in 1.

The structure (Fig. 1) contains double chains of strongly-distorted $MoO_5(H_2O)$ octahedra sharing edges; Mo-O = 1.65-2.37 Å. The chains are joined by hydrogen bonds, 2.76-3.14 Å.

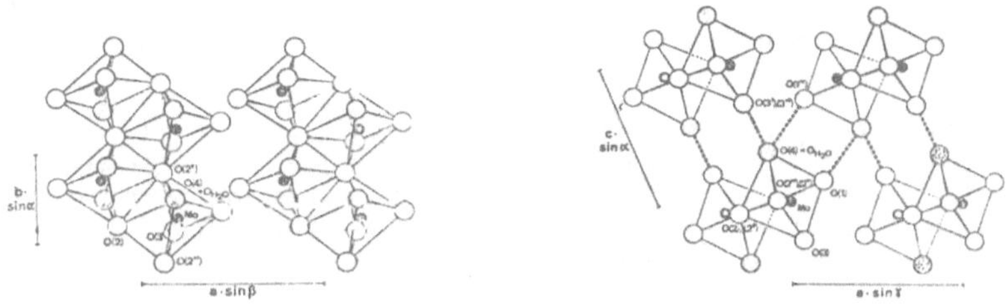

Fig. 1. Structure of α-$MoO_3.H_2O$; broken lines are hydrogen bonds.

1. Structure Reports, 40A, 178.

CERIUM OXIDE

Ce_7O_{12}

I. S.P. RAY and D.E. COX, 1975. J. Solid State Chem., 15, 333-343.

II. S.P. RAY, A.S. NOWICK and D.E. COX, 1975. Ibid., 15, 344-351.

Rhombohedral, $R\bar{3}$, a = 6.80 [not 8.60] Å, α = 99.4° (hexagonal cell has a = 10.37, c = 9.67 Å), Z = 1. Neutron diffraction data, R = 0.144 for 79 and 24 reflexions (two data sets at different wavelengths).

Atomic positions (hexagonal axes)

		x	y	z
Ce(1) in	18(f)	0.4135	0.1258	-0.0134
Ce(2)	3(a)	0	0	0
O(1)	18(f)	0.4445	0.1473	-0.2685
O(2)	18(f)	0.4572	0.1629	-0.7716
n O(3)	6(c)	0	0	- 1/4

Isostructural with UY_6O_{12} (1). Some oxygen is disordered in position 6(c). Ce-O = 2.24-2.64 Å.

1. Structure Reports, 31A, 162.

PRASEODYMIUM OXIDE

Pr_7O_{12}

R.B. von DREELE, L. EYRING, A.L. BOWMAN and J.L. YARNELL, 1975. Acta Cryst., B31, 971-974.

Rhombohedral, $R\bar{3}$, a = 6.7431 Å, α = 99.306°, D_m not given, Z = 1. Neutron powder data with profile fitting.

Isostructural with Tb_7O_{12} (1) and UY_6O_{12} (2). The structure (Fig. 1) is derived from the MO_2 fluorite structure and has a pair of vacancies in the oxygen array along [111], which result in M_4O_6 defect clusters which occur in pairs with one metal atom in common. Pr atoms are 6- and 7-coordinate, Pr-O = 2.22-2.40 Å.

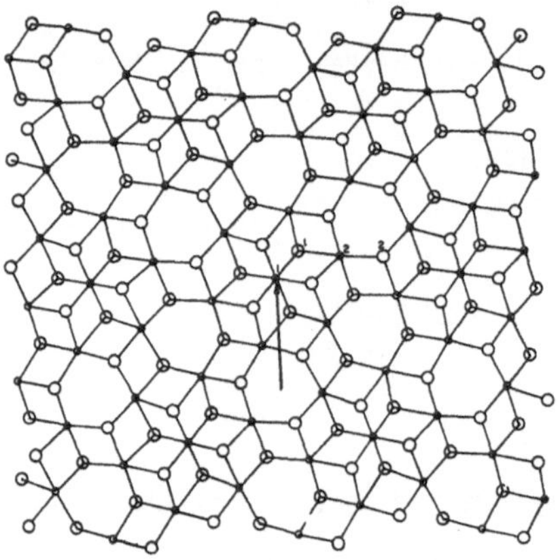

Fig. 1. Structure of Pr_7O_{12}.

1. Structure Reports, 26, 364.
2. Ibid., 31A, 162.

NEODYMIUM OXIDE

Nd_2O_3

J.X. BOUCHERLE and J. SCHWEIZER, 1975. Acta Cryst., B31, 2745-2746.

Trigonal, $P\bar{3}m1$, unit cell parameters not given, Z = 1. Neutron powder data, R = 0.02 for 68 reflexions. Nd in 2(d): z = 0.2462; O(1) in 2(d): z = 0.6466; O(2) in 1(a).

The structure is as previously described (1).

1. Strukturbericht, 1, 244, 261, 745.

CALCIUM SODIUM ALUMINATE

$Ca_{8.5}NaAl_6O_{18}$

F. NISHI and Y. TAKÉUCHI, 1975. Acta Cryst., B31, 1169-1173.

Orthorhombic, Pbca, a = 10.875, b = 10.859, c = 15.105 Å, D_m not given, Z = 4. Mo radiation, R = 0.074 and 0.035 for 2006 and 1223 reflexions (two data sets).

The structure contains Al_6O_{18} rings of six corner-sharing AlO_4 tetrahedra (Fig. 1), Al-O = 1.722-1.770(4) Å. Ca ions have irregular 6-, 7-, and 8-coordination, one of the five non-equivalent Ca ions being split into two half-atoms separated by 0.56 Å. Na ions are partly distributed over a set of Ca positions, and partly at the centre of the tetrahedral ring.

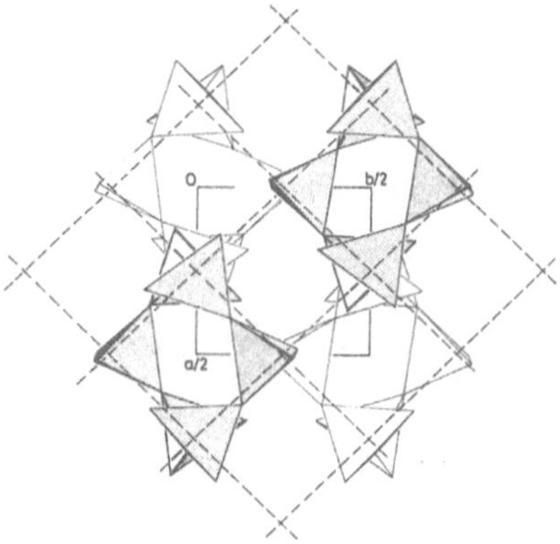

Fig. 1. Structure of $Ca_{8.5}NaAl_6O_{18}$.

TRICALCIUM ALUMINATE

$Ca_3Al_2O_6$ $3CaO.Al_2O_3$

P. MONDAL and J.W. JEFFERY, 1975. Acta Cryst., B31, 689-697.

Cubic, Pa3, a = 15.263 Å, D_m = 3.016, Z = 24. Cu radiation, R = 0.051 for 1191 reflexions (films, photometer and visual intensities).

The structure (Fig. 1a) consists of rings of six AlO_4 tetrahedra, Al_6O_{18} (Fig. 1b), eight to a unit cell, surrounding holes of radius 1.47 Å at 1/8, 1/8, 1/8 and its symmetry-related positions, with Ca^{2+} ions holding the rings together. The Al tetrahedra are fairly regular, Al-O = 1.729-1.768(3) Å, O-Al-O = 101-124°, but the Ca octahedra are considerably distorted, Ca-O = 2.258-3.075(3) Å. The compound is one of the main components of Portland cement, and many previous attempts have been made to determine the structure (1, 2).

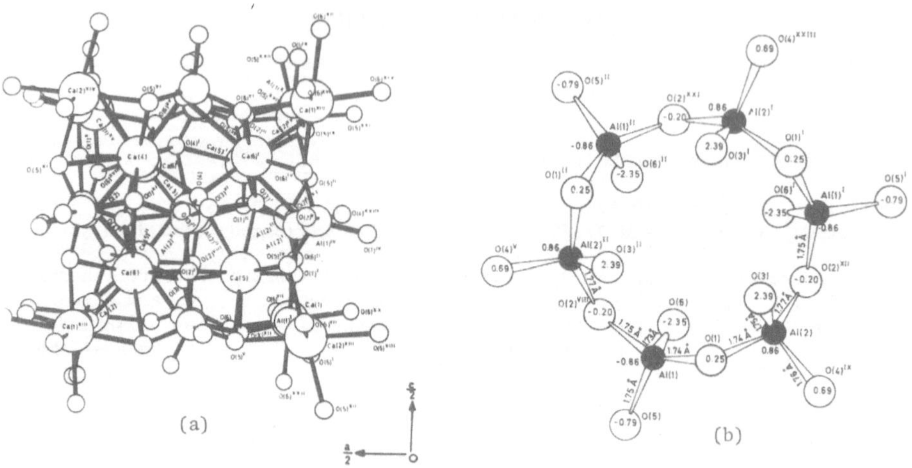

Fig. 1. $Ca_3Al_2O_6$; (a) one-eighth of the unit cell, and (b) the Al_6O_{18} ring viewed along [$\bar{1}\bar{1}\bar{1}$] (bond lengths and distances from the Al mean plane are given).

1. Strukturbericht, 2, 60, 344; 5, 40; Structure Reports, 8, 164.
2. A.E. MOORE, 1966. Mag. Concr. Res., 18 (55), 59.

STRONTIUM ALUMINATE

$SrAl_{12}O_{19}$ $SrO.6Al_2O_3$

A.J. LINDOP, C. MATTHEWS and D.W. GOODWIN, 1975. Acta Cryst., B31, 2940-2941.

Hexagonal, $P6_3/mmc$, a = 5.562, c = 21.972 Å, D_m not given, Z = 2. Mo radiation, R = 0.060 for 609 reflexions.

The material is isostructural with magnetoplumbite (1) and related substances (e.g. 2), the structure being very close to that of $CaO.6Al_2O_3$ (3).

1. Strukturbericht, 6, 74.
2. Structure Reports, 32A, 290.
3. Ibid., 33A, 274.

YTTRIUM ALUMINATE

$YAlO_3$

R. DIEHL and G. BRANDT, 1975. Mater. Res. Bull., 10, 85-90.

Orthorhombic, Pnma, a = 5.330, b = 7.375, c = 5.180 Å (in agreement with 1), D_m = 5.36, Z = 4. Mo radiation, R = 0.075 for 880 reflexions.

Isostructural with $GdFeO_3$ (2); Al-O = 1.901-1.921, Y-O = 2.237-3.262 Å.

1. Structure Reports, 20, 274.
2. Ibid., 20, 273.

STRONTIUM LANTHANUM ALUMINATE

STRONTIUM GADOLINIUM ALUMINATE

$SrLa_2Al_2O_7$
$SrGd_2Al_2O_7$

J. FAVA and G. Le FLEM, 1975. Mater. Res. Bull., 10, 75-80.

Tetragonal, I4/mmm, a = 3.776, 3.707, c = 20.21, 19.80 Å, for La, Gd compounds, respectively, Z = 2. Powder data.

Isostructural with $Sr_3Ti_2O_7$ (1). In the Gd compound, Sr and Gd occupy the 12- and 9-coordinate sites respectively; in the La compound, Sr and La are disordered.

1. S.N. RUDDLESEN and P. POPPER, 1958. Acta Cryst., 11, 54.

SODIUM GALLATE

Na_5GaO_4

D. FINK and R. HOPPE, 1975. Z. anorg. Chem., 414, 193-202.

Orthorhombic, Pbca, a = 10.26, b = 5.95, c = 18.06 Å, D_m = 3.04, Z = 8. Mo radiation, R not given for 885 reflexions.

The structure contains isolated GaO_4^{5-} tetrahedra, connected by Na^+ ions. Ga-O = 1.85-1.88 Å.

YTTRIUM GALLATE

$YGaO_3$ (high-temperature)

S. GELLER, J.B. JEFFRIES and P.J. CURLANDER, 1975. Acta Cryst., B31, 2770-2774.

Hexagonal, $P6_3cm$, a = 6.065, c = 11.615 Å, D_m not given, Z = 6. Mo radiation, R = 0.040 for 266 reflexions (sample quenched from 1950°C; difficulties with convergence of the least-squares refinement).

Atomic positions

		x	y	z
Y(1)	in 2(a)	0	0	0.2716
Y(2)	4(b)	1/3	2/3	0.2347
Ga	6(c)	0.3376	0	0
O(1)	6(c)	0.3104	0	0.1578
O(2)	6(c)	0.6394	0	0.3419
O(3)	2(a)	0	0	0.4837
O(4)	4(b)	1/3	2/3	0.0161

Isostructural with $LuMnO_3$ (1). The structure contains six layers of loosely packed oxygen ions in an ABCACB stacking sequence. Y ions have 7-coordination, Y-O = 2.28-2.54 Å, and the Ga ion has fivefold trigonal bipyramidal coordination, Ga-O = 1.84-2.06 Å.

1. Structure Reports, 28, 138.

CADMIUM INDATE

$CdIn_2O_4$

I. RASINES, 1974. Z. Kristallogr., 140, 410-413.

Cubic, Fd3m, a = 9.166 Å, D_m = 6.92, Z = 8. Cu radiation, R(I) = 0.05 for 38 reflexions (powder data). M in 8(a) and 16(d); O in 32(e): x = 0.390 (origin at $\bar{4}3m$).

Spinel structure; the cation distribution is uncertain. M-O = 2.22 (tetrahedral), 2.35 Å (octahedral).

LEAD GERMANATE

$Pb_5Ge_3O_{11}$

M.I. KAY, R.E. NEWNHAM and R.W. WOLFE, 1975. Ferroelectrics, 9, 1-6.

Trigonal, P3, a = 10.190, c = 10.624 Å, [Z = 3]. Neutron diffracation data, R = 0.053 for 304 reflexions.

Space group P3 (1) is confirmed, rather than P$\bar{6}$ (2). The structure is quite similar to that previously described (2), except for the removal of the mirror plane.

1. Y. IWATA, H. KOIZUMI, N. KOYANO, I. SHIBUYA and N. NIIZEKI, 1973. J. Phys. Soc. Japan, 35, 314; Structure Reports, 39A, 364.
2. Structure Reports, 39A, 214.

SCANDIUM OXYGERMANATE

Sc_2GeO_5 $Sc_2O_3.GeO_2$

Ju.A. GORBUNOV, B.A. MAKSIMOV, Ju.A. KHARITONOV and N.V. BELOV, 1974. Kristallografija, 19, 1081-1083 [Soviet Physics - Crystallography, 19, 669-670].

Monoclinic, B2/b, a = 10.927, b = 10.656, c = 10.486 Å, γ = 93.5°, D_m not given, Z = 12. Mo radiation, R = 0.045 for 2320 reflexions.

The structure (Fig. 1) contains tetrahedral GeO_4^{2-} ions, O^{2-} ions, and Sc^{3+} ions with octahedral coordination. Ge-O = 1.69-1.79, Sc-O = 1.98-2.21 Å.

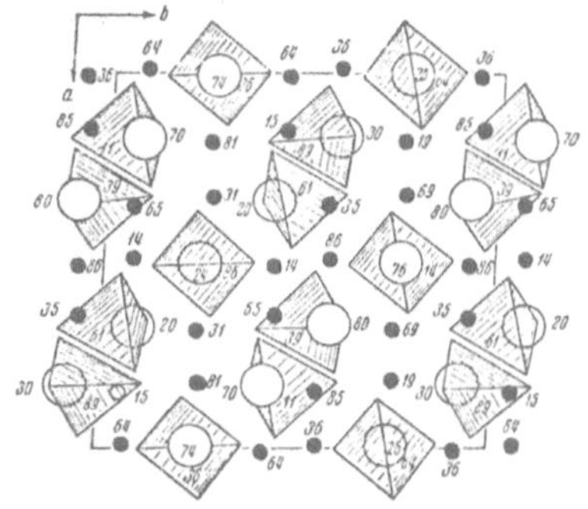

Fig. 1. Structure of Sc_2GeO_5; large open circles are O^{2-}, small filled circles are Sc^{3+}

URANYL GERMANATE DIHYDRATE

$(UO_2)_2GeO_4.2H_2O$

J.P. LEGROS and Y. JEANNIN, 1975. Acta Cryst., B31, 1140-1143.

Orthorhombic, Fddd, a = 8.179, b = 11.515, c = 19.379 Å, D_m = 5.15, Z = 8. R = 0.068 for 285 reflexions.

U coordination pentagonal bipyramids share edges to form chains, which are linked by edge-sharing with GeO_4 tetrahedra (Fig. 1).

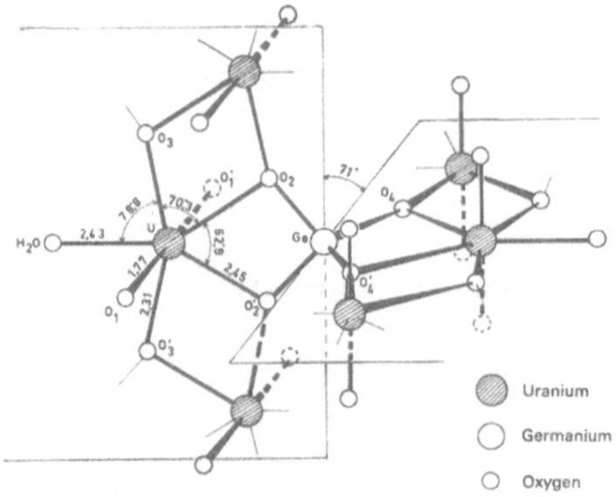

Fig. 1. Structure of uranyl germanate dihydrate.

COPPER URANYL GERMANATE

$[Cu(H_2O)_4](UO_2HGeO_4)_2.2H_2O$

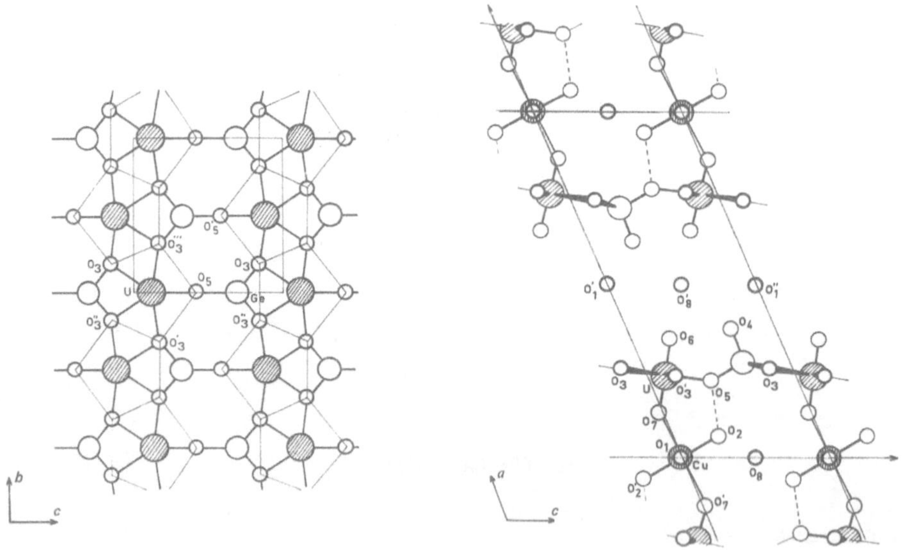

Fig. 1. The (UO_2HGeO_4) sheet (left side) and view of the structure along b (right side) of copper uranyl germanate.

J.P. LEGROS and Y. JEANNIN, 1975. Acta Cryst., B31, 1133-1139.

Monoclinic, C2/m, a = 17.66, b = 7.148, c = 6.817 Å, β = 112.8°, D_m = 4.18, Z = 2. Mo radiation, R = 0.109 for 606 reflexions.

The structure contains infinite $(UO_2HGeO_4^-)_n$ sheets, between which are water molecules and $Cu(H_2O)_4$ groups (Fig. 1). U has pentagonal bipyramidal coordination, U-O = 1.75 (uranyl), 2.24-2.44 Å; Ge has tetrahedral coordination, Ge-O = 1.70-1.75 Å; and Cu has distorted octahedral geometry, Cu-4 H_2O = 1.94 (x 2), 1.98 (x 2), Cu-2 O(uranyl) = 2.41 Å. The structure also contains hydrogen bonds.

POTASSIUM ORTHOSTANNATE

K_4SnO_4

R. MARCHAND, Y. PIFFARD and M. TOURNOUX, 1975. Acta Cryst., B31, 511-514.

Triclinic, P$\bar{1}$, a = 6.48, b = 6.51, c = 9.70 Å, α = 71.82, β = 99.89, γ = 113.13°, D_m = 3.12, Z = 2. Mo radiation, R = 0.029 for 4006 reflexions.

The structure (Fig. 1) contains discrete tetrahedral SnO_4^{4-} ions, Sn-O = 1.947-1.960(3) Å, O-Sn-O = 105.7-114.4°. K ions have irregular 4- or 5-coordination, K-O = 2.61-2.94 Å.

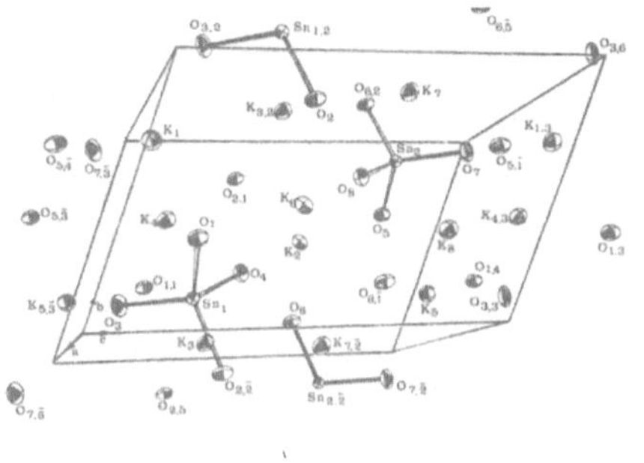

Fig. 1. Structure of K_4SnO_4.

STRONTIUM PLUMBATE(IV)

$SrPbO_3$

H.-L. KELLER, K.-H. MEIER and H. MÜLLER-BUSCHBAUM, 1975. Z. Naturforsch., 30B, 277-278.

Orthorhombic, Pnma, a = 5.964, b = 8.320, c = 5.860 Å, D_m not given, Z = 4. R = 0.061 for 348 reflexions.

Atomic positions

			x	y	z
Sr	in	4(c)	0.455	1/4	0.988
Pb		4(a)	0	0	0
O(1)		8(d)	0.296	0.947	0.205
O(2)		4(c)	0.035	1/4	0.113

Isostructural with $LaErO_3$ (1). In the perovskite-type structure, Pb has octahedral coordination (Pb-O = 2.16-2.19 Å) and Sr has 8-coordination (Sr-O = 2.39-2.98 Å).

1. Structure Reports, 37A, 265.

MAGNESIUM ANTIMONITE

$MgSb_2O_4$

C. GIROUX-MARAINE and G. PEREZ, 1975. Rev. Chim. Minér., 12, 427-432.

Tetragonal, $P4_2/mbc$, a = 8.476, c = 5.938 Å, D_m = 5.09, Z = 4. Mo radiation, R = 0.042 for 315 reflexions.

Atomic positions

			x	y	z
Sb	in	8(h)	0.3246	0.3351	0
Mg		4(d)	0	1/2	1/4
O(1)		8(h)	0.1019	0.3650	0
O(2)		8(g)	0.1775	0.6775	1/4

The structure contains chains of edge-sharing MgO_6 octahedra, connected by Sb atoms with trigonal pyramidal coordination. Mg-O = 2.06, 2.13, Sb-O = 1.91, 2.00 Å.

THALLIUM(I) ANTIMONATE(V)

$TlSbO_3$
Tl_5SbO_5

I. M. BOUCHAMA and M. TOURNOUX, 1975. Rev. Chim. Minér., 12, 80-92.

II. Idem, 1975. Ibid., 12, 93-101.

$TlSbO_3$-2Ha
Trigonal, $P\bar{3}1c$, a = 5.31, c = 14.25 Å, Z = 4. Mo radiation, R = 0.074 for 303 reflexions. Tl in 4(f): z = 0.0666; Sb(1) in 2(a); Sb(2) in 2(c); O in 12(i): (0.9481, 0.2832, 0.1733).

$TlSbO_3$-2Hb
Hexagonal, $P6_322$, a = 5.31, c = 14.25 Å, Z = 4. Mo radiation, R = 0.090 for 72 reflexions (film data). Tl in 4(f): z = 0.068; Sb(1) in 2(b); Sb(2) in 2(c); O in 12(i): (0.963, 0.297, 0.181).

Tl_5SbO_5
Orthorhombic, $Cmc2_1$, a = 10.74, b = 11.78, c = 13.58 Å, Z = 8. Mo radiation, R = 0.098 for 618 reflexions.

The $TlSbO_3$ polytypes contain sheets of corner-sharing SbO_6 distorted octahedra, Sb-O = 1.98; Tl has trigonal-pyramidal coordination, Tl-O = 2.54 Å, with three other O at 3.8 Å. The two polytypes differ in the stacking of the sheets.

The Tl_5SbO_5 structure contains isolated Sb_2O_{10} groups, which consist of two SbO_6 octahedra sharing an edge, Sb-O = 1.93-2.27 Å. The six independent Tl ions exhibit 2-, 3-, 4-, and 5-coordinations, all of which suggest stereochemical activity for the lone electron-pair; Tl-O = 2.28-3.33 Å.

RUTILE-LIKE ANTIMONATES

$MSbO_4$ (M = Cr, Fe, Rh, Al, Ga)
MSb_2O_6 (M = Mg, Co, Ni, Cu, Zn)

J.D. DONALDSON, A. KJEKSHUS, D.G. NICHOLSON and T. RAAKE, 1975. Acta Chem. Scand., A29, 803-809.

These materials have been examined by X-ray powder, and other methods. The first group (Cr...Ga) have random-rutile structures, with tetragonal a and c averaging 4.57 and 3.05 Å, whilst the second group (Mg...Zn) are tri-rutiles, with respective cell parameters averaging 4.65 and 9.26 Å. However $CuSb_2O_6$ is slightly distorted to the monoclinic system.

BISMUTH IRON OXIDE

BISMUTH ZINC OXIDE

$Bi_{25}FeO_{40}$
$Bi_{38}ZnO_{60}$

D.C. CRAIG and N.C. STEPHENSON, 1975. J. Solid State Chem., 15, 1-8.

Cubic, I23, a = 10.179, 10.194 Å, Z = 1, 2/3. Cu radiation, R = 0.021 for the Fe compound, 0.027 for 204 reflexions for the Zn compound.

Atomic positions

$Bi_{25}FeO_{40}$			x	y	z
24 Bi^{3+}	in	24(f)	0.17635	0.31796	0.01409
1 Fe^{3+} + 1 Bi^{5+}		2(a)	0	0	0
8 O(1)		8(c)	0.6885	0.6885	0.6885
24 O(2)		24(f)	0.6346	0.7521	0.9887
8 O(3)		8(c)	0.8926	0.8926	0.8926

$Bi_{38}ZnO_{60}$					
24 Bi^{3+}	in	24(f)	0.82352	0.68178	0.98603
2/3 Zn + 4/3 Bi^{5+}		2(a)	0	0	0
8 O(1)		8(c)	0.3117	0.3117	0.3117
24 O(2)		24(f)	0.3661	0.2496	0.0118
8 O(3)		8(c)	0.1069	0.1069	0.1069

The materials are isostructural (the specimens studied are optical enantiomorphs). Bi^{3+} is bonded to five oxygen atoms at 2.07-2.60 Å with the lone electron-pair completing an octahedron. Twenty-four octahedra share corners to form a $Bi_{24}O_{40}$ cage within which are two tetrahedral sites occupied by Bi^{5+}/Fe^{3+} and Bi^{5+}/Zn^{2+}; the tetrahedra are regular, M-O = 1.89 Å.

CALCIUM STRONTIUM SCANDATE

$Ca_2SrSc_6O_{12}$ $Ca_{0.67}Sr_{0.33}Sc_2O_4$

H. MÜLLER-BUSCHBAUM and W. MUSCHICK, 1975. Z. anorg. Chem., 412, 209-214.

Hexagonal, $P6_3/m$, a = 9.659, c = 3.136 Å, Z = 1. Mo radiation, R = 0.044 for 1144 reflexions.

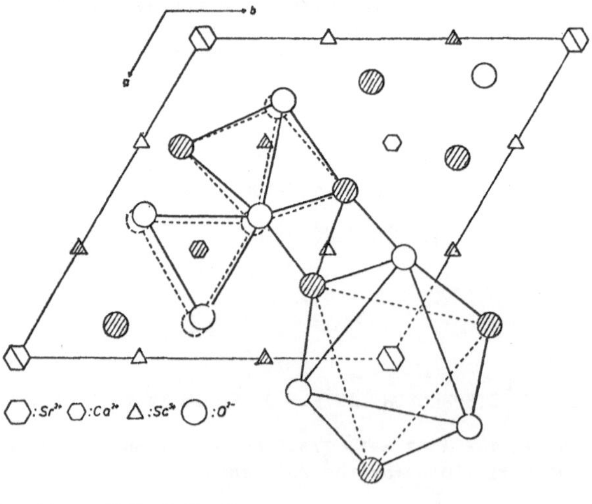

Fig. 1. Structure of $Ca_2SrSc_6O_{12}$.

Atomic positions

		x	y	z
6 Sc	in 6(h)	0.3460	-0.0030	1/4
2 Ca	2(c)	1/3	2/3	1/4
1 Sr	2(b)	0	0	0
6 O(1)	6(h)	0.195	0.309	1/4
6 O(2)	6(h)	0.530	0.398	1/4

The structure (Fig. 1) contains octahedrally coordinated Sc and Sr, Sc-O = 2.06-2.21, Sr-O = 2.73. Ca has six oxygen neighbours in a trigonal prism at 2.33 Å, with three further oxygens above the rectangular faces at 3.02 Å.

EUROPIUM SCANDATE

$EuScO_3$

M. FAUCHER and P. CARO, 1975. Mater. Res. Bull., 10, 1-8.

Orthorhombic, [Pbnm], a = 5.502, b = 5.752, c = 7.954 Å, D_m not given, Z = 4. Powder data, 19 reflexions.

Isostructural with $GdFeO_3$ (1).

1. Structure Reports, 20, 273; 30A, 323.

THALLIUM(I) TITANIUM MAGNESIUM OXIDE

$Tl_x(Mg,Ti)O_2$

A. VERBAÈRE, M. DION and M. TOURNOUX, 1975. Rev. Chim. Minér., 12, 156-174.

Tetragonal, I4/m, a = 10.181, c = 2.975 Å for $Tl_{0.2}(Mg_{0.1}Ti_{0.9})O_2$, Z = 8. Mo radiation, R = 0.041 for 228 reflexions. (Ti,Mg), O(1), O(2) in 8(h): x = 0.1666, 0.2066, 0.1646, y = 0.3503, 0.1550, 0.5394; Tl in fractionally-occupied 4(e) positions: z = 0.047, 0.34, 0.22.

Hollandite-type structure (1). Another composition, $Tl_{0.4}(Mg_{0.2}Ti_{0.8})O_2$, is orthorhombic, isostructural with $Rb_x(Mn,Ti)O_2$ (2).

1. Structure Reports, 13, 190.
2. Ibid., 33A, 294, 303.

ARMALCOLITE

$(Mg,Fe)Ti_2O_5$

J.R. SMYTH, 1974. Earth Planet. Sci. Letters, 24, 262-270.

Orthorhombic, Bbmm, a = 9.743, b = 10.001, c = 3.728 Å, D_m not given, Z = 4. Mo radiation, R = 0.060 for 582 reflexions, lunar material.

Pseudobrookite, Fe_2TiO_5, structure (1), with Ti in 8(f) and (Fe,Mg) in 4(c).

1. Strukturbericht, 2, 53, 336; Structure Reports, 21, 279; 22, 312.

PSEUDORUTILE

$Fe_2Ti_3O_9$

I.E. GREY and A.F. REID, 1975. Amer. Min., 60, 898-906.

Hexagonal, $P6_322$, a = 14.375, c = 4.615 Å. Mo radiation, R = 0.069 (films, visual intensities).

The structure contains hexagonal close-packed oxygens, as previously proposed (1), with metal atoms randomized with two-thirds occupancy over half the available octahedral sites and fully ordered in the remaining sites. The ordered atoms form chains of alternately filled and empty octahedral sites along c; poor correlation between these chains leads to diffuse streaking of the metal-ordering reflexions.

1. G. TEUFER and A.K. TEMPLE, 1966. Nature, 211, 179.

LANTHANUM DITITANATE

(Orthorhombic)

$La_2Ti_2O_7$

K. SCHEUNEMANN and H. MÜLLER-BUSCHBAUM, 1975. J. Inorg. Nucl. Chem., 37, 1879-1881.

Orthorhombic, $Pna2_1$, a = 25.745, b = 7.810, c = 5.547 Å, D_m not given, Z = 8. Mo radiation, R = 0.070 for 694 reflexions.

Isostructural with $Ca_2Nb_2O_7$ (1) and $Sr_2Nb_2O_7$ (2). Ti-O = 1.80-2.21, La-O = 2.42-3.84 Å.

1. Structure Reports, 40A, 194.
2. This volume, p. 241.

LANTHANUM DITITANATE

(Monoclinic)

$La_2Ti_2O_7$

M. GASPERIN, 1975. Acta Cryst., B31, 2129-2130.

Monoclinic, $P2_1$, a = 7.800, b = 13.011, c = 5.546 Å, γ = 98.60°, D_m not given, Z = 4. Mo radiation, R = 0.075 for 1544 reflexions.

The structure (Fig. 1) is the same as that of monoclinic $Ca_2Nb_2O_7$ (1), and is closely related to that of one of the orthorhombic forms of $Ca_2Nb_2O_7$ (2). It contains slabs of four distorted TiO_6 octahedra, linked by La ions; two of the latter have 12 oxygen neighbours, and the other two have 7- and 8-coordination, with further neighbours at greater distances.

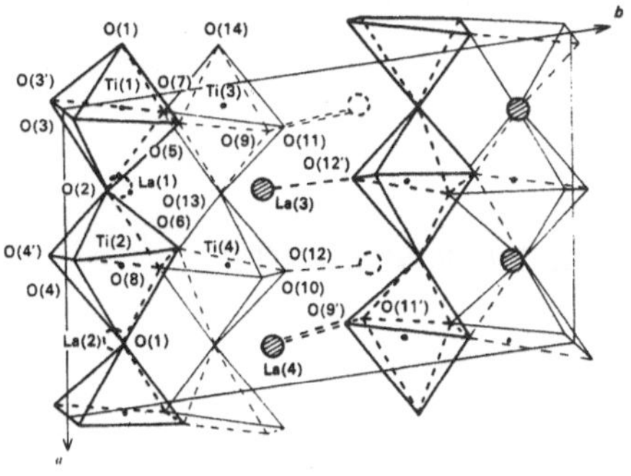

Fig. 1. Structure of lanthanum dititanate.

1. J.K. BRANDON and H.D. MEGAW, 1970. Phil. Mag., 21, 189.
2. Structure Reports, 40A, 194.

NEODYMIUM TITANATE

$Nd_2Ti_2O_7$

K. SCHEUNEMANN and H. MÜLLER-BUSCHBAUM, 1975. J. Inorg. Nucl. Chem., 37, 2261-2263.

Monoclinic, $P2_1$, a = 7.677, b = 5.465, c = 26.013 Å, β = 98.4°, D_m not given, Z = 8. Mo radiation, R not given for 3429 reflexions.

The structure is quite similar to that of the orthorhombic $Ca_2Nb_2O_7$ (1), $Sr_2Nb_2O_7$ (2), and $La_2Ti_2O_7$ (3), and contains perovskite-like blocks. Ti has octahedral coordination, Ti-O = 1.82-2.18 Å, and most Nd atoms have twelve coordination.

1. Structure Reports, 40, 194.
2. This volume, p. 241.
3. Preceding report.

STRONTIUM DIZIRCONATE

$Sr_3Zr_2O_7$

V. LONGO and D. MINICHELLI, 1974. Ceramurgia, 4, 25-28.

Orthorhombic, Pmmm, a = 4.113, b = 4.102, c = 20.941 Å, D_m = 5.15, Z = 2. Cu radiation, powder data, R = 0.18 for 163 reflexions (50 observed).

The structure is assumed to be similar to that of tetragonal $Sr_3Ti_2O_7$ (1), but some disorder is suspected and further work is proposed.

1. Structure Reports, 22, 308; 24, 440.

CALCIUM HAFNIUM OXIDES

$CaHf_4O_9$
$Ca_6Hf_{19}O_{44}$

I. J.G. ALLPRESS, H.J. ROSSELL and H.G. SCOTT, 1975. J. Solid State Chem., 14, 264-273.

$CaHf_4O_9$
Monoclinic, C2/c, a = 17.698, b = 14.500, c = 12.021 Å, β = 119.47°, D_m not given, Z = 16. Cu radiation, R = 0.07 for 116 reflexions (powder data).

$Ca_6Hf_{19}O_{44}$
Rhombohedral, R$\bar{3}$c, a = 12.059 Å, α = 98.31°, Z = 2 (hexagonal cell, a = 18.244, c = 17.613 Å, Z = 6). Cu radiation, R = 0.12 for 106 reflexions (powder data).

Both phases are superstructures derived from the defect fluorite structure by ordering of cations and of anion vacancies. Ca ions are 8-coordinate, and Hf ions 6-, 7-, or 8-coordinate. It is pointed out that the structure determinations are incomplete, and the proposed structures are not necessarily correct.

$Ca_2Hf_7O_{16}$

II. H.J. ROSSELL and H.G. SCOTT, 1975. J. Solid State Chem., 13, 345-350.

Rhombohedral, R$\bar{3}$, a = 9.527 Å, α = 38.80° (hexagonal cell has a = 6.329, c = 26.396 Å), D_m = 8.346, Z = 1. Cu radiation, R = 0.07 for 88 reflexions (powder data).

The rhombohedral cell has a volume 2.25 times that of a fluorite subcell. The cations are ordered on the cation sites of the fluorite structure, with Ca ions segregated into discrete layers parallel to (111) of fluorite. There is some evidence that the formal anion vacancies are also ordered, but the anion coordinates have not been accurately determined (σ ∿ 0.1 Å).

SODIUM METAVANADATE

$NaVO_3$

K. RAMANI, A.M. SHAIKH, B. SWAMINATHA REDDY and M.A. VISWAMITRA, 1975. Ferroelectrics, 9, 49-56.

Monoclinic, Cc, a = 10.494, b = 9.434, c = 5.863 Å, β = 108°48', D_m = 2.92, Z = 8, for the room-temperature ferroelectric phase (above 380°C, the paraelectric phase has space group C2/c). Cu radiation, R = 0.077 for 375 reflexions (films, visual intensities).

Isostructural with $LiVO_3$ (1), except for displacements from C2/c in the ferroelectric phase. The structure contains alternate chains of VO_4 tetrahedra and Na coordination octahedra; V-O = 1.62-1.63 (terminal), 1.78-1.81 (bridging), Na-O = 2.27-2.59(2) Å.

1. Structure Reports, 39A, 222.

SODIUM VANADATE

α'-NaV_2O_5

A. CARPY and J. GALY, 1975. Acta Cryst., B31, 1481-1482.

Orthorhombic, $P2_1mn$, a = 11.318, b = 3.611, c = 4.797 Å, Z = 2. Mo radiation, R = 0.079 for 117 reflexions (film data).

The structure is essentially as previously described (1), except that Na is now considered to have eight- rather than seven-coordination.

1. Structure Reports, 31A, 140.

POTASSIUM VANADATE(V)

POTASSIUM CHROMATE(V)

POTASSIUM MANGANATE(V)

K_3VO_4
K_3CrO_4
K_3MnO_4

R. OLAZCUAGA, J.-M. REAU, G. LeFLEM and P. HAGENMULLER, 1975. Z. anorg. Chem., 412, 271-280.

γ-Forms
Cubic, $P2_13$, a = 8.304, 8.307, 8.251 Å, for V, Cr, Mn, Z = 4. Powder data suggest an Na_2CaSiO_4-type structure (1).

β-Forms
Tetragonal, $I\bar{4}2m$, a = 5.93, 5.91, 5.90, c = 8.11, 8.11, 8.01 Å, Z = 2. Powder data suggest a structure similar to that of $K_3Cr(O_2)_4$ (2), with oxygen atoms replacing peroxide groups.

1. Strukturbericht, 2, 159, 557.
2. Structure Reports, 8, 159; 22, 330; 24, 448; 28, 224; 38A, 266.

POTASSIUM VANADYL PYROVANADATE

$K_2V_3O_8$ $K_2(VO)[V_2O_7]$

J. GALY and A. CARPY, 1975. Acta Cryst., B31, 1794-1795.

Tetragonal, P4bm, a = 8.870, c = 5.215 Å, D_m not given, Z = 2. Mo radiation, R = 0.038 for 396 reflexions.

Isostructural with fresnoite, $Ba_2(TiO)[Si_2O_7]$ (1). The structure (Fig. 1) contains $(V_3O_8^{2-})_n$ layers parallel to (001), consisting of corner-sharing VO_5 square pyramids and pairs of VO_4 tetrahedra. The layers are joined by K ions, which have ten-fold pentagonal antiprismatic coordination. V-O (VO_5, V^{4+}) = 1.58, 1.95 (x 4), V-O (V_2O_7, V^{5+}) = 1.63-1.79(1), K-O = 2.77-3.51 Å.

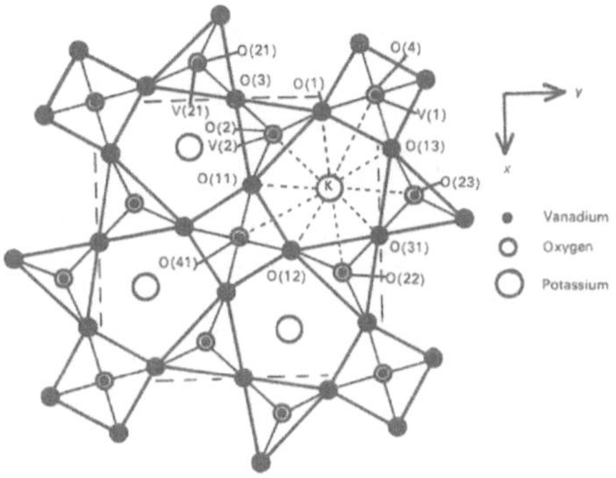

Fig. 1. Structure of $K_2V_3O_8$.

1. Structure Reports, 32A, 442.

CALCIUM DIVANADATE DIHYDRATE

$Ca_2V_2O_7.2H_2O$

J.A. KONNERT and H.T. EVANS, 1975. Acta Cryst., B31, 2688-2690.

Triclinic, $P\bar{1}$, a = 8.134, b = 8.202, c = 6.868 Å, α = 96.24, β = 113.36, γ = 106.19°, D_m = 2.81, Z = 2. Mo radiation, R = 0.070 for 1730 reflexions.

The structure (Fig. 1) contains layers of packed $V_2O_7^{4-}$ ions and H_2O molecules, interleaved with Ca^{2+} ions. The V_2O_7 is 18° from an eclipsed conformation, V-O = 1.80 (bridging), 1.69 Å (terminal), V-O-V = 139°. Ca ions have 7- and 8-coordination, Ca-O = 2.34-2.70 Å, and the water molecules are involved in hydrogen bonds, 2.64-2.87 Å.

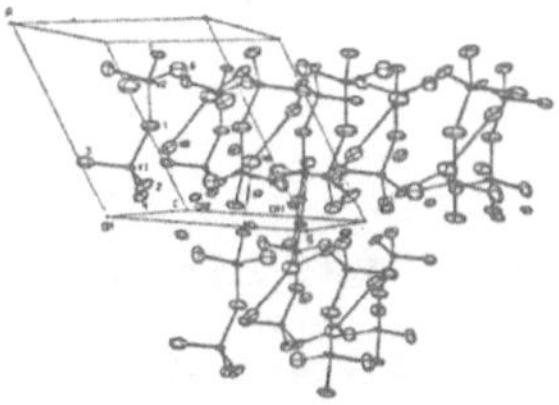

Fig. 1. Structure of calcium divanadate dihydrate.

INDIUM VANADATE(IV)

INDIUM TITANATE(IV)

In_2VO_5
In_2TiO_5

J. SENEGAS, J.-P. MANAUD and J. GALY, 1975. Acta Cryst., B31, 1614-1618.

In_2VO_5
Orthorhombic, Pnma, a = 7.232, b = 3.468, c = 14.82 Å, D_m = 6.52, Z = 4. Mo radiation, R = 0.046 for 1263 reflexions.

In_2TiO_5
Orthorhombic, Pnma, a = 7.237, b = 3.429, c = 14.86 Å, D_m = 6.25, Z = 4.

In the In_2VO_5 structure (Fig. 1) the two independent In atoms both have distorted octahedral coordination, the octahedra sharing edges to form ribbons along b with groups of four octahedra. The ribbons are joined by V atoms which have trigonal bipyramidal coordination. In-O = 2.14-2.24, V-O = 1.76-2.03(1) Å. In_2TiO_5 is isostructural.

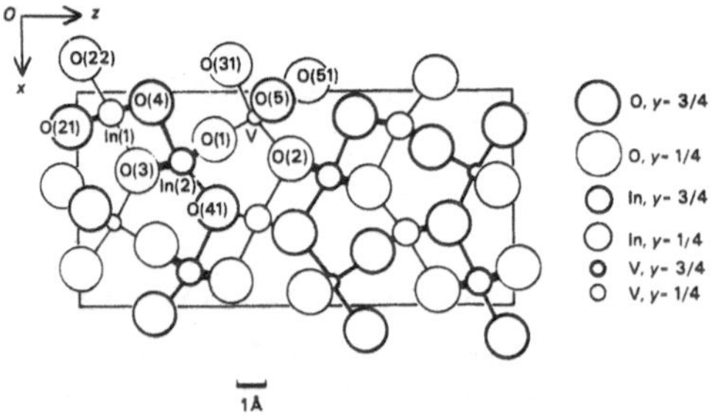

Fig. 1. Structure of In_2VO_5.

COPPER(II) DIVANADATE

α-$Cu_2V_2O_7$

C. CALVO and R. FAGGIANI, 1975. Acta Cryst., B31, 603-605.

Orthorhombic, Fdd2, a = 20.645, b = 8.383, c = 6.442 Å, D_m = 3.969, Z = 8. Mo radiation, R = 0.028 for 488 reflexions.

The structure is as previously described in an independent study (1); the present results are much more accurate. V-O = 1.648-1.743(5), Cu-O = 1.880-1.972, 2.542(5) Å.

1. Structure Reports, 39A, 228.

COPPER VANADATE

$Cu_4V_{2.15}O_{9.38}$

H.-P. CHRISTIAN and H. MÜLLER-BUSCHBAUM, 1975. Z. Naturforsch., 30B, 175-178.

Orthorhombic, $P2_12_12_1$, a = 15.021, b = 8.564, c = 6.055 Å, D_m not given, Z = 4. R = 0.078 for 830 reflexions.

The structure (Fig. 1) contains layers of edge-sharing CuO_6 distorted octahedra (Cu-O = 1.94-2.65 Å). The layers are linked by VO_4 tetrahedra (V-O = 1.63-1.79 Å) and two types of trigonal bipyramid, one type occupied by Cu (Cu-O = 1.88-2.23 Å) and the other type partially occupied by V (V-O = 1.62-2.65 Å, with a sixth oxygen at 2.82 Å).

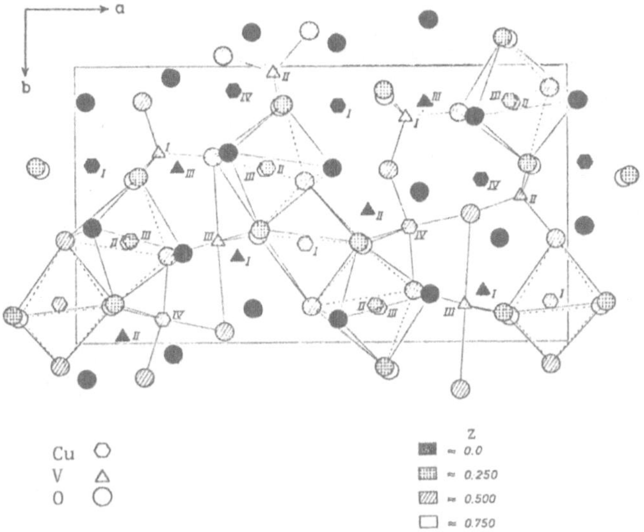

Fig. 1. Structure of $Cu_4V_{2.15}O_{9.38}$.

SODIUM POTASSIUM NIOBATE

$Na_{1-x}K_xNbO_3$ (x = 0.1)

M. AHTEE and A.W. HEWAT, 1975. Acta Cryst., A31, 846-850.

Monoclinic, Pm, a = 7.878, b = 7.792, c = 7.863 Å, β = 90.59°, for x = 0.1, Z = 4. Neutron powder data.

Materials with x = 0.02 and 0.10 (phase Q) have distorted perovskite structures, space group Pm, but close to $P2_1ma$ (1). Nb atoms are displaced from the centres of octahedra, which are regular, but slightly tilted about a and c.

1. M. WELLS and H.D. MEGAW, 1961. Proc. Phys. Soc., 78, 1258.

STRONTIUM DINIOBATE

$Sr_2Nb_2O_7$

I. N. ISHIZAWA, F. MARUMO, T. KAWAMURA and M. KIMURA, 1975. Acta Cryst., B31, 1912-1915.

Orthorhombic, $Cmc2_1$, a = 3.933, b = 26.726, c = 5.683 Å, D_m not given, Z = 4. Mo radiation, R = 0.081 for 676 reflexions.

The structure (Fig. 1) contains distorted perovskite-type slabs of NbO_6 octahedra parallel to (010), and Sr ions; Nb-O = 1.84-2.31; Sr-O = 2.43-3.08 Å (7- and 12-coordination).

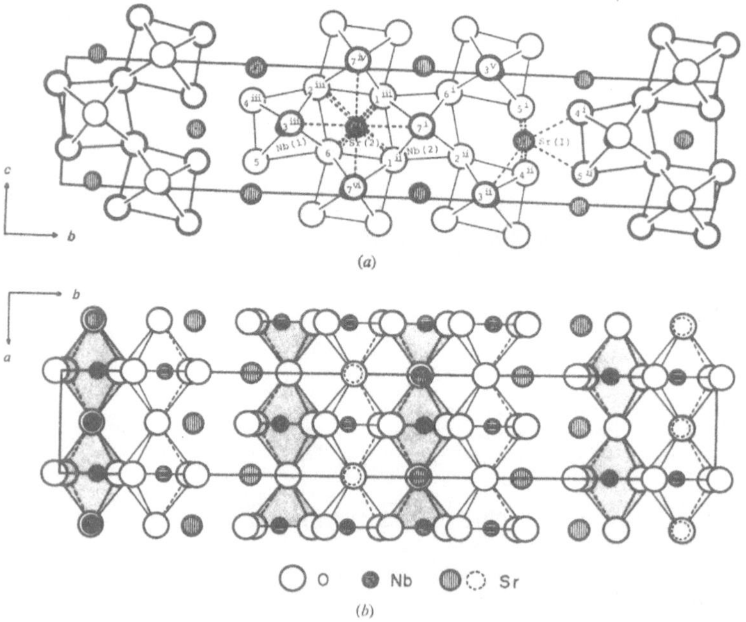

Fig. 1. Structure of $Sr_2Nb_2O_7$ viewed (a) along a, and (b) along c.

II. K. SCHEUNEMANN and H. MÜLLER-BUSCHBAUM, 1975. J. Inorg. Nucl. Chem., 37, 1679-1680.

Orthorhombic, $Pna2_1$, a = 26.647, b = 7.936, c = 5.705 Å, D_m not given, Z = 8. Mo radiation, R = 0.068 for 853 reflexions.

Isostructural with $Ca_2Nb_2O_7$ (1). Nb-O = 1.87-2.22, Sr-O = 2.42-3.92 Å.

1. Structure Reports, 40A, 194.

GERMANIUM NIOBIUM OXIDE

$GeNb_{18}O_{47}$ $GeO_2.9Nb_2O_5$

J.S. ANDERSON, D.J.M. BEVAN, A.K. CHEETHAM, R.B. von DREELE, J. HUTCHISON and J. STRÄHLE, 1975. Proc. Roy. Soc., A, 346, 139-156.

Tetragonal, I4/m, a = 15.73, c = 3.83 Å, Z = 1. Mo radiation, R = 0.034 for 1224 reflexions, and neutron powder data.

Isostructural with $P_2O_5.9Nb_2O_5$ (1); the anion lattice is essentially complete, with excess cations in an extra site with distorted octahedral coordination.

1. Structure Reports, 29, 422; 30A, 336.

PYROCHLORES

$Eu_2M_2O_7$ (M = Sn, Ti, Zr)

I. M. FAUCHER and P. CARO, 1975. J. Solid State Chem., 12, 1-11.

Powder data suggest a pyrochlore structure for M = Sn and Ti, and a mixture of a disordered and a pyrochlore phase for M = Zr.

$CdM_2O_6.H_2O$
$PbM_2O_6.H_2O$
$A_xH_{2-2x}M_2O_6.H_2O$
A = Ca, Sr, Ba; M = Nb, Ta

II. D. GROULT, C. MICHEL and B. RAVEAU, 1975. J. Inorg. Nucl. Chem., 37, 2203-2205.

Cubic, Fd3m, a ∿ 10.5 Å, Z = 8, powder data. Pyrochlore structures, Cd, Pb, and A in 16(d); H_2O in 16(d) in the Cd and Pb compounds, 8(b) in the other compounds; Nb (Ta) in 16(c); O in 48(f): x ∿ 0.32.

$AMWO_6.H_2O$ (A = Li, Na, Ag; M = Nb, Ta, Sb)

III. C. MICHEL, D. GROULT and B. RAVEAU, 1975. J. Inorg. Nucl. Chem., 37, 247-250.

IV. C. MICHEL, D. GROULT, A. DESCHANVRES and B. RAVEAU, 1975. Ibid., 37, 251-255.

Cubic, Fd3m, a = 10.25-10.39 Å, Z = 8. Powder data. Pyrochlore structures, but some weak reflexions suggest space group F23 for the NaNb compound.

TIN TANTALATE

$Sn_{1.75}(Ta,Sn)_2O_{6.5}$

V. T. BIRCHALL and A.W. SLEIGHT, 1975. J. Solid State Chem., 13, 118-130.

Cubic, Fd3m, a = 10.591 Å, Z = 8. Cu radiation, R = 0.016 for 46 reflexions (powder data).

Pyrochlore-type structure, but with Sn^{2+} displaced by 0.4 Å from the $\bar{3}m$ site.

CALCIUM TETRATANTALATE

$CaTa_4O_{11}$

M. ISOBE, F. MARUMO. S. IWAI and M. KIMURA, 1975. Acta Cryst., B31, 908-910.

Hexagonal, $P6_322$, a = 6.2173, c = 12.271 Å, D_m not given, Z = 2. Mo radiation, R = 0.032 for 291 reflexions.

The structure (Fig. 1) contains two independent Ta atoms; Ta(1) has pentagonal bipyramidal seven-coordination, and Ta(2) is octahedral. Ca has distorted-cubic eight-coordination. Ta(1) polyhedra edge-share to form layers; Ta(2) and Ca polyhedra form mixed edge-sharing layers, and the two types of layer are joined by edge-sharing between Ta(1) and Ca polyhedra. Ta-O = 1.96-2.43(2), Ca-O = 2.48 and 2.61(3) Å.

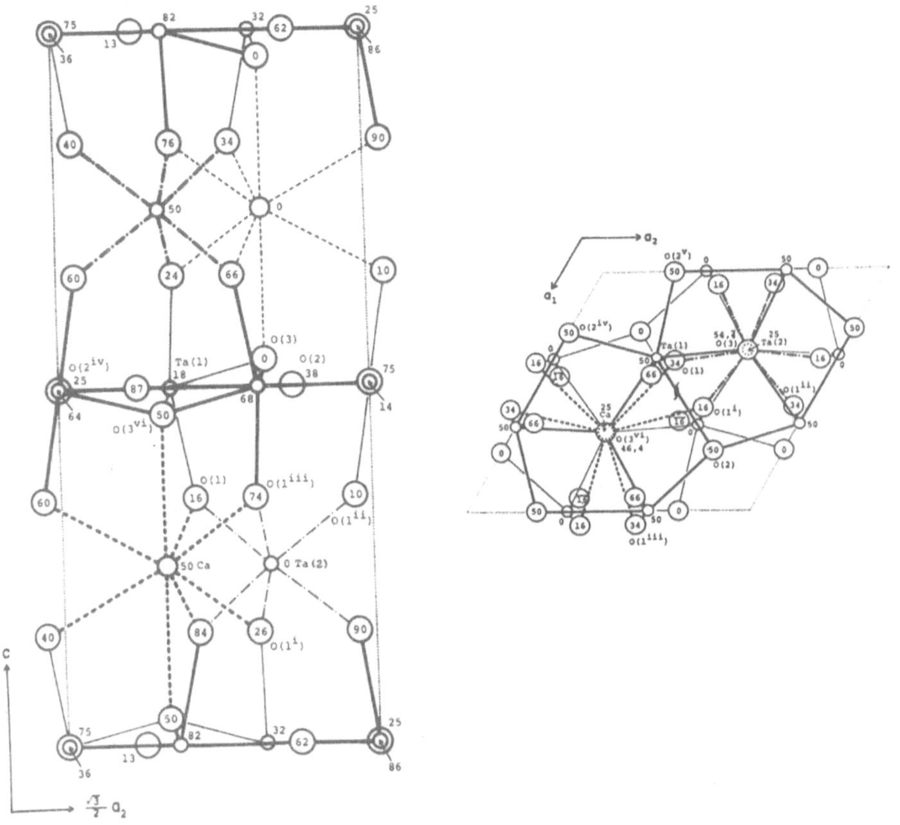

Fig. 1. Structure of $CaTa_4O_{11}$ viewed along a and c (in the latter projection Ta(2) at (2/3, 1/3, 3/4) and Ca at $(\overline{1}/3, \overline{2}/3, 3/4)$ are omitted for clarity).

THALLIUM(I) DITANTALATE

$Tl_2Ta_2O_6$

M. GANNE and M. TOURNOUX, 1975. Mater. Res. Bull., 10, 1313-1318.

Cubic, Fd3m, a = 10.66 Å, Z = 8. Mo radiation, R = 0.051 for 93 reflexions. 16 Ta in 16(c); 48 O in 48(f): z = 0.3060; 8 Tl(1) in 16(d); 8 Tl(2) in 32(e): x = 0.5259 (origin at $\overline{3}$m).

The displacement of some of the Tl^+ ions from the 16(d) position suggests that the lone pair is sterically active.

EUROPIUM TANTALATE

Eu_3TaO_6

K. SATO, G. ADACHI and J. SHIOKAWA, 1975. Mater. Res. Bull., 10, 113-116.

Cubic, Fm3m, a = 8.309 Å, D_m = 8.45, Z = 4. Cu radiation, R = 0.024 for 16 reflexions (powder data).

$(NH_4)_3FeF_6$-type structure (1) assumed (with x(O) = 0.25). Magnetic measurements suggest the formulation $Eu(II)_2Eu(III)TaO_6$.

1. Strukturbericht, 1, 437, 449.

LITHIUM CHROMATE

Li_2CrO_4

I.D. BROWN and R. FAGGIANI, 1975. Acta Cryst., B31, 2364-2365.

Rhombohedral, R$\bar{3}$, a = 14.005, c = 9.405 Å, D_m = 2.426, Z = 18 (hexagonal cell). Mo radiation, R = 0.075 for 825 reflexions.

Atomic positions

	x	y	z
Cr	0.4563	0.3194	0.5838
O(1)	0.5618	0.4462	0.5812
O(2)	0.3424	0.3276	0.5856
O(3)	0.4627	0.2576	0.7303
O(4)	0.4573	0.2511	0.4405
Li(1)	0.4581	0.3149	0.9196
Li(2)	0.4541	0.3135	0.2522

Phenacite structure (1). Li-O = 1.96(2), Cr-O = 1.653(7) Å.

1. Strukturbericht, 1, 356, 402; 2, 517; Structure Reports, 37A, 342.

CHROMIUM SILVER OXIDE

$AgCrO_2$

E. GEHLE and H. SABROWSKY, 1975. Z. Naturforsch., 30B, 659-661.

Rhombohedral, R$\bar{3}$m, a = 2.985, c = 18.51 Å, D_m = 6.63, Z = 3. Cu radiation, R = 0.086 for 302 reflexions. Ag in 3(a); Cr in 3(b); O in 6(c): z = 0.1119.

Delafossite-type structure (1); Ag-O = 2.04, Cr-O = 1.89 Å. Previous work in 2.

1. Structure Reports, 37A, 262.
2. Ibid., 21, 304.

MERCURY(II) CHROMATE HEMIHYDRATE

$HgCrO_4.0{\cdot}5H_2O$

K. AURIVILLIUS and C. STÅLHANDSKE, 1975. Z. Kristallogr., 142, 129-141.

Monoclinic, C2/c, a = 11.832, b = 5.262, c = 14.637 Å, β = 121.01°, D_m = 5.51, Z = 8. Neutron diffraction data, R = 0.030 for 870 reflexions.

The structure is as previously determined by X-ray methods (1). The H atom is at (0.0725, 0.5076, 0.2878); O-H = 0.945(4) Å, H-O-H = 103.7°.

1. Structure Reports, 38A, 265.

DODECAMOLYBDOPHOSPHORIC ACID HYDRATE

$H_3Mo_{12}PO_{40}.29\text{-}31H_2O$

R. STRANDBERG, 1975. Acta Chem. Scand., A29, 359-364.

Tetragonal, $I4_1/amd$, a = 16.473, c = 23.336 Å, D_m = 2.42, Z = 4. Mo radiation, R = 0.047 for 3114 reflexions.

The $Mo_{12}PO_{40}$ unit (Fig. 1) has its central P atom at a site of symmetry $\bar{4}2m$. These units are linked into infinite chains by hydrogen bonded O...(H_2O)...O bridges, which involve six of the water molecules. The other water molecules could not be located, and may be non-structural. The locations of the three acidic protons could not be defined from the results of this analysis. The Mo-O bonds fall into three groups, with distances averaging 1.70, 1.92, and 2.43 Å.

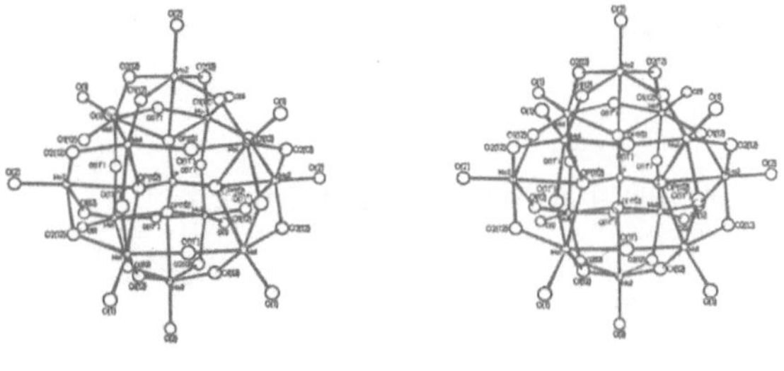

Fig. 1. Stereoview of the $H_3Mo_{12}PO_{40}$ molecule.

AMMONIUM DIMOLYBDATE

$(NH_4)_2Mo_2O_7$

A.W. ARMOUR, M.G.B. DREW and P.C.H. MITCHELL, 1975. J. Chem. Soc., Dalton, 1493-1496.

Triclinic, $P\bar{1}$, a = 7.937, b = 7.305, c = 7.226 Å, α = 93.88, β = 114.33, γ = 82.58°, D_m = 2.95, Z = 2. Mo radiation, R = 0.061 for 990 reflexions.

Isostructural with potassium dimolybdate (1). The structure contains NH_4^+ and $[Mo_2O_7^{2-}]_n$ ions. The latter are infinite chains, made up of pairs of edge-sharing MoO_6 distorted octahedra, adjacent pairs being linked by corner-sharing with two MoO_4 distorted tetrahedra; Mo-O = 1.71-2.31(1) Å.

1. Structure Reports, 37A, 248.

AMMONIUM HEPTAMOLYBDATE(VI) TETRAHYDRATE

POTASSIUM HEPTAMOLYBDATE(VI) TETRAHYDRATE

$(NH_4)_6Mo_7O_{24}.4H_2O$
$K_6Mo_7O_{24}.4H_2O$

H.T. EVANS, B.M. GATEHOUSE and P. LEVERETT, 1975. J. Chem. Soc., Dalton, 505-514.

Ammonium salt
Monoclinic, $P2_1/c$, a = 8.393, b = 36.170, c = 10.472 Å, β = 115.96°, D_m = 2.871, Z = 4. Mo radiation, R = 0.076 for 8197 reflexions (film data).

Potassium salt
Monoclinic, $P2_1/c$, a = 8.15, b = 35.68, c = 10.30 Å, β = 115.2°, D_m = 3.23, Z = 4. Cu radiation, R = 0.12 for 2178 reflexions (film data).

Structure as in 1. The heptamolybdate ion contains seven MoO_6 octahedra condensed by edge-sharing; Mo-O = 1.71-2.42 Å.

1. Structure Reports, 13, 263; 32A, 296.

AMMONIUM 9-MOLYBDOMANGANATE HYDRATE

$(NH_4)_6[MnMo_9O_{32}].xH_2O$ (x = 6-8)

R. ALLMANN and H. d'AMOUR, 1975. Z. Kristallogr., 141, 342-353.

Rhombohedral, R32, a = 10.075 Å, α = 104.32° (hexagonal cell has a = 15.914, c = 12.401 Å), D_m = 3.01, Z = 1. Mo radiation, R = 0.035 for 752 reflexions.

The structure is as previously described (1). The polyanion (Fig. 1) has D_3 symmetry and consists of ten condensed octahedra; mean Mn-O = 1.90, Mo-O = 1.70 (unshared O), 1.97 (O bonded to 2 Mo), 2.22 (O bonded to 3 Mo and 1 Mn), Mn...Mo = 3.172 and 3.191, Mo...Mo = 3.238 and 3.356 Å. The ammonium ions lie on twofold axes, N...O = 2.65-3.21 Å; only six of the eight presumed water molecules could be found, and these are distributed over twelve positions.

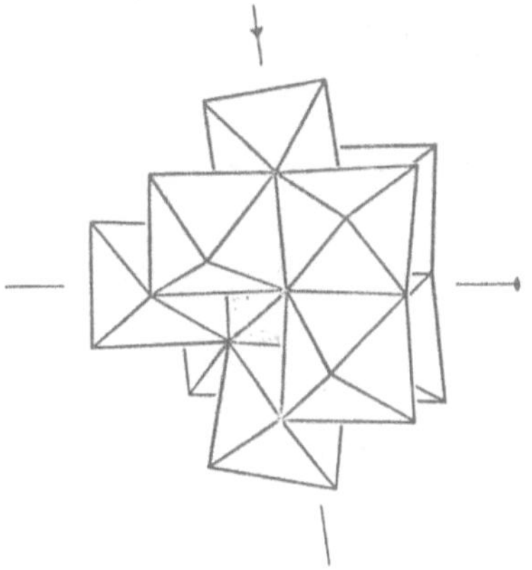

Fig. 1. The $[MnMo_9O_{32}]^{6-}$ polyanion.

1. Structure Reports, 18, 534.

LITHIUM TETRAMOLYBDATE

$Li_2Mo_4O_{13}$ (high-temperature form)

B.M. GATEHOUSE and B.K. MISKIN, 1975. J. Solid State Chem., 15, 274-282.

Triclinic, P$\bar{1}$, a = 8.612, b = 11.562, c = 8.213 Å, α = 94.45, β = 96.38, γ = 111.24°, D_m not given, Z = 3. Mo radiation, R = 0.031 for 2883 reflexions.

Like the low-temperature form (1) H-$Li_2Mo_4O_{13}$ is a derivative structure of V_6O_{13} (Fig. 1), and can be related to L-$Li_2Mo_4O_{13}$ by movement of one of the shear planes in the structure. Interatomic distances are similar in the two forms.

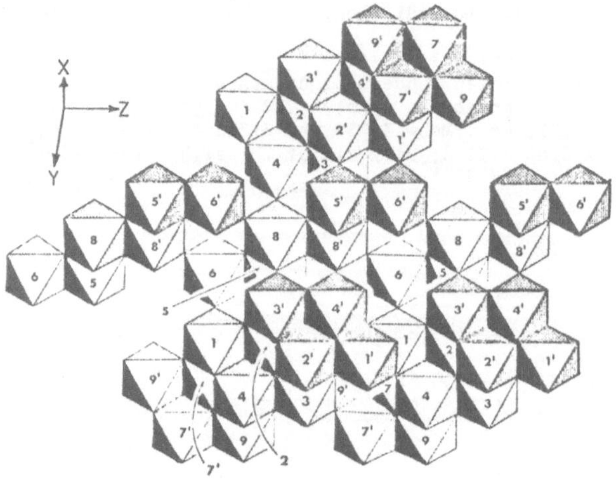

Fig. 1. Idealized structure of H-$Li_2Mo_4O_{13}$.

1. Structure Reports, 40A, 199.

SODIUM MOLYBDATE DIHYDRATE

$Na_2MoO_4.2H_2O$

K. MATSUMOTO, A. KOBAYASHI and Y. SASAKI, 1975. Bull. Chem. Soc. Japan, 48, 1009-1013.

Orthorhombic, Pbca, a = 8.463, b = 10.552, c = 13.827 Å, D_m = 2.51, Z = 8. Mo radiation, R = 0.037 for 1497 reflexions.

The structure contains alternate layers of MoO_4^{2-} tetrahedra and water molecules, linked by Na ions and hydrogen bonds. Mo-O = 1.75-1.79(1), Na-O = 2.39-2.48 (octahedral) and 2.30-2.41 (trigonal bipyramidal), O-H...O = 2.77-2.82 Å.

HEXASODIUM 18-MOLYBDODIPHOSPHATE HYDRATE

$Na_6Mo_{18}P_2O_{62}.24H_2O$

R. STRANDBERG, 1975. Acta Chem. Scand., A29, 350-358.

Monoclinic, C2/c, a = 23.091, b = 13.481, c = 23.157 Å, β = 100.35°, D_m = 3.10, Z = 4. Mo radiation, R = 0.040 for 7260 reflexions.

The $Mo_{18}P_2O_{62}{}^{6-}$ anion has symmetry C_2 and consists of two Mo_9PO_{34} units (1), sharing six oxygen atoms. Between edge-sharing MoO_6 octahedra the Mo...Mo distances range from 3.35 to 3.39 Å; between corner-sharing octahedra, from 3.66 to 3.84 Å. Two Na ions take part in O...Na...O links between separate anions, whilst the third Na is part of an O...Na...(H_2O)...Na...O chain. [More than 300 interatomic distances and angles are recorded.]

1. Structure Reports, 40A, 200.

CAESIUM PENTAMOLYBDATE

CAESIUM HEPTAMOLYBDATE

$Cs_2Mo_5O_{16}$
$Cs_2Mo_7O_{22}$

B.M. GATEHOUSE and B.K. MISKIN, 1975. Acta Cryst., B31, 1293-1299.

$Cs_2Mo_5O_{16}$
Monoclinic, C2/c, a = 21.44, b = 5.559, c = 14.338 Å, β = 122.74°, D_m = 4.62, Z = 4. Mo radiation, R = 0.041 for 2999 reflexions.

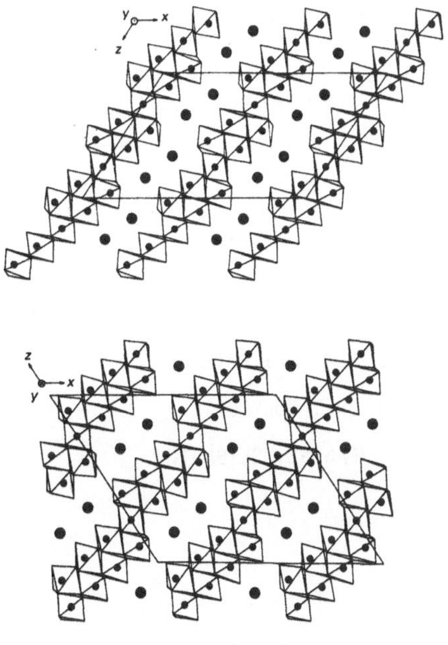

Fig. 1. Structures of $Cs_2Mo_5O_{16}$ (top) and $Cs_2Mo_7O_{22}$ (bottom).

$Cs_2Mo_7O_{22}$
Monoclinic, C2/c, a = 21.54, b = 5.537, c = 18.91 Å, β = 122.71°, D_m = 4.55, Z = 4. Mo radiation, R = 0.039 for 1762 reflexions.

The two structures are very similar; both are composed of sheets of linked Mo-O octahedra joined by Cs-O polyhedra (Fig. 1). Mo-O = 1.69-2.51, Cs-O = 2.94-3.99 Å.

RUBIDIUM BISMUTH MOLYBDATE

α-$RbBi(MoO_4)_2$

R.F. KLEVCOVA, L.P. SOLOV'EVA and V.A. VINOKUROV, 1975. Kristallografija, 20, 270-275 [Soviet Physics - Crystallography, 20, 164-167].

Monoclinic, $P2_1/c$, a = 11.63, b = 12.09, c = 5.28 Å, β = 92.5°, D_m = 5.44, Z = 4. Mo radiation, R = 0.12 for 1105 reflexions (film data).

The structure (Fig. 1) contains MoO_4 tetrahedra which share corners with edge-sharing eight-coordinate Rb and Bi polyhedra. Mo-O = 1.73-1.91, Bi-O = 2.25-2.63, Rb-O = 2.77-3.12 Å.

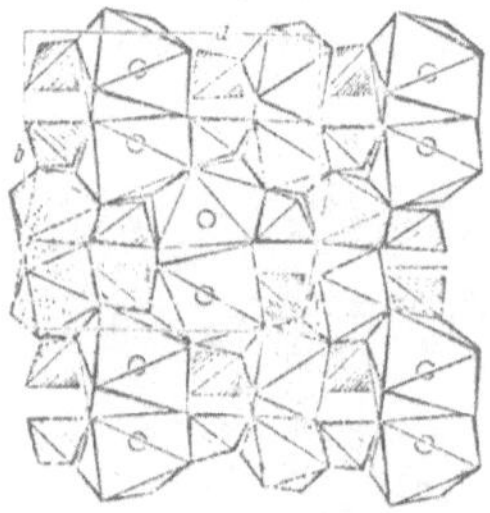

Fig. 1. Structure of α-$RbBi(MoO_4)_2$; circles are Rb atoms.

IRON MOLYBDATE

COBALT MOLYBDATE

MANGANESE MOLYBDATE

$Fe_2Mo_3O_8$
$Co_2Mo_3O_8$
$Mn_2Mo_3O_8$

D. BERTRAND and H. KERNER-CZESKLEBA, 1975. J. Phys., Fr., 36, 379-390.

Hexagonal, $P6_3mc$, a = 5.777, 5.767, 5.799, c = 10.057, 9.916, 10.268 Å, Z = 2. Neutron powder data.

Isostructural with $Zn_2Mo_3O_8$ (1).

1. Structure Reports, 21, 333; 31A, 155.

SODIUM ZINC MOLYBDATE

SODIUM SCANDIUM TUNGSTATE

$Na_{2.2}Zn_{0.9}(MoO_4)_2$
$Na_5Sc(WO_4)_4$

V.A. EFREMOV, Ju.A. VELIKODNIJ and V.K. TRUNOV, 1970. Kristallografija, 20, 287-292 [Soviet Physics - Crystallography, 20, 176-179].

Monoclinic, $C2/c$, a = 12.62, 12.85, b = 13.66, 13.92, c = 7.16, 7.28 Å, β = 112°16', 112°59', D_m not given, Z = 6, 3. Mo radiation, R = 0.12 and 0.11 for 1200 and 560 reflexions, respectively (film data).

The compounds are isostructural, the structure (Fig. 1) containing a three-dimensional framework of MoO_4 or WO_4 tetrahedra linked by MO_6 octahedra, M = (5.4 Zn + 2.6 Na)/8 in the Zn compound, and (3 Sc + 5 Na)/8 in the Sc compound. The other Na ions have 6-, 10- and 12-coordination [including some rather long Na-O distances].

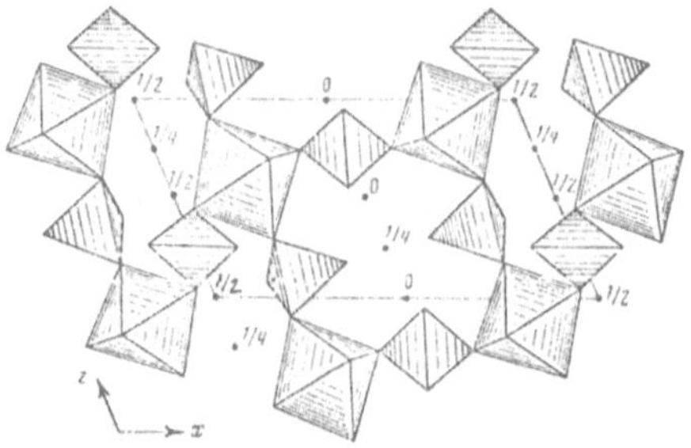

Fig. 1. Structures of $Na_{2.2}Zn_{0.9}(MoO_4)_2$ and $Na_5Sc(WO_4)_4$; black dots are Na ions.

SODIUM SCANDIUM MOLYBDATE

$Na_5Sc(MoO_4)_4$

R.F. KLEVCOVA, L.P. KOZEEVA and P.V. KLEVCOV, 1975. Kristallografija, 20, 925-930 [Soviet Physics - Crystallography, 20, 571-574].

Monoclinic, C2/c, a = 12.89, b = 13.88, c = 7.26 Å, β = 113°, D_m = 3.30, Z = 3. Mo radiation, R = 0.097 for 1364 reflexions (films, visual intensities).

The structure [see 1] contains two types of slightly-distorted MoO_4 tetrahedra, Mo-O = 1.77 Å. The cations are considerably disordered, with a (Na,Sc) octahedron (average M-O = 2.28 Å) and three types of Na polyhedron (average Na-O = 2.42, 2.57, 2.63 Å). Edge- and corner-sharing result in a three-dimensional structure.

1. Preceding report.

POTASSIUM ZINC MOLYBDATE

$K_2Zn_2(MoO_4)_3$

C. GICQUEL-MAYER and G. PEREZ, 1975. Rev. Chim. Minér., 12, 537-545.

Monoclinic, $P2_1/c$, a = 7.00, b = 9.01, c = 20.68 Å, β = 111.15° [given in the text as 111.65°], D_m = 3.80, Z = 4. Mo radiation, R = 0.057 for 2315 reflexions.

The structure contains MoO_4 tetrahedra, a group of four edge-sharing ZnO_6 octahedra, and K^+ ions with two different six coordinations. Mo-O = 1.71-1.82, Zn-O = 2.05-2.45, K-O = 2.64-2.99 Å.

LITHIUM LANTHANUM MOLYBDATE

α-$LiLa(MoO_4)_2$

R.F. KLEVCOVA, 1975. Kristallografija, 20, 746-750 [Soviet Physics - Crystallography, 20, 456-458].

Orthorhombic, Pbca, a = 10.09, b = 9.92, c = 13.58 Å, Z = 8. Mo radiation, R = 0.087 for 1103 reflexions (films, visual intensities).

The structure contains 9-coordinate La (La-O = 2.41-2.84 Å), two types of MoO_4 tetrahedra (Mo-O = 1.71-1.84 Å), and 5-coordinate Li (trigonal bipyramid, Li-O = 1.86-2.39 Å, next nearest O neighbours at 3.12-3.33 Å). These polyhedra are joined by common faces, edges, and corners to form two layers parallel to (100), one containing La dimers and Mo tetrahedra, and the other Mo tetrahedra and Li polyhedra.

SAMARIUM MOLYBDATE

α-Sm_2MoO_6

P.V. KLEVCOV, L.Ju. KHARČENKO and R.F. KLEVCOVA, 1975. Kristallografija, 20, 571-578 [Soviet Physics - Crystallography, 20, 349-353].

Monoclinic, I2/c, a = 15.79, b = 11.28, c = 5.47 Å, β = 91.2°, D_m = 6.72, Z = 8. Mo radiation, R = 0.118 for 923 reflexions (films, visual intensities).

Isostructural with Nd_2WO_6 (1). Mo has distorted tetrahedral coordination, Mo-O = 1.75-1.82 Å, with a fifth oxygen at 2.22 Å; Sm ions have eight-coordination, Sm-O = 2.21-2.81 Å.

1. Structure Reports, 35A, 239.

POTASSIUM SAMARIUM MOLYBDATE

$K_5Sm(MoO_4)_4$

P.V. KLEVCOV, L.P. KOZEEVA, V.I. PROTASOVA, L.Ju. KHARČENKO, L.A. GLINSKAJA, R.F. KLEVCOVA and V.V. BAKAKIN, 1975. Kristallogrifija, 20, 57-62 [Soviet Physics - Crystallography, 20, 31-33].

Rhombohedral, R$\bar{3}$m, a = 5.98, c = 20.75 Å, D_m = 3.78, Z = 1.5. Mo radiation, R = 0.14 for 360 reflexions (film data). (0.5 K + 0.5 Sm) in 3(a); Mo, K, O(1) in 6(c): z = 0.4002, 0.196, 0.314; O(2) in 18(h): (0.157, 0.314, 0.574).

The compound is $K_2(K_{0.5}Sm_{0.5})(MoO_4)_2$, and is isostructural with palmierite, $K_2Pb(SO_4)_2$ (1). Mo has tetrahedral, (K,Sm) octahedral, and K 10-coordination; Mo-O = 1.72, 1.79, (K,Sm)-O = 2.66, K-O = 2.45-3.14 Å.

1. Structure Reports, 10, 143; 11, 390; 13, 326; 17, 499.

LITHIUM DITUNGSTATE

$Li_2W_2O_7$

K. OKADA, H. MORIKAWA, F. MARUMO and S. IWAI, 1975. Acta Cryst., B31, 1451-1454.

Triclinic, P$\bar{1}$, a = 8.283, b = 7.050, c = 5.037 Å, α = 85.40, β = 102.13, γ = 110.29°, D_m not given, Z = 2. Mo radiation, R = 0.069 for 2995 reflexions.

The structure (Fig. 1) is built up from distorted WO_6 octahedra and LiO_4 tetrahedra. The WO_6 octahedra share edges to form $W_2O_7^{2-}$ double chains along c (Fig. 2). W-O = 1.74-2.30(2), Li-O = 1.90-2.11(5) Å.

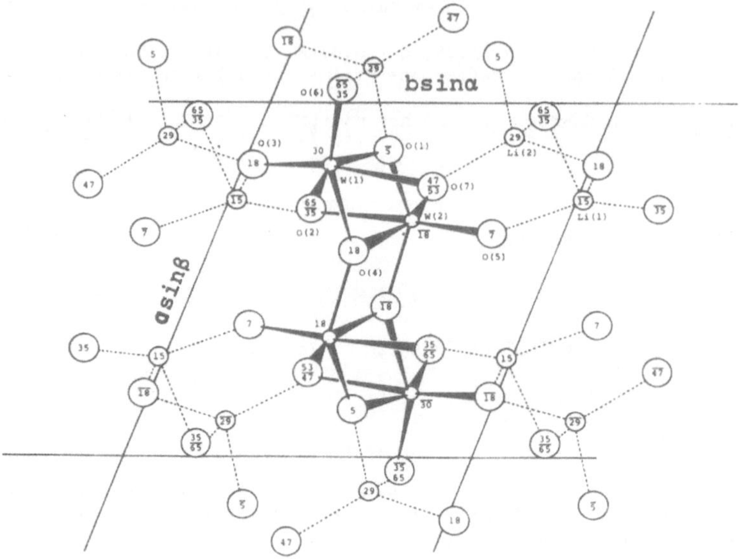

Fig. 1. Structure of $Li_2W_2O_7$.

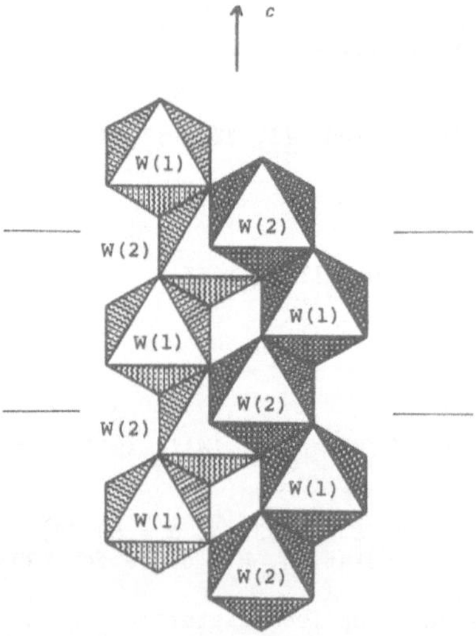

Fig. 2. Idealized double chain in $Li_2W_2O_7$.

SODIUM DITUNGSTATE

$Na_2W_2O_7$

K. OKADA, H. MORIKAWA, F. MARUMO and S. IWAI, 1975. Acta Cryst., B31, 1200-1201.

Orthorhombic, Cmca, a = 7.216, b = 11.899, c = 14.716 Å, D_m not given, Z = 8. Mo radiation, R = 0.054 for 2051 reflexions.

Isostructural with $Na_2Mo_2O_7$ (1). W atoms have distorted octahedral and nearly regular tetrahedral coordinations (Fig. 1), W-O = 1.73-2.25 and 1.76-1.81(1) Å, respectively. Na ions have six-coordination (octahedron and pentagonal pyramid).

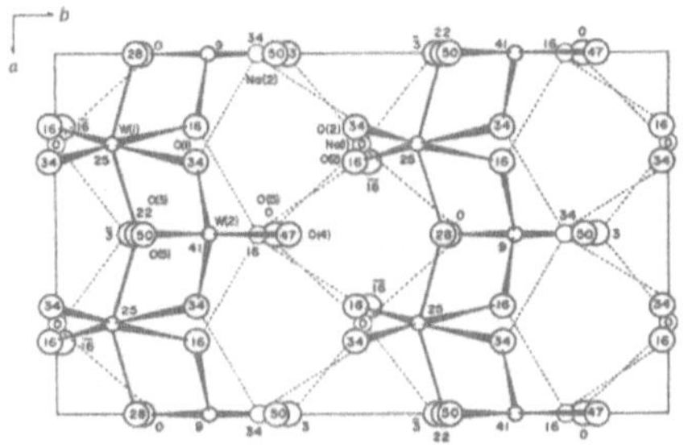

Fig. 1. Structure of $Na_2W_2O_7$.

1. Structure Reports, 13, 260; 24, 449; 32A, 298.

SODIUM PHOSPHOTUNGSTATE

$NaH_2[PW_{12}O_{40}].xH_2O$ (x = 12-14)

R. ALLMANN and H. d'AMOUR, 1975. Z. Kristallogr., 141, 161-173.

Triclinic, $P\bar{1}$, a = 14.06, b = 14.07, c = 14.03 Å, α = 93.2, β = 112.7, γ = 118.1°, D_m = 4.69, Z = 2. Mo radiation, R = 0.13 for 5375 reflexions.

The $PW_{12}O_{40}{}^{3-}$ ion has pseudo T_d symmetry, with a structure similar to that in phosphotungstic acid (1); it contains a central PO_4 tetrahedron surrounded by twelve WO_6 octahedra. Average bond lengths are P-O = 1.55, W-O = 1.69, 1.91, 2.42 Å. The Na ion has octahedral coordination, Na-O = 2.31-2.44 Å, and the water molecules are zeolitic, with $H_5O_2{}^+$ and $H_7O_3{}^+$ groupings being distinguishable.

1. Strukturbericht, 3, 113, 463; 4, 56, 183.

POTASSIUM 12-TUNGSTOSILICATE

β-$K_4SiW_{12}O_{40}.9H_2O$

K.Y. MATSUMOTO, A. KOBAYASHI and Y. SASAKI, 1975. Bull. Chem. Soc. Japan, 48, 3146-3151.

Orthorhombic, Pnma, a = 20.62, b = 15.57, c = 12.95 Å, D_m = 4.88, Z = 4. Mo radiation, R = 0.138 for 2007 reflexions.

The structure contains a discrete β-$SiW_{12}O_{40}^{4-}$ polyanion (C_s symmetry), which is a geometrical isomer of the well-known Keggin ion (1), α-$SiW_{12}O_{40}^{4-}$ (T_d symmetry). The α and β structures are related by a 60° rotation of one of the four W_3O_{13} units which make a cage around the central SiO_4 tetrahedron. The water appears to be zeolitic.

1. Strukturbericht, 3, 113, 463.

BARIUM 12-TUNGSTOSILICATE

α-$Ba_2SiW_{12}O_{40}.16H_2O$

A. KOBAYASHI and Y. SASAKI, 1975. Bull. Chem. Soc. Japan, 48, 885-888.

Monoclinic, C2/c, a = 25.09, b = 12.10, c = 18.79 Å, β = 122.5°, D_m = 4.95, Z = 4. Mo radiation, R = 0.13 for 2952 reflexions.

The structure contains the well-known Keggin ion (1),$SiW_{12}O_{40}^{4-}$, with an SiO_4 tetrahedron in the centre. These ions are linked by Ba ions and by water molecules (four of which could be not be located and are probably zeolitic).

1. Strukturbericht, 3, 113, 463.

POTASSIUM POLYVANADOTUNGSTATE

$K_7V_5W_8O_{40}.12H_2O$

K. NISHIKAWA, A. KOBAYASHI and Y. SASAKI, 1975. Bull. Chem. Soc. Japan, 48, 3152-3155.

Cubic, P$\bar{4}$3m, a = 10.62 Å, D_m = 4.0, Z = 1. Mo radiation, R = 0.057 for 256 reflexions.

The structure contains a Keggin-type polyanion (1), $VM_{12}O_{40}^{7-}$ (M_{12} = 4V + 8W, randomly mixed), with a central VO_4 tetrahedron. 3 K ions are in the 3(d) sites, and 4 are distributed in the 6(g) positions; both types have eight-coordination.

1. Strukturbericht, 3, 113, 463.

BARIUM TUNGSTEN LITHIUM OXIDE

$Ba_5W_3Li_2O_{15}$

E.F. JENDREK, A.D. POTOFF and L. KATZ, 1975. J. Solid State Chem., 14, 165-171.

Hexagonal, $P6_3/mmc$, a = 5.761, c = 23.719 Å, D_m not given, Z = 2. Mo radiation, R = 0.064 for 426 reflexions.

Atomic positions

			x	y	z
Ba(1)	in	2(d)	1/3	2/3	3/4
Ba(2)		4(f)	1/3	2/3	0.0489
Ba(3)		4(e)	0	0	0.1529
W,Li(1)		4(f)	1/3	2/3	0.2004
W(2)		4(f)	1/3	2/3	0.9031
O(1)		12(k)	0.5061	0.0122	0.1479
O(2)		6(h)	0.1740	0.3480	1/4
O(3)		12(k)	0.1740	0.3480	0.5538

The compound has a ten-layer structure as previously determined (1, 2) with a |(5)|(5)| stacking of BaO_3 layers. Pairs of face-sharing MO_6 octahedra (M = W or Li) are linked through corner-sharing by strings of three corner-sharing octahedra (2); in the latter the two outer octahedra are occupied by W, but no evidence was found of Li in the middle one, which appears to be empty. However, the independent neutron powder study (2) indicates Li in 2(a), so that the present material may be $Ba_5W_{3.2}Li_{0.8}O_{15}$ or $Ba_5W_3LiO_{14.5}$.

1. Structure Reports, 38A, 221.
2. Ibid., 40A, 205.

ALUMINUM TUNGSTATE(V)

$AlWO_4$

J.P. DOUMERC, M. VLASSE, M. POUCHARD and P. HAGENMULLER, 1975. J. Solid State Chem., 14, 144-151.

Monoclinic, C2/m, a = 9.069, b = 5.705, c = 4.541 Å, β = 92.29°, D_m = 7.69, Z = 4. Mo radiation, R = 0.055 for 974 reflexions.

Atomic positions

			x	y	z
Al	in	4(i)	0.2590	0	0.4858
W		4(g)	0	0.7716	0
O(1)		8(j)	0.1504	0.2449	0.3012
O(2)		4(i)	0.1141	0	0.7791
O(3)		4(i)	0.4023	0	0.1975

Isostructural with one of the forms of $(Cr,V)O_2$ (1). The structure contains a rutile-like framework, with W(V) stabilized by W-W interactions, 2.61 Å. W-O = 1.90-2.02, Al-O = 1.88-1.92 Å, both octahedral coordination.

1. Structure Reports, 38A, 242.

POTASSIUM GALLIUM TUNGSTATE

$KGa_{0.33}W_{5.66}O_{18}$

Ju.A. VELIKODNIJ and V.K. TRUNOV, 1975. Kristallografija, 20, 1043-1044 [Soviet Physics - Crystallography, 20, 638].

Hexagonal, $P6_3/mcm$, a = 7.305, c = 7.570 Å, Z = 1. Mo radiation, R = 0.12 for 165 reflexions (films, visual intensities).

Atomic positions

			x	y	z
1 K	in	2(b)	0	0	0
0.33 Ga + 5.67 W		6(g)	0.4790	0	1/4
12 O(1)		12(j)	0.4460	0.2230	1/4
6 O(2)		6(f)	1/2	0	0

Hexagonal tungsten bronze structure (1); W-O = 1.76-1.96(2), K-O = 3.26-3.65(6) Å.

1. Structure Reports, 17, 401.

COPPER(II) TUNGSTATE

$CuWO_4$

S. KLEIN and H. WEITZEL, 1975. J. Appl. Cryst., 8, 54-59.

Refinement with neutron powder data confirms previous results (1).

1. Structure Reports, 35A, 226.

SILVER TUNGSTATE

Ag_2WO_4 (high-temperature)

P.M. SKARSTAD and S. GELLER, 1975. Mater. Res. Bull., 10, 791-800.

Orthorhombic, Pn2n, a = 10.89, b = 12.03, c = 5.92 Å, D_m = 7.88, Z = 8. Ag radiation, R = 0.071 for 524 reflexions.

The structure contains $W_4O_{16}^{8-}$ ions which consist of four condensed edge-sharing WO_6 octahedra. These anions are linked by Ag^+ ions which exhibit four types of coordination: distorted trigonal prismatic, octahedral, tetrahedral, and linear. Difficulty was encountered in refining the oxygen parameters, and standard deviations are large (0.1 Å).

LITHIUM MANGANATE(III)

SODIUM MANGANATE(III)

$LiMnO_2$
β-$NaMnO_2$

R. HOPPE, G. BRACHTEL and M. JANSEN, 1975. Z. anorg. Chem., 417, 1-10.

Orthorhombic, Pmnm, a = 2.81, 2.86, b = 5.76, 6.34, c = 4.57, 4.79 Å, D_m = 4.2, 4.2, Z = 2. Mo radiation, R = 0.067 and 0.120 for 98 and 157 reflexions, respectively. Mn, Li(Na) in 2(a); O(1), O(2) in 2(b): y = 0.6347, 0.126, 0.144, 0.602 for the Li compound, 0.6265, 0.133, 0.172, 0.582 for the Na compound.

The compounds are isostructural. The Mn coordination octahedra are distorted as a result of the Jahn-Teller effect, Mn-O = 1.89, 1.96, 2.29 Å in the Li compound, 1.91, 1.95, 2.41 Å in the Na compound. Li and Na have octahedral coordination, Li-O = 2.10-2.29, Na-O = 2.30-2.41 Å.

MAGNESIUM MANGANITE

$MgMn_2O_4$

N.K. RADHAKRISHNAN and A.B. BISWAS, 1975. Z. Kristallogr., 142, 117-120.

Tetragonal, space group not given, a = 8.070, c = 9.281 Å, Z not given. Neutron diffraction powder data, R(I) = 0.020 for 16 reflexions.

Tetragonally-distorted spinel structure (1), with degree of inversion about 20%, i.e. $(Mg_{0\cdot8}Mn_{0\cdot2})_{tetr}[Mg_{0\cdot2}Mn_{1\cdot8}]_{oct}$.

1. Structure Reports, 21, 330.

COBALT(II) MANGANESE(IV) OXIDE

$Co_2Mn_3O_8$

A. RIOU and A. LECERF, 1975. Acta Cryst., B31, 2487-2490.

Orthorhombic, $Pmn2_1$, a = 5.743, b = 4.915, c = 9.361 Å, D_m = 5.13, Z = 2. Mo radiation, R = 0.065 for 184 reflexions (films, densitometer intensities).

The structure (Fig. 1) contains $[Mn_3O_8{}^{4-}]_n$ sheets, connected by Co^{2+} ions. Mn has octahedral coordination, and Co has tetrahedral and distorted octahedral coordinations; Mn-O = 1.81-2.00, Co-O = 1.91-2.02 (tetrahedral), 2.05-2.37 Å (octahedral).

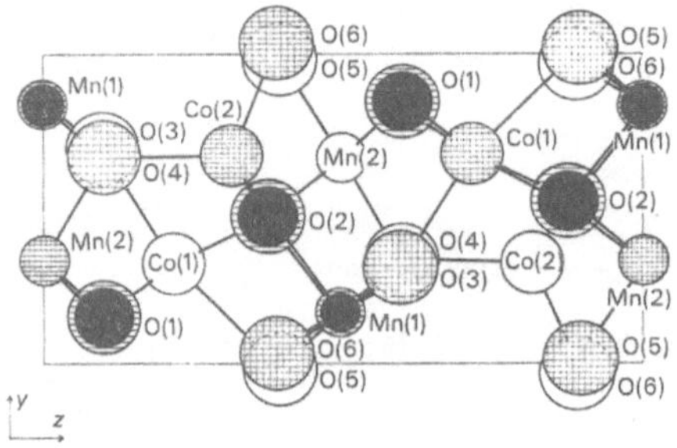

Fig. 1. Structure of $Co_2Mn_3O_8$.

CALCIUM COPPER(II) MANGANATE(IV)

$CaCu_3Mn_4O_{12}$

J. CHENAVAS, J.C. JOUBERT, M. MAREZIO and B. BOCHU, 1975. J. Solid State Chem., 14, 25-32.

Cubic, Im3, a = 7.241 Å, D_m not given, Z = 2. Mo radiation, R = 0.020 for 92 reflexions. Ca in 2(a); Cu in 6(b); Mn in 8(c); O in 24(g): x = 0.3033, y = 0.1822.

Perovskite-like structure, similar to that of $NaMn_3Mn_4O_7$ (1). Ca-O = 2.56 (x 12), Cu-O = 1.94 (x 4), Mn-O = 1.92 (x 6) Å.

1. Structure Reports, 39A, 247.

POTASSIUM PERRHENATE

$KReO_4$

C.J.L. LOCK and G. TURNER, 1975. Acta Cryst., B31, 1764-1765.

Tetragonal, $I4_1/a$, a = 5.660, c = 12.667 Å, D_m = 4.68, Z = 4. Mo radiation, R = 0.020 for 236 reflexions. Origin at $\bar{1}$, Re in 4(a); K in 4(b); O in 16(f): x = 0.1177, y = 0.0288, z = 0.2022.

The structure is essentially as previously described (1). The ReO_4^- ion is tetrahedral, Re-O = 1.723(4) Å. K-O = 2.78 and 2.85 Å.

1. Structure Reports, 24, 340.

BARIUM MANGANESE RHENIUM OXIDE

Ba_2MnReO_6

C.P. KHATTAK, D.E. COX and F.F.Y. WANG, 1975. J. Solid State Chem., 13, 77-83.

Cubic, Fm3m, a = 8.166 Å, Z = 4. Neutron powder data; Ba in 8(c), Mn in 4(a), Re in 4(b), O in 24(e): x = 0.2648.

Ordered perovskite-type structure.

STRONTIUM FERRITE

$Sr_2Fe_2O_5$

C. GREAVES, A.J. JACOBSON, B.C. TOFIELD and B.E.F. FENDER, 1975. Acta Cryst., B31, 641-646.

Orthorhombic, Icmm, a = 5.6727, b = 15.582, c = 5.5303 Å, Z = 4. Neutron powder data with profile analysis.

The structure is described in space group Icmm, with disorder of one Fe and one O atom. An ordered domain would have space group Ibm2, like brownmillerite (1). Electron diffraction patterns seem to rule out Pcmn ($Ca_2Fe_2O_5$ structure, 2).

1. Structure Reports, 37A, 259.
2. Ibid., 23, 387; 37A, 258.

BARIUM STRONTIUM FERRITE

$BaSrFe_4O_8$

I. D. HERMANN-RONZAUD and M. BACMANN, 1975. Acta Cryst., B31, 665-668.

II. M.C. CADÉE, 1975. Ibid., B31, 2012-2015.

Trigonal, P$\bar{3}$1m (compare P$\bar{6}$m2 in 1), a = 5.450, 5.446, c = 8.101, 8.082 Å, in I, II, respectively, Z = 1. I: X-ray and neutron powder data, R = 0.09 and 0.024 for 18 and 12 reflexions; II: Cu radiation, powder data, R = 0.04. Ba in 1(a); Sr in 1(b); Fe in 4(h): z = 0.23, 0.229; O(1) in 2(c); O(2) in 6(k): x = 0.35, 0.347, z = 0.29, 0.296 in I, II respectively.

The material is isostructural with $BaCaFe_4O_8$ (2) and does not have space group P$\bar{6}$m2 as previously suggested (1). In the structure (Fig. 1), Fe has tetrahedral coordination, with (from I) Fe-O = 1.83(3) Å, Ba and Sr have octahedral coordination, Ba-O = 2.95(5), Sr-O = 2.60(5); Ba has six further oxygen neighbours at 3.15 Å.

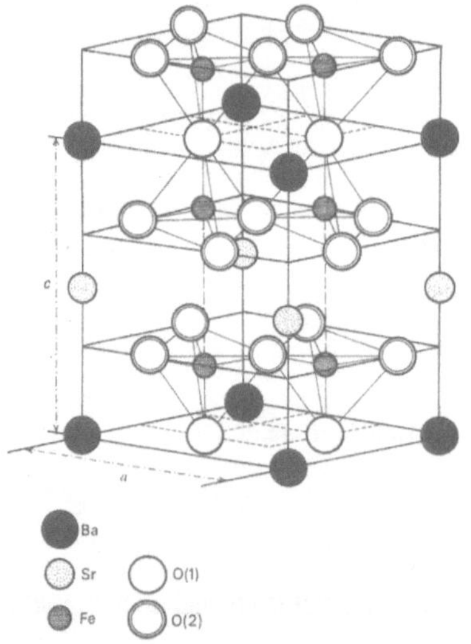

Fig. 1. Structure of $BaSrFe_4O_8$.

1. Structure Reports, 39A, 250.
2. Ibid., 37A, 259.

TITANOMAGNETITE

$Fe(Fe,Ti)_2O_4$

M.Z. STOUT and P. BAYLISS, 1975. Canad. Miner., 13, 86-88.

Cubic, Fd3m, a = 8.3976 Å, Z = 8. Mo radiation, R = 0.037 for 114 reflexions; x(O) = 0.26277.

Spinel structure (1), with Fe^{2+} in the tetrahedral site, (Fe^{2+},Fe^{3+},Ti^{4+}) in the octahedral site.

1. Strukturbericht, 1, 350.

NICKEL TIN FERRITE

$Fe_{2-2x}Ni_{1+x}Sn_xO_4$

C. DJEGA-MARIADASSOU, F. BASILE and P. POIX, 1975. J. Solid State Chem., 13, 99-106.

Cubic, [Fd3m], a = 8.337-8.395 Å, for x = 0-0.3, [Z = 8]. Co radiation, R = 0.06 for 20 reflexions (powder data); x(O) = 0.380.

Spinel structure, with Fe^{3+} in the A-sites.

YTTRIUM IRON GARNET

$Y_3Fe_5O_{12}$

M. BONNET, A. DELAPALME, H. FUESS and M. THOMAS, 1975. Acta Cryst., B31, 2233-2240.

Cubic, Ia3d, a = 12.376 Å, Z = 8. Mo radiation, R = 0.032 for 503 reflexions.

Garnet structure, as previously described (1). Oxygen parameters are (-0.0271, 0.0567, 0.1504).

1. Structure Reports, 20, 280; 21, 294; 27, 510.

STRONTIUM EUROPIUM FERRATE

$SrEu_2Fe_2O_7$

M. DROFENIK and L. GOLIČ, 1975. Cryst. Struct. Comm., 4, 589-592.

Tetragonal, $P4_2/mnm$, a = 5.507, c = 19.876 Å, D_m = 6.65, Z = 4. Mo radiation, R = 0.027 for 334 reflexions.

Atomic positions

	x	y	z
Eu	0.2299	0.2299	0.1827
Sr	0.2415	0.2415	0
Fe	0.2438	0.2438	0.4026
O(1)	0	1/2	0.3925
O(2)	0.3013	0.3013	0.2920
O(3)	0	0	0.0891
O(4)	0	0	0.3797
O(5)	0.2877	-0.2877	0

The structure (Fig. 1) can be considered as a stacking of layers $(SrO)(Fe_2O_4)$-$(Eu_4O_4)(Fe_2O_4)(SrO)$. Fe has distorted octahedral, Eu seven-, and Sr eight-coordination; Fe-O = 1.95-2.24, Eu-O = 2.24-2.58, Sr-O = 2.58-2.93 Å.

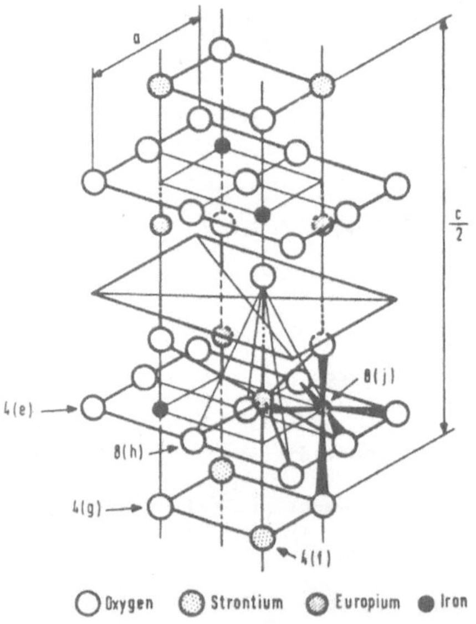

Fig. 1. Structure of $SrEu_2Fe_2O_7$.

YTTERBIUM FERRATE

$YbFe_2O_4$

K. KATO, I. KAWADA, N. KIMIZUKA and T. KATSURA, 1975. Z. Kristallogr., 141, 314-320.

Rhombohedral, R$\bar{3}$m, a = 3.455, c = 25.054 Å, D_m not given, Z = 3. Mo radiation, R = 0.056 for 424 reflexions. Yb in 3(a); Fe, O(1), O(2) in 6(c): z = 0.2150, 0.2925, 0.1292.

The structure contains alternating $YbO_{3/2}$ and $(Fe^{2+},Fe^{3+})_2O_{5/2}$ layers. Yb has octahedral and Fe trigonal bipyramidal coordination, Yb-O = 2.24, Fe-O = 1.94, 2.01, 2.15(1) Å.

YTTERBIUM EUROPIUM FERRATE

$Yb_{0.5}Eu_{0.5}Fe_2O_4$

B. MALAMAN, O. EVRARD, N. TANNIERES, J. AUBRY, A. COURTOIS and J. PROTAS, 1975. Acta Cryst., B31, 1310-1312.

Rhombohedral, R$\bar{3}$m, a = 3.486, c = 24.92 Å, D_m = 6.60, Z = 3. Mo radiation, R = 0.059 for 192 reflexions. (Yb,Eu) in 3(a); (Fe^{2+},Fe^{3+}), O(1) and O(2) in 6(c): z = 0.2141, 0.2914, and 0.1295, respectively.

The rare-earth ions have octahedral coordination, and Fe ions are trigonal bipyramidal (Fig. 1).

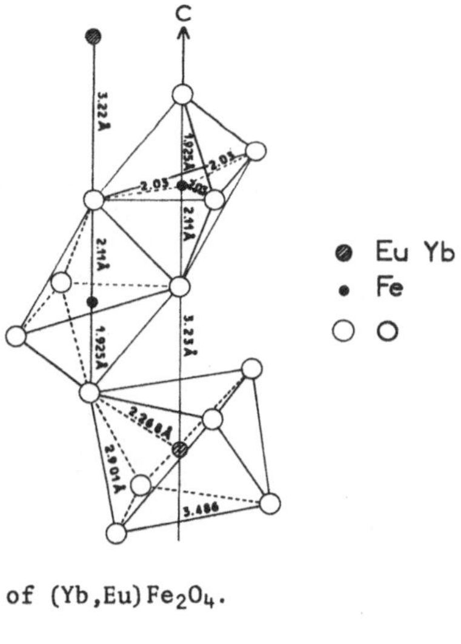

Fig. 1. Structure of $(Yb,Eu)Fe_2O_4$.

SODIUM FERRATE RUTHENATE

$NaFeRuO_4$

J. DARRIET and A. VIDAL, 1975. Bull. Soc. Fr. Minér. Crist., 98, 374-377.

Orthorhombic, Pnma, a = 9.218, b = 2.956, c = 10.848 Å, D_m = 5.46, Z = 4. Cu radiation, R = 0.074 for 144 reflexions (films, visual intensities).

Isostructural with $CaFe_2O_4$ (1), with disorder of Fe and Ru. $NaRu_2O_4$ is also isostructural.

1. Structure Reports, 21, 290; 37A, 257.

BISMUTH RUTHENIUM OXIDE

$Bi_3Ru_3O_{11}$

F. ABRAHAM, D. THOMAS and G. NOWOGROCKI, 1975. Bull. Soc. Fr. Minér. Crist., 98, 25-29.

Cubic, Pn3, a = 9.302 Å, D_m = 8.8, Z = 4. Mo radiation, R = 0.069 for 572 reflexions.

Atomic positions

		x	y	z
Bi(1)	in 4(b)	0	0	0
Bi(2)	8(e)	0.3838	0.3838	0.3838
Ru	12(g)	0.3897	3/4	1/4
O(1)	12(f)	0.590	1/4	1/4
O(2)	8(e)	0.152	0.152	0.152
O(3)	24(h)	0.599	0.247	0.547

Isostructural with $Bi_3GaSb_2O_{11}$ (1). Pairs of RuO_6 octahedra share an edge to give Ru_2O_{10} units; these share corners to form a three-dimensional structure, with voids which accommodate Bi_8O_4 groups formed from four Bi_4O tetrahedra. The Ru...Ru distance in the Ru_2O_{10} group (2.60 Å) suggests metal-metal interaction. Ru-O = 1.89-2.03, Bi-O = 2.21-2.82 Å.

1. T.P. SLEIGHT, C.R. HARE and A.W. SLEIGHT, 1968. Mater. Res. Bull., 3, 437.

LITHIUM COBALTATE(II)

LITHIUM ZINCATE

Li_4CoO_3
Li_4ZnO_3

M. JANSEN, P. KASTNER and R. HOPPE, 1975. Z. anorg. Chem., 414, 69-75.

Rhombohedral, R3m, a = 13.10, 13.14, c = 7.98, 7.99 Å, for Co and Zn compounds. Mo radiation, R = 0.10 and 0.11 for 130 and 235 reflexions, respectively.

The unit cells are pseudo-cubic, a = 9 Å, with cubic close-packing of oxygen atoms. The structures (and stoichiometries) have not been fully established.

SODIUM COBALTATE(IV)

Na_4CoO_4

M. JANSEN, 1975. Z. anorg. Chem., 417, 35-40.

Triclinic, P1, a = 8.65, b = 5.70, c = 6.40 Å, α = 123.9, β = 98.1, γ = 99.2°, D_m = 2.87, Z = 2. R = 0.053 for 1380 reflexions.

The structure contains tetrahedral CoO_4 groups, Co-O = 1.76-1.85 Å, O-Co-O = 105-114°, and Na ions with 4- and 5-coordinations, Na-O = 2.27-2.92 Å.

POTASSIUM COBALTATE(III)

RUBIDIUM COBALTATE(III)

$KCoO_2$
$RbCoO_2$

M. JANSEN and R. HOPPE, 1975. Z. anorg. Chem., 417, 31-34.

$KCoO_2$
Tetragonal, P4/nmm, a = 3.81, c = 7.91 Å, D_m = 3.65, Z = 2. Powder data. Co, K, O(1) in 2(c): z = 0.08, 0.66, 0.32; O(2) in 2(a).

$RbCoO_2$
Orthorhombic, [Pbca], a = 5.66, b = 11.35, c = 16.35 Å, D_m = 4.46, Z = 16. Powder data. Parameters of $KGaO_2$ (1) assumed.

In $KCoO_2$, Co has five-coordination, Co-O = 1.91, 1.99 (x 4) Å.

1. E. VIELHABER and R. HOPPE, 1969. Z. anorg. Chem., 369, 14; this volume, p. 419.

POTASSIUM COBALTATE(III)

$KCoO_2$

C. DELMAS, C. FOUASSIER and P. HAGENMULLER, 1975. J. Solid State Chem., 13, 165-171.

Powder data indicate two forms, both of which have structures related to that of cristobalite.

POTASSIUM IRIDATE(V)

$KIrO_3$

R. HOPPE and K. CLAES, 1975. J. Less-Common Metals, 43, 129-142.

Cubic, Pn3, a = 9.487 Å, D_m = 6.50, Z = 12. Mo radiation, R = 0.091 for 376 reflexions.

Atomic positions

			x	y	z
K(1)	in	4(b)	0	0	0
K(2)		8(e)	0.4065	0.4065	0.4065
Ir		12(g)	0.5922	3/4	1/4
O(1)		12(f)	0.616	1/4	1/4
O(2)		24(h)	0.250	0.591	0.550

Isostructural with $KSbO_3$ (1). Ir has octahedral coordination, Ir-O = 1.91 and 1.98 Å; K has three O at 2.66 and three at 2.88 Å.

1. Structure Reports, 11, 443.

LITHIUM CUPRATE(III)

Li_3CuO_3

H.-N. MIGEON, A. COURTOIS, M. ZANNE, C. GLEITZER and J. AUBRY, 1975. Rev. Chim. Minér., 12, 203-209.

Tetragonal, $P4_2/mnm$, a = 8.71, c = 3.58 Å, D_m = 3.17, Z = 4. Powder data, R = 0.11. Cu, Li(2) in 4(f): x = 0.382, 0.199; Li(1), O(1) in 8(i): x = 0.600, 0.170, y = 0.170, 0.367; O(2) in 4(g): x = 0.389.

Isostructural with Li_3AuO_3 (1). The structure contains $Cu_2O_6^{6-}$ groups, in which Cu has square-planar coordination, Cu-O = 1.84, 2.00 Å. [One very short Li(2)-O distance, 1.49 Å, casts some doubt on this Li position.]

1. Structure Reports, 35A, 214.

MAGNESIUM OXOCUPRATE

$MgCu_2O_3$

H. DRENKHAHN and H. MÜLLER-BUSCHBAUM, 1975. Z. anorg. Chem., 418, 116-120.

Orthorhombic, Pmmn, a = 4.00, b = 9.35, c = 3.19 Å, Z = 2. R = 0.083 for 282 reflexions.

Atomic positions

		x	y	z
Mg	in 2(a)	1/4	1/4	0.369
Cu	4(e)	3/4	0.086	0.820
O(1)	2(b)	3/4	1/4	0.453
O(2)	4(e)	1/4	0.094	0.869

Isostructural with $CaCu_2O_3$ (1). Cu has distorted octahedral coordination, Cu-O = 1.94-2.01, 2.53, 2.76 Å, and Mg has octahedral coordination, Mg-O = 2.02-2.16 Å.

1. Structure Reports, 34A, 248.

NEODYMIUM CUPRATE

Nd_2CuO_4

H. MÜLLER-BUSCHBAUM and W. WOLLSCHLÄGER, 1975. Z. anorg. Chem., 414, 76-80.

Tetragonal, I4/mmm, a = 3.945, c = 12.171 Å, D_m not given, Z = 2. R = 0.06 for 344 reflexions. Cu in 2(a); Nd in 4(e): z = 0.3513; O(1) in 4(c); O(2) in 4(d).

The compound does not have the K_2NiF_4 structure-type (which has O(2) in 4(e), 1), but represents a new structure-type (Fig. 1). Cu has square-planar coordination, Cu-O = 1.97 Å, and Nd has eight oxygen neighbours, Nd-O = 2.32 and 2.68 Å.

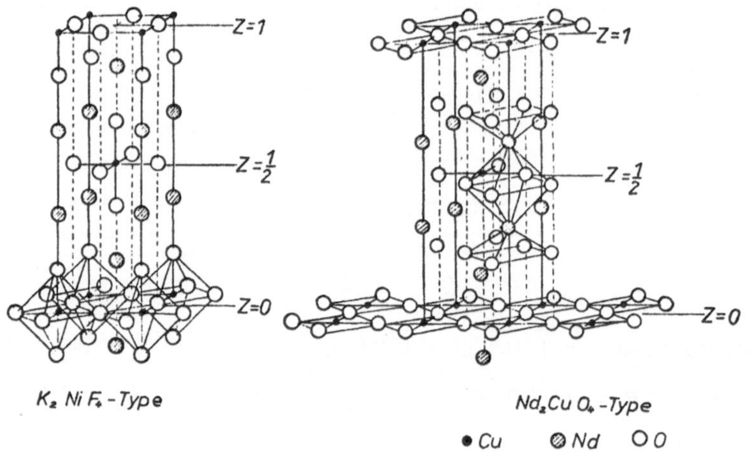

Fig. 1. K_2NiF_4 and Nd_2CuO_4 structure-types.

1. Structure Reports, 17, 332; 19, 323.

LITHIUM SILVER OXIDE

$LiAg_3O_2$

M. JANSEN, 1975. Z. Naturforsch., 30B, 854-858.

Orthorhombic, Ibam, a = 5.974, b = 9.945, c = 5.694 Å, D_m = 7.04, Z = 4. Mo radiation, R = 0.052 for ℓ = 2n reflexions.

Atomic positions

			x	y	z
Li	in	4(b)	1/2	0	1/4
Ag(1)		4(c)	0	0	0
Ag(2)		8(e)	1/4	1/4	1/4
O		8(j)	0.3268	0.0945	0

The structure contains branched O-Ag-O-Ag chains, in which Ag atoms have linear coordination, Ag-O = 2.16 Å, Ag-O-Ag = 83 and 97°. Li has distorted tetrahedral coordination, Li-O = 1.99 Å; the tetrahedra share edges.

RARE-EARTH OXIDES

$RbLnO_2$ (Ln = La, Nd, Sm, Eu, Gd)
$CsNdO_2$

H. BRUNN and R. HOPPE, 1975. Z. anorg. Chem., 417, 213-220.

$RbLnO_2$
Rhombohedral, $R\bar{3}m$, a = 3.74-3.55, c = 19.64-19.53 Å, D_m = 5.14-6.10, Z = 3. Powder data; α-$NaFeO_2$-type structure (1), Rb in 3(a); Nd in 3(b); O in 6(c): z = 0.22.

$CsNdO_2$
Hexagonal, $P\bar{6}m2$, a = 3.67, c = 13.72 Å, D_m = 6.28, Z = 2. Powder data; β-$RbScO_2$-type structure (2), Cs in 1(a) and 1(f); Nd in 2(h): z = 0.25; O in 2(i): z = 0.18, and 2(g): z = 0.32.

1. Strukturbericht, 3, 75, 392
2. R. HOPPE and H. SABROWSKY, 1965. Z. anorg. Chem., 339, 144; this volume, p. 420.

CAESIUM URANATE

γ-$Cs_2U_4O_{12}$

A.B. van EGMOND, 1975. J. Inorg. Nucl. Chem., 37, 1929-1931.

Cubic, Fd3m, a = 11.230 Å, Z = 4. Powder data, U in 16(c); Cs in 8(b); O in 48(f): x = -0.051 (origin at centre).

U has distorted dodecahedral coordination, U-O = 2.07 Å. Cs has 6 O neighbours at 3.64 and 12 at 4.06 Å; since these distances are rather large, the Cs ions are probably disordered.

U and Cs positions are given for the α (rhombohedral) and β (monoclinic) forms.

BISMUTH URANATE

Bi_2UO_6

A.S. KOSTER, J.P.P. RENAUD and G.D. RIECK, 1975. Acta Cryst., B31, 127-131.

At 1000°C
Trigonal, $P\bar{3}$, a = 4.045, c = 9.90 Å, Z = 1. Cr radiation, R = 0.05 for 46 reflexions (powder data).

At 20°C
Monoclinic, C2, a = 6.872, b = 4.009, c = 9.690 Å, β = 90.16°, D_m = 9.20, Z = 2.
X-ray and neutron powder data, R = 0.10 for 70 reflexions.

The structures are closely related (Fig. 1), and both can be regarded as stackings of layers of interlocked UO_8 polyhedra and layers of a Bi-O network. In the high-temperature phase, U has hexagonal bipyramidal coordination, U-O = 2.40(5) (equatorial), 2.1(1) Å (axial). In the 20°C phase, four of the equatorial oxygens are closer to U (2.01(3) Å) than the other two (2.34 Å); U-O(axial) = 1.91(1) Å. Bi-O = 2.3-3.4 Å.

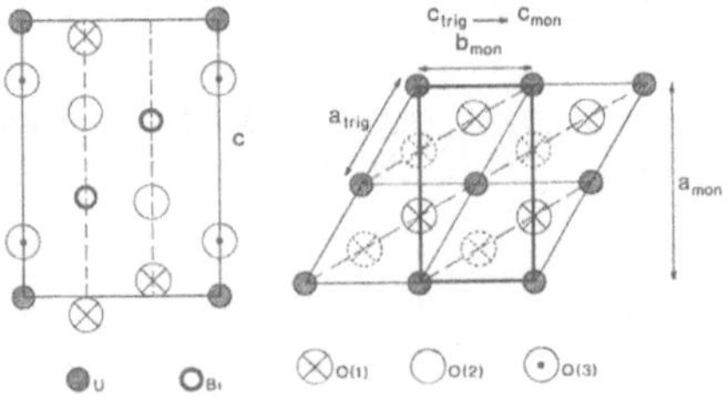

Fig. 1. Structure of Bi_2UO_6 (1000°C). Left, a section parallel to (11$\bar{2}$0); right, projection along c.

LANTHANON URANIUM OXIDES

$(Ln,U)O_2$ (Ln = Y, La, Nd, Ho, Lu)

H. WEITZEL and C. KELLER, 1975. J. Solid State Chem., 13, 136-141.

Cubic, [Fm3m], a = 5.293-5.507 Å, [Z = 4]. Neutron powder data. Fluorite structure.

LITHIUM POTASSIUM HYDROXIDE

2LiOH.KOH

N.V. RANNEV, T.A. DEMIDOVA and V.E. ZAVODNIK, 1974. Kristallografija, 19, 998-1001 [Soviet Physics - Crystallography, 19, 618-619].

Monoclinic, $P2_1/m$, a = 6.134, b = 5.810, c = 5.197 Å, β = 103°12', D_m = 1.85, Z = 2. Mo radiation, R = 0.097 for 760 reflexions.

Atomic positions

	x	y	z
K	0.2647	0.1519	3/4
Li	0.1852	0.6511	0.0044
O(1)	0.0745	0.3833	1/4
O(2)	0.2059	0.9097	1/4
O(3)	0.4125	0.6692	3/4

The structure contains $Li(OH)_4$ tetrahedra which share edges to form bands along c, between which are the K^+ ions with 12-coordination; Li-O = 1.91-2.14(1), K-O = 2.83-3.77 Å.

BARIUM HYDROXYGERMANATE

$Ba_5[(Ge,C)O_4]_3(OH)$

Ju.A. MALINOVSKIJ, E.A. POBEDIMSKAJA and N.V. BELOV, 1975. Kristallografija, 20, 644-646 [Soviet Physics - Crystallography, 20, 395-396].

Hexagonal, $P6_3/m$, a = 10.207, c = 7.734 Å, D_m not given, Z = 2. Mo radiation, R = 0.057 for 369 reflexions.

Hydroxyapatite structure (1, 2) (z(OH) = 0.168).

1. Structure Reports, 22, 411; 29, 369.
2. K. SUDARSANAN and R.A. YOUNG, 1969. Acta Cryst., B25, 1534; this volume, p. 423.

CADMIUM HYDROXYGERMANATE

$Cd_2Ge_3O_7(OH)_2$

E.L. BELOKONEVA, M.I. SIROTA, M.A. SIMONOV and N.V. BELOV, 1975. Kristallografija, 20, 42-45 [Soviet Physics - Crystallography, 20, 21-23].

Triclinic, $P\bar{1}$, a = 7.081, b = 8.059, c = 7.559 Å, α = 108.90, β = 84.46, γ = 117.40°, D_m = 5.3, Z = 2. Mo radiation, R = 0.080 for 1125 reflexions.

The structure contains $[Ge_3O_7(OH)_2]_n$ bands (Fig. 1), which contain six-membered rings of corner-sharing Ge coordination octahedra and tetrahedra, which edge- and corner-share with neighbouring rings; Ge-O = 1.75 (tetrahedral), 1.89 Å (octahedral). Cd has octahedral and seven coordination, Cd-O = 2.19-2.43 Å.

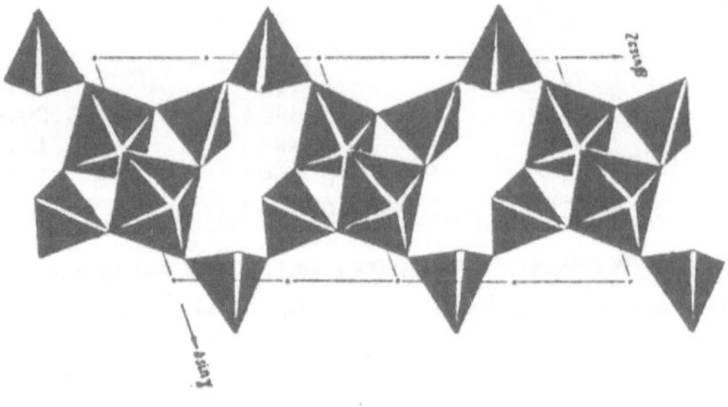

Fig. 1. The $[Ge_3O_7(OH)_2]_n$ band.

SODIUM HEXAHYDROXYANTIMONATE(V)

$NaSb(OH)_6$

T. ASAI, 1975. Bull. Chem. Soc. Japan, 48, 2677-2679.

Tetragonal, $P4_2/n$, a = 8.029, c = 7.894 Å, D_m = 3.20, Z = 4. Mo radiation, R = 0.063 for 465 reflexions.

Atomic positions

	x	y	z
Na	3/4	3/4	3/4
Sb	1/4	1/4	1/4
O(1)	0.1942	0.0274	0.3482
O(2)	0.0292	0.3320	0.3228
O(3)	0.1605	0.1815	0.0281

The structure is essentially as previously described (1), but with significant shifts in oxygen parameters. It contains $Sb(OH)_6$ octahedra, Sb-O = 1.98, $Na(OH)_6$ octahedra, Na-O = 2.39, and O-H...O hydrogen bonds, 2.74-3.15 Å.

1. Strukturbericht, 6, 26, 118.

COPPER HYDROXYCHROMATE

$Cu_{11}(OH)_{14}(CrO_4)_4$

A. RIOU, 1974. Bull. Soc. Fr. Minér. Crist., 97, 405-410.

Triclinic, $P\bar{1}$, a = 5.850, b = 10.859, c = 9.919 Å, α = 93.95, β = 100.44, γ = 104.87°, D_m = 3.92, Z = 1. Mo radiation, R = 0.12 for 550 reflexions (film data).

The structure (Fig. 1) contains 7 Cu atoms in octahedral layers; between these layers are 4 more Cu atoms also octahedrally coordinated, and tetrahedrally coordinated Cr atoms. The six independent Cu octahedra all exhibit Jahn-Teller distortions, Cu-O = 1.78-2.14 and 2.22-2.56 Å. Cr-O = 1.55-1.90(8) Å.

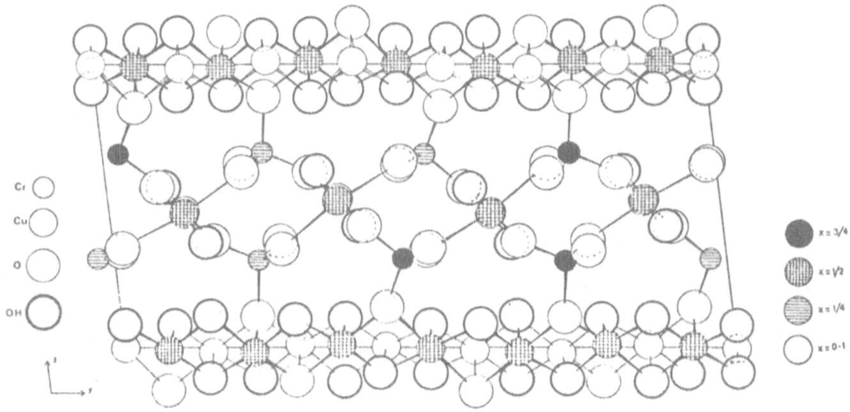

Fig. 1. Structure of $Cu_{11}(OH)_{14}(CrO_4)_4$.

LITHIUM HEXAHYDROXYPLATINATE(IV)

SODIUM HEXAHYDROXYPLATINATE(IV)

$Li_2Pt(OH)_6$
$Na_2Pt(OH)_6$

M. TRÖMEL and E. LUPPRICH, 1975. Z. anorg. Chem., 414, 160-168.

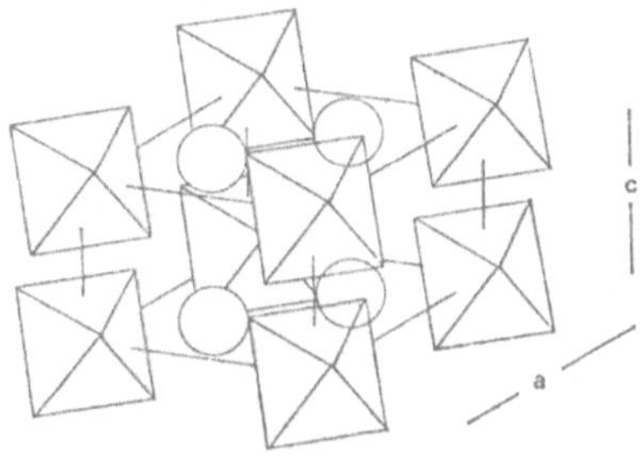

Fig. 1. Structure of $Li_2Pt(OH)_6$.

Trigonal, P$\bar{3}$1m, a = 5.362, 5.831, c = 4.647, 4.755 Å, Z = 1. Mo radiation, R = 0.039 for 296 reflexions for Li compound; Cu radiation, powder data, R = 0.048 for 30 reflexions for Na compound. Pt in 1(a); Li or Na in 2(c); O in 6(k): x = 0.325, 0.287, z = 0.254, 0.249.

The structure (Fig. 1) is related to that of Li_2ZrF_6 (1), except that the alkali metal ions are in 2(c) (rather than 2(d)). Pt-O = 2.11, 2.05, Li-O = 2.16, Na-O = 2.41 Å.

1. Structure Reports, 24, 277; 39A, 154.

CALCIUM HEXAHYDROXYPLATINATE(IV)

$CaPt(OH)_6$

M. TRÖMEL and E. LUPPRICH, 1975. Z. anorg. Chem., 414, 169-175.

Trigonal, P$\bar{3}$1c, a = 5.890, c = 9.591 Å, D_m = 3.85, Z = 2. Cu radiation, powder data. Ca in 2(c); Pt in 2(b); O in 12(i): (0.23, 0.32, 0.12).

The structure (Fig. 1) contains octahedral $Pt(OH)_6^{2-}$ ions, Pt-O = 2.04 Å; Ca-O = 2.20 Å. Hydrogen positions are calculated and possible hydrogen bonding is discussed.

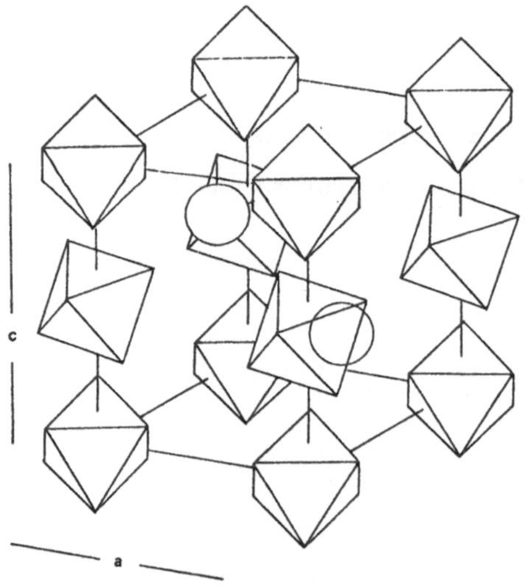

Fig. 1. Structure of $CaPt(OH)_6$.

POTASSIUM ANTIMONY OXYSULPHIDE

$K_3SbS_3.2Sb_2O_3$

H.A. GRAF and H. SCHÄFER, 1975. Z. anorg. Chem., 414, 220-230.

Hexagonal, $P6_3$, a = 14.256, c = 5.621 Å, D_m = 4.16, Z = 2. Mo radiation, R = 0.090 for 2200 reflexions.

The structure contains tubes along c of corner-sharing SbO_3 trigonal pyramids, within which are the K ions, with SbS_3 pyramids between the tubes (Fig. 1). Sb-O = 1.93-2.03, Sb-S = 2.36, K-O = 2.77-3.00 (6 distances), K-S = 3.24 Å.

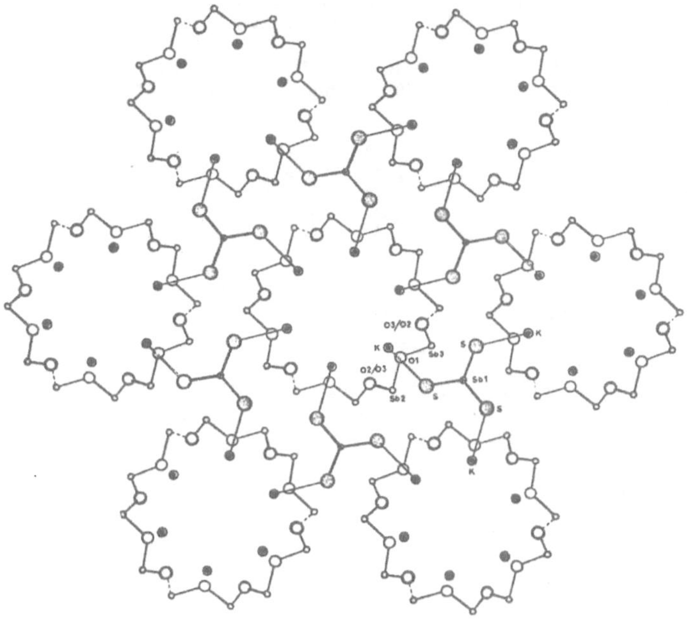

Fig. 1. Structure of $K_3SbS_3.2Sb_2O_3$.

CALIFORNIUM OXYSULPHIDE

Cf_2O_2S

W.H. ZACHARIASEN, 1975. J. Inorg. Nucl. Chem., 37, 1441-1442.

The hexagonal phase described as a form of Cf metal (1) is Cf_2O_2S, with a Ce_2O_2S-type structure (2). The cubic phase (1) is CfS.

1. R.G. HAIRE and R.D. BAYBARZ, 1974. J. Inorg. Nucl. Chem., 36, 1295; Structure Reports, 40A, 104.
2. Structure Reports, 12, 174.

HYDRAZINIUM(1+) HYDROGEN SULPHIDE

N_2H_5HS

F. LAZARINI and M. VARDJAN-JAREC, 1975. Acta Cryst., B31, 2355-2356.

Monoclinic, $P2_1/c$, a = 7.175, b = 7.824, c = 12.318 Å, β = 99.66°, D_m = 1.26, Z = 8. Cu, Mo radiation, R = 0.058 for 872 reflexions (films, visual intensities).

The structure (Fig. 1) contains $N_2H_5^+$ and HS^- ions, joined by weak N-H...S hydrogen bonds which involve only the NH_3 groups. N-N = 1.44(1) Å; N-H...S = 3.23-3.29 Å, 146-165°.

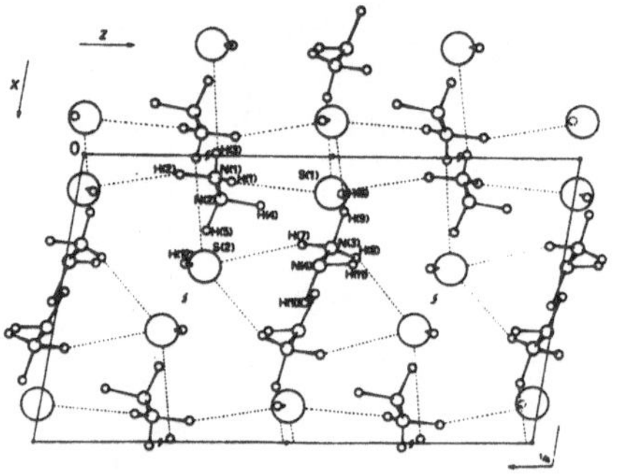

Fig. 1. Structure of N_2H_5HS.

POTASSIUM METATHIOSTANNATE DIHYDRATE

$K_2SnS_3.2H_2O$

W. SCHIWY, C. BLUTAU, D. GÄTHJE and B. KREBS, 1975. Z. anorg. Chem., 412, 1-10.

Orthorhombic, Pnma, a = 6.429, b = 15.621, c = 10.569 Å, D_m = 2.06, Z = 4. Mo radiation, R = 0.040 for 1303 reflexions.

The structure (Fig. 1) contains $(SnS_3^{2-})_n$ polyanions, which consist of SnS_4 tetrahedra sharing corners; Sn-S = 2.35 (terminal), 2.42 and 2.46 Å (bridging). K ions have 6 and 7 neighbours, K...S = 3.31-3.33, K...O = 2.74-3.10 Å, and the water molecules are involved in hydrogen bonding, O-H...S = 3.25-3.34, O-H...O = 2.77 Å.

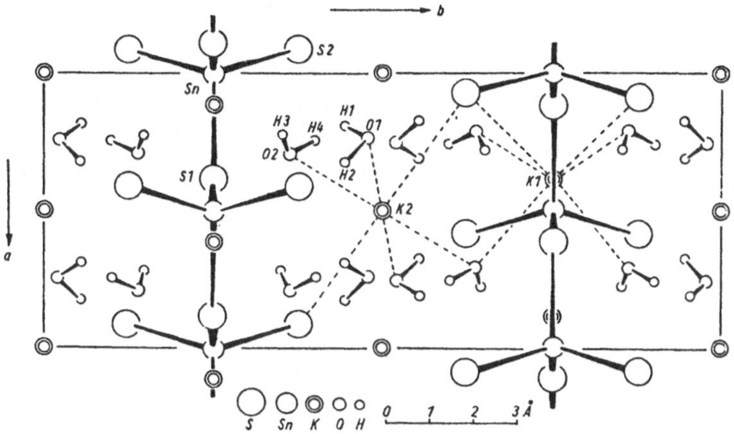

Fig. 1. Structure of $K_2SnS_3.2H_2O$

TITANIUM SULPHIDE AMMONIA

TANTALUM SULPHIDE AMMONIA

$TiS_2.NH_3$
$TaS_2.NH_3$

R.R. CHIANELLI, J.C. SCANLON, M.S. WHITTINGHAM and F.R. GAMBLE, 1975. Inorg. Chem., 14, 1691-1696.

$TiS_2.NH_3$
Rhombohedral, R3m, a = 3.427, c = 26.55 Å, Z = 3. Mo radiation, powder data. Ti, N, and 2S in 3(a): z = 0, 0.167, ±0.385, with additional disorder across z = 0.

$TaS_2.NH_3$
Hexagonal, a = 3.320, c = 18.16 Å, Z = 2. Powder data. Structure not fully established.

The materials are three- and two-layer structures, respectively. N in $TiS_2.NH_3$ is midway between S layers.

SODIUM BORATE HYDRATE

$Na_3[B_3O_5(OH)_2]$ $0{\cdot}5[3Na_2O.3B_2O_3.2H_2O]$

E. CORAZZA, S. MENCHETTI and C. SABELLI, 1975. Acta Cryst., B31, 1993-1997.

Orthorhombic, Pnma, a = 8.923, b = 7.152, c = 9.548 Å, D_m = 2.2, Z = 4. Mo radiation, R = 0.029 for 829 reflexions.

The basic structural unit is the isolated $[B_3O_5(OH)_2]^{3-}$ anion (Fig. 1), which lies on a mirror plane; mean B-O = 1.48 (tetrahedral), 1.38 Å (triangular). The anions are joined by O(1)-H...O(6) hydrogen bonds, 2.77 Å, and by the Na^+ ions, which have octahedral and seven coordination, Na-O = 2.37-2.84 Å.

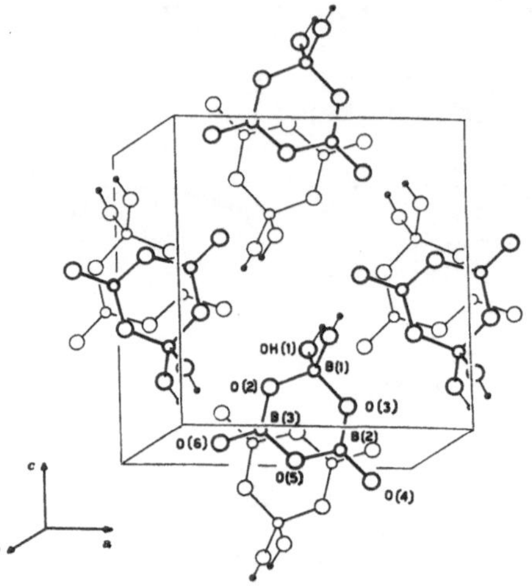

Fig. 1. The $[B_3O_5(OH)_2]^{3-}$ anions.

AMEGHINITE

$Na[B_3O_3(OH)_4]$ $0\cdot5(Na_2O.3B_2O_3.4H_2O)$

A. DAL NEGRO, J.M. MARTIN POZAS and L. UNGARETTI, 1975. Amer. Min., 60, 879-883.

Monoclinic, C2/c, a = 18.428, b = 9.882, c = 6.326 Å, β = 104°23', D_m = 2.030, Z = 4. Mo radiation, R = 0.038 for 1024 reflexions.

The structure contains isolated $B_3O_3(OH)_4^-$ anions, which consist of a six-membered ring composed of one tetrahedron and two triangles sharing corners, B-O = 1.44-1.51 and 1.34-1.40 Å. The anions are linked by hydrogen bonds and by the sodium ions, Na-O = 2.40-2.67 Å (distorted octahedra, which share an edge to form centrosymmetric pairs).

NASINITE

$Na_2[B_5O_8(OH)].2H_2O$

E. CORAZZA, S. MENCHETTI and C. SABELLI, 1975. Acta Cryst., B31, 2405-2410.

Orthorhombic, $Pna2_1$, a = 12.015, b = 6.518, c = 11.173 Å, D_m = 2.12, Z = 4. Mo radiation, R = 0.027 for 1320 reflexions.

The structure closely resembles that of the synthetic K compound (1). It contains $[B_5O_8(OH)^{2-}]_n$ polyanion sheets in the bc plane, joined by eight-coordinate Na ions and hydrogen bonds (Fig. 1). B-O = 1.44-1.50 (tetrahedral), 1.35-1.38 (triangular), Na-O = 2.38-3.10 Å.

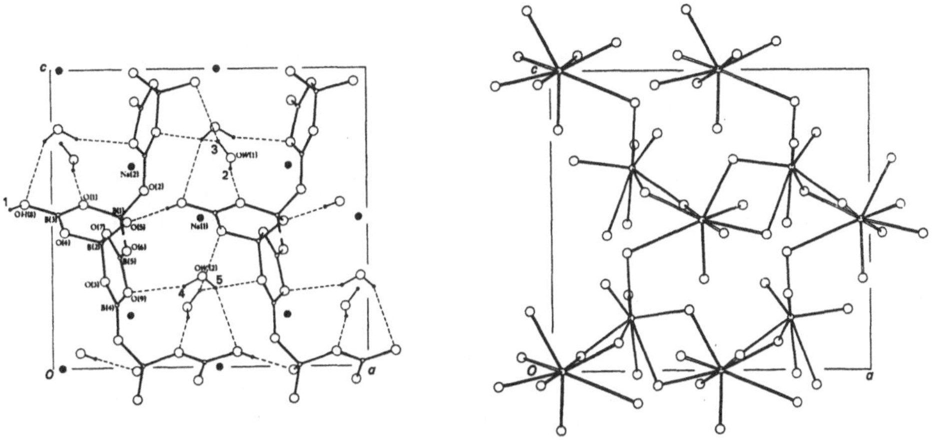

Fig. 1. Structure of nasinite (left) and the Na coordination (right).

1. Structure Reports, 34A, 350.

POTASSIUM METABORATE HYDRATE

RUBIDIUM METABORATE HYDRATE

$K_3[B_3O_4(OH)_4].2H_2O$ $3[KBO_2.4/3H_2O]$

$Rb_3[B_3O_4(OH)_4].2H_2O$ $3[RbBO_2.4/3H_2O]$

I. I.I. ZVIEDRE, Ja.K. OZOLS and A.F. IEVIN'Š, 1974. Latv. PSR Zinat. Akad. Vest., Kim. Ser., No. 4, 387-394.

II. I.I. ZVIEDRE and A.F. IEVIN'Š, 1974. Ibid., No. 4, 395-400.

Orthorhombic, $Pna2_1$, a = 7.81, 7.90, b = 13.71, 13.95, c = 8.80, 9.28 Å, for K, Rb salts, respectively, Z = 4. R(K) = 0.12.

The compounds are isostructural. The anions contain a ring of two tetrahedra and one triangle sharing corners. K and Rb ions have 8-coordination. [The K salt has been described previously by the same authors (1).]

1. Structure Reports, 38A, 386.

CAESIUM METABORATE TETRAHYDRATE

α-$Cs[B(OH)_4].2H_2O$ α-$CsBO_2.4H_2O$

I.I. ZVIEDRE and A.F. IEVIN'Š, 1974. Latv. PSR Zinat. Akad. Vest., Kim. Ser., No. 4, 401-405.

Tetragonal, $I\bar{4}$, a = 6.08, c = 8.58 Å, D_m = 2.51, Z = 2.

The structure contains tetrahedral $B(OH)_4^-$ ions, 8-coordinate Cs^+ ions, and water molecules. Each OH group is coordinated to two cations and hydrogen bonded to two water molecules. [The structure has been described previously by the same authors (1).]

1. Structure Reports, 39A, 263.

RUBIDIUM BERYLLIUM BORATE FLUORIDE

CAESIUM BERYLLIUM BORATE FLUORIDE

$RbBe_2(BO_3)F_2$
$CsBe_2(BO_3)F_2$

I.A. BAJDINA, V.V. BAKAKIN, L.P. BACANOVA, N.A. PAL'ČIK, N.V. PODBEREZSKAJA and L.P. SOLOV'EVA, 1975. Ž. Strukt. Khim., 16, 1050-1053.

Monoclinic, C2, a = 7.695, 7.695, b = 4.441, 4.446, c = 7.087, 7.544 Å, β = 111.2, 109.7°, Z = 2. R = 0.16 and 0.09 for about 350 reflexions.

Atomic positions

	Rb			Cs		
	x	y	z	x	y	z
Rb/Cs	0	0	0	0	0	0
B	0	0.512	1/2	0	0.484	1/2
Be	0.194	-0.010	0.582	0.192	-0.011	0.582
O(1)	0	0.182	1/2	0	0.197	1/2
O(2)	0.345	0.143	0.503	0.346	0.161	0.501
F	0.226	0.498	0.182	0.233	0.473	0.204

Isostructural with the K compound (1).

1. Structure Reports, 35A, 188.

MAGNESIUM FLUOROBORATE

α-Mg_2BO_3F

A.A. BROVKIN and L.V. NIKIŠOVA, 1975. Kristallografija, 20, 740-745 [Soviet Physics - Crystallography, 20, 452-455].

Orthorhombic, $Pna2_1$, a = 20.44, b = 4.530, c = 11.80 Å, D_m = 3.07, Z = 16. Mo radiation, R = 0.105 for 1171 reflexions (films, visual intensities).

The material studied had composition $Mg_2(BO_3)_{1.14}F_{0.58}$. The structure contains chains of edge-sharing MgO_6 octahedra along c, with blocks of six additional octahedra in contact with the chains. Neighbouring blocks are drawn together by three BO_3 triangles, which are arranged around a F atom with local threefold symmetry. BO_3/3F isomorphism is found.

MAGNESIUM BORATE FLUORIDE

$Mg_5(BO_3)_3F$

A.A. BROVKIN and L.V. NIKIŠOVA, 1975. Kristallografija, 20, 415-418 [Soviet Physics - Crystallography, 20, 252-254].

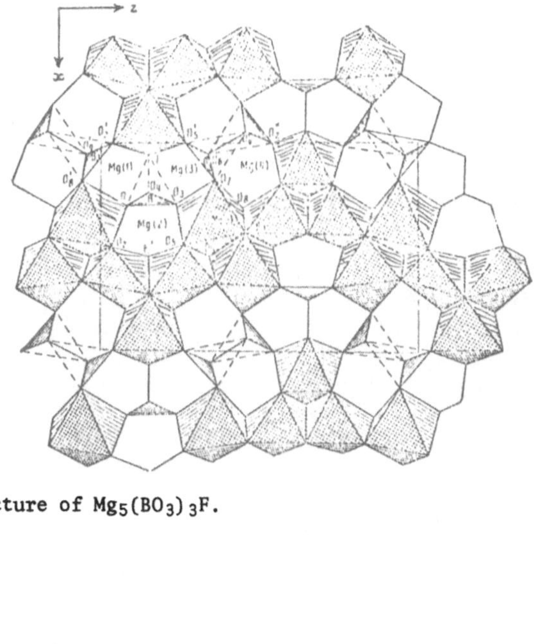

Fig. 1. Structure of $Mg_5(BO_3)_3F$.

Orthorhombic, $Pna2_1$, a = 10.061, b = 4.536, c = 14.865 Å, D_m = 3.12, Z = 4. Cu, Mo radiations, R = 0.103 for 464 reflexions (films, visual intensities).

The structure (Fig. 1) contains Mg coordination octahedra joined by edge-sharing and via triangular BO_3 groups. B-O = 1.33-1.43, Mg-F = 1.98, Mg-O = 2.01-2.24 Å. There is some relation to the structure of leucophoenicite (1).

1. Structure Reports, 35A, 466.

FLUOBORITE

$Mg_3(F,OH)_3(BO_3)$

A. DAL NEGRO and C. TADINI, 1974. Miner. (Tschermaks) Petrogr. Mitt., 21, 94-100.

Hexagonal, $P6_3/m$, a = 8.827, c = 3.085 Å, D_m = 2.94, Z = 2. Mo radiation, R = 0.029 for 403 reflexions. B in 2(c); Mg, O, F(OH) in 6(h): x = 0.3668, 0.4495, 0.2982, y = 0.0297, 0.8456, 0.2089.

The structure is as previously described (1). In the BO_3 group, B-O = 1.387(1) Å; in the $MgO_3(F,OH)_3$ octahedron, Mg-O = 2.086, 2.112 (x 2), Mg-(F,OH) = 1.955, 1.977 (x 2) Å.

1. Structure Reports, 13, 349.

SZAIBELYITE

(ASCHARITE)

$Mg_2(OH)[B_2O_4(OH)]$ $2MgHBO_3$

Y. TAKÉUCHI and Y. KUDOH, 1975. Amer. Min., 60, 273-279.

The paper is essentially identical with a previous publication by the same authors (1).

1. Structure Reports, 39A, 263.

CALCIUM ORTHOBORATE

$Ca_3(BO_3)_2$

A. VEGAS, F.H. CANO and S. GARCIÁ-BLANCO, 1975. Acta Cryst., B31, 1416-1419.

Rhombohedral, R$\bar{3}$c, a = 8.6377, c = 11.849 Å, D_m = 3.05, Z = 6. Mo radiation, R = 0.016 for 544 reflexions.

The structure (Fig. 1) is as previously described (1), containing planar BO_3^{3-} ions, B-O = 1.384(1) Å, and CaO_8 distorted square-antiprisms, Ca-O = 2.347-2.732 Å.

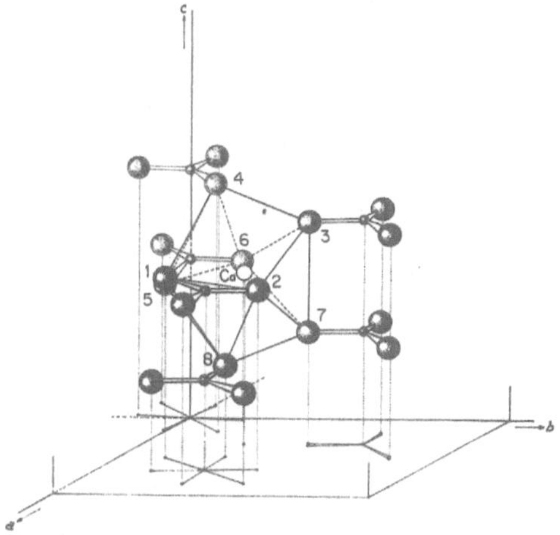

Fig. 1. Structure of $Ca_3(BO_3)_2$.

1. Structure Reports, 34A, 356.

STRONTIUM BORATE APATITE

$(Sr,Na)_{10}(PO_4)_6BO_2$

C. CALVO, R. FAGGIANI and N. KRISHNAMACHARI, 1975. Acta Cryst., B31, 188-192.

Trigonal, P$\bar{3}$, a = 9.734, c = 7.279 Å, D_m not given, Z = 1. Mo radiation, R = 0.065 for 960 reflexions.

The structure (Fig. 1) is apatite-like. Ninefold-coordinated Sr and sixfold-coordinated (Sr,Na,vacancy) cations alternate in columns parallel to c. The PO_4 groups are rotated from their position in apatite and a linear O-B-O group is present. The remaining Sr ions have seven-coordination. P-O = 1.535, B-O = 1.25(1) Å.

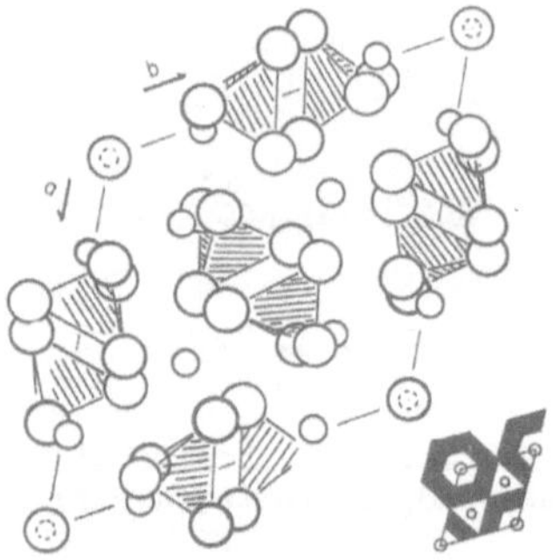

Fig. 1. Structure of $(Sr,Na)_{10}(PO_4)_6BO_2$. B is dotted; the inset shows the hexagonal column arrangement.

STRONTIOBORITE

$SrB_8O_{11}(OH)_4$ $SrO.4B_2O_3.2H_2O$

A.A. BROVKIN, N.V. ZAJAKINA and V.S. BROVKINA, 1975. Kristallografija, 20, 911-916 [Soviet Physics - Crystallography, 20, 563-566].

Monoclinic, $P2_1$, a = 9.909, b = 8.130, c = 7.623 Å, β = 108.4°, D_m = 2.40, Z = 2. Cu radiation, R = 0.11 for 722 reflexions (films, visual intensities).

The structure contains layers of $B_8O_{11}(OH)_4{}^{2-}$ polyanions, which consist of three tetrahedra and five triangles sharing corners; B-O = 1.36-1.57 (tetrahedra), 1.30-1.46(5) Å (triangles). The layers are joined by Sr^{2+} ions (9-coordination, Sr-O = 2.52-2.72 Å) and hydrogen bonds (O-H...O = 2.66-2.99 Å).

RUBIDIUM STRONTIUM OCTABORATE

$Rb_2Sr[B_4O_5(OH)_4]_2.8H_2O$ $Rb_2O.SrO.4B_2O_3.12H_2O$

N.P. IVČENKO and E.N. KURKUTOVA, 1975. Kristallografija, 20, 533-537 [Soviet Physics - Crystallography, 20, 326-328].

Orthorhombic, $P2_12_12_1$, a = 11.61, b = 12.67, c = 16.72 Å, D_m = 2.1, Z = 4. Cu radiation, R = 0.15 for 2240 reflexions (films, visual intensities).

The structure contains isolated $[B_4O_5(OH)_4]^{2-}$ anions, consisting of rings of two tetrahedra and two triangles, similar to those in borax (1). Rb ions have eight-coordination, and the Sr ion has distorted octahedral coordination. Mean B-O = 1.47 (tetrahedra), 1.40 (triangles); Rb-O = 2.78-3.95, Sr-O = 2.34-2.47 Å.

1. Structure Reports, 20, 376.

THALLIUM(I) PYROSILICATE

THALLIUM(I) ORTHOBORATE

$Tl_6Si_2O_7$
Tl_3BO_3

Y. PIFFARD, R. MARCHAND and M. TOURNOUX, 1975. Rev. Chim. Minér., 12, 210-217.

$Tl_6Si_2O_7$
Trigonal, P$\bar{3}$, a = 9.697, c = 7.827 Å, D_m = 7.26, Z = 2. Mo radiation, R = 0.045 for 1910 reflexions.

Tl_3BO_3
Hexagonal, P6_3/m, a = 9.275, c = 3.775 Å, Z = 2. R = 0.051 for 124 reflexions. Tl in 6(h): x = 0.2963, y = 0.3551; B in 2(d); O in 6(h): x = 0.5695, y = 0.1593.

In both structures Tl^+ ions are bonded to three oxygen atoms all on one side, with sterically-active lone electron-pairs; Tl-O = 2.41-2.81 Å. Isolated BO_3 groups are stacked along c, B-O = 1.40 Å, and the pyrosilicate structure can be derived by replacement of two BO_3 groups by one Si_2O_7 group.

CHROMIUM CHLORIDE BORACITE

$Cr_3B_7O_{13}Cl$

R.J. NELMES and F.R. THORNLEY, 1974. J. Phys., C, 7, 3855-3874.

Cubic, F$\bar{4}$3c, a = 12.132 Å, Z = 8. Mo radiation, R = 0.025 for 215 reflexions. Cr in 24(c); Cl in 8(b); B(1) in 24(d); B(2) in 32(e): x = 0.080; O(1) in 8(a); O(2) in 96(h): x = 0.181, y = 0.020, z = 0.097.

Isostructural with the Mg compound (1).

1. Structure Reports, 15, 282; 39A, 264.

MANGANESE TETRABORATE NONAHYDRATE

$Mn[B_4O_5(OH)_4].7H_2O$ $MnB_4O_7.9H_2O$

I. I.R. BERZINJA, Ja.K. OZOLS and A.F. IEVIN'Š, 1974. Latv. PSR Zinat. Akad. Vest., Kim, Ser., No. 6, 648-650.

II. Idem, 1975. Kristallografija, 20, 419-422 [Soviet Physics - Crystallography, 20, 255-256].

Triclinic, $P\bar{1}$, a = 10.76, b = 9.90, c = 7.98 Å, α = 57°51', β = 103°40', γ = 107°50', D_m = 1.80, Z = 2. Cu radiation, R = 0.103 for 1700 reflexions (films, visual intensities).

The structure (Fig. 1) contains isolated $[B_4O_5(OH)_4]^{2-}$ ions which consist of two tetrahedra and two triangles sharing corners. Mn has octahedral coordination to 5 H_2O and 1 OH, the octahedra being linked to the borate anions by sharing the OH group and by hydrogen bonding. Mean B-O = 1.48 (tetrahedra), 1.38 (triangles), Mn-O = 2.18 Å.

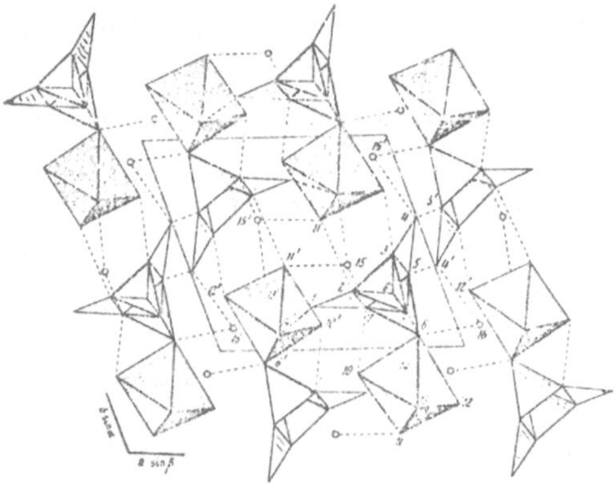

Fig. 1. Structure of $MnB_4O_7.9H_2O$.

GAUDEFROYITE

$Ca_4Mn_3O_3(BO_3)_3(CO_3)$

O.V. JAKUBOVIČ, M.A. SIMONOV and N.V. BELOV, 1975. Kristallografija, 20, 152-155 [Soviet Physics - Crystallography, 20, 87-88].

Hexagonal, $P6_3$, a = 10.606, c = 5.879 Å, Z = 2. Mo radiation, R = 0.076 for 1235 reflexions.

The structure contains Ca polyhedra (7- and 9-coordination) and Mn octahedra, linked by triangular CO_3 groups (which lie within empty oxygen octahedra) and by triangular BO_3 groups (Fig. 1). B-O = 1.34-1.43, C-O = 1.31, Mn-O = 1.86-2.29, Ca-O = 2.34-2.67 Å. Previous study in 1.

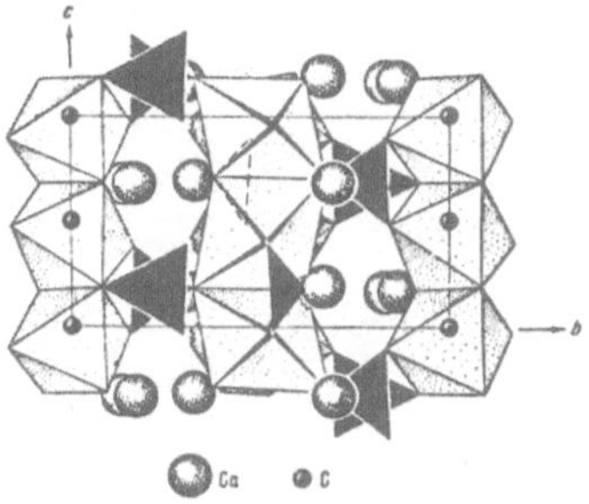

Fig. 1. Structure of gaudefroyite.

1. M.-M. GRANGER and J. PROTAS, 1965. C.R. Acad. Sci., Paris, 260, 4553.

IRON BORATE

Fe_3BO_6

R. DIEHL and G. BRANDT, 1975. Acta Cryst., B31, 1662-1665.

Orthorhombic, Pnma, a = 10.048, b = 8.531, c = 4.466 Å, D_m = 4.78, Z = 4. Mo radiation, refinement of the structure (1), R = 0.031 for 1241 reflexions.

Isostructural with norbergite (2), as previously reported (1). B has tetrahedral coordination, B-O = 1.44-1.50(1) Å, and both independent Fe atoms have distorted octahedral geometry, Fe-O = 1.86-2.23 Å.

1. Structure Reports, 30A, 417.
2. Strukturbericht, 2, 121, 516; Structure Reports, 34A, 366.

ALUMINUM COBALT BORATE

$Al_4Co(BO_4)_2O_2$

J.J. CAPPONI and M. MAREZIO, 1975. Acta Cryst., B31, 2440-2443.

Monoclinic, $P2_1/c$, a = 7.221, b = 4.371, c = 9.534 Å, β = 108.45°, D_m not given, Z = 2. Mo radiation, R = 0.020 for 392 reflexions (twinned crystal).

The material is synthetic and is an anhydrous boron chondrodite. The structure is similar to that of chondrodite (1) and contains layers of octahedral chains linked by boron tetrahedra (Fig. 1). Average interatomic distances are Al-O = 1.94, Co-O = 2.08, B-O = 1.49 Å.

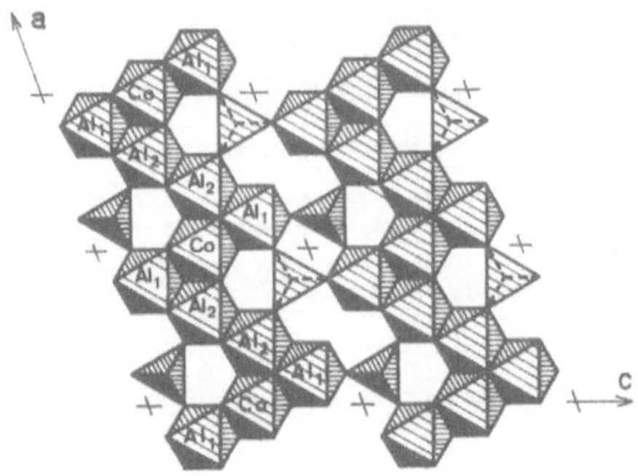

Fig. 1. Structure of $Al_4Co(BO_4)_2O_2$.

1. Strukturbericht, 2, 119, 517; Structure Reports, 35A, 469.

COBALT LANTHANUM METABORATE

$CoLa(BO_2)_5$

G.K. ABDULLAEV, K.S. MAMEDOV and G.G. DŽAFAROV, 1975. Ž. Strukt. Khim., 16, 71-76 [J. Struct. Chem., 16, 61-65].

Monoclinic, $P2_1/n$, a = 8.85, b = 7.63, c = 9.66 Å, β = 92.0°, D_m = 4.2, Z = 4. Cu radiation, R = 0.14 for 630 reflexions (films, visual intensities).

The structure contains layers parallel to (101) of infinite $[B_5O_{10}{}^{5-}]_n$ ions, which contain 3 B atoms with tetrahedral and 2 B atoms with triangular coordination. The layers are joined by Co (octahedral coordination) and La (10-coordination) ions. B-O = 1.50 (tetrahedra), 1.38 (triangles), Co-O = 2.05-2.23, La-O = 2.47-2.91(4) Å.

SAMARIUM METABORATE

GADOLINIUM METABORATE

$Sm(BO_2)_3$
$Gd(BO_2)_3$

G.K. ABDULLAEV, K.S. MAMEDOV and G.G. DŽAFAROV, 1975. Kristallografija, 20, 265-269 [Soviet Physics - Crystallography, 20, 161-163].

Monoclinic, I2/c, a = 6.33, 6.28, b = 8.06, 8.02, c = 7.85, 7.80 Å, β = 93, 93°, D_m not given, Z = 4. Cu radiation, R = 0.12 and 0.13 for 220 and 200 reflexions, respectively (film data).

Isostructural with the La (1) and Nd (2) compounds. The structure contains $(B_6O_{12}{}^{6-})_n$ chains of two tetrahedra and four triangles, linked by Ln ions with 10-coordination. B-O = 1.47-1.51 (tetrahedra), 1.34-1.38 (triangles), Sm-O = 2.31-2.71, Gd-O = 2.28-2.70(2) Å.

1. Structure Reports, 35A, 282.
2. Ibid., 38A, 298.

SODIUM CARBONATE MONOHYDRATE

(THERMONATRITE)

$Na_2CO_3.H_2O$

K.K. WU and I.D. BROWN, 1975. Acta Cryst., B31, 890-892.

Orthorhombic, $P2_1ab$, a = 6.472, b = 10.724, c = 5.259 Å, D_m = 2.255, Z = 4. Neutron diffraction data, R = 0.059 for 612 reflexions.

The structure is as determined by X-ray methods (1, 2), except that the hydrogen positions lie between those postulated in 2 and those predicted in 3.

1. Strukturbericht, 4, 42, 157.
2. Structure Reports, 40A, 305.
3. W.H. BAUR, 1972. Acta Cryst., B28, 1456.

NORTHUPITE

$Na_3Mg(CO_3)_2Cl$

A. DAL NEGRO, G. GIUSEPPETTI and C. TADINI, 1975. Miner. (Tschermaks) Petrogr. Mitt., 22, 158-163.

Cubic, Fd3, a = 14.069 Å, D_m = 2.38, Z = 16. Mo radiation, R = 0.023 for 357 reflexions. Origin at centre; Na in 48(f): x = 0.8977; Mg in 16(d); C in 32(e): x = 0.2831; Cl in 16(c); O in 96(g): (0.2627, 0.2301, 0.3546).

The structure is essentially as previously described (1, 2). Na has distorted octahedral coordination, Na-O = 2.411, 2.437(1), Na-Cl = 2.873(1) Å (each x 2), and Mg has more regular octahedral coordination, Mg-O = 2.073(1) Å (x 6). The carbonate group is planar, C-O = 1.284 Å.

1. Strukturbericht, 2, 80, 399.
2. T. WATANABÉ, 1933. Sci. Papers Phys. Chem. Res. (Tokyo), 21, 40.

CALCIUM CARBONATE

$CaCO_3$-II (high-pressure phase)

L. MERRILL and W.A. BASSETT, 1975. Acta Cryst., B31, 343-349.

Monoclinic, $P2_1/c$, a = 6.334, b = 4.948, c = 8.033 Å, β = 107.9°, D_x = 2.77, Z = 4. Mo radiation, R = 0.078 for 112 reflexions at 15 kbar.

The structure (Fig. 1) is related to that of calcite, the transition from calcite to the high-pressure phase involving rotation of carbonate groups by 11° and a small displacement of adjacent planes of Ca ions parallel to a calcite $(10\bar{1}4)$ plane.

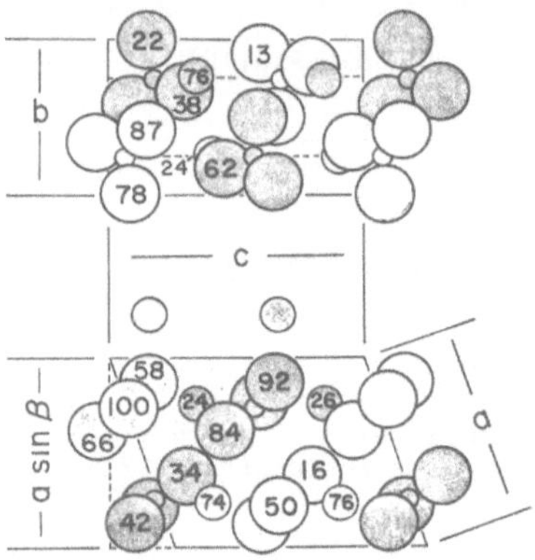

Fig. 1. Structure of $CaCO_3$-II.

STRONTIUM AND BARIUM CARBONATES
(High-temperature Forms)

$SrCO_3$
$BaCO_3$

K.O. STRØMME, 1975. Acta Chem. Scand., A29, 105-111.

Disorder affecting high-temperature (rhombohedral or cubic) phases has been studied by configurational-entropy calculations, whose results are compared with calorimetric data. The type of disorder postulated may also affect such structures as $AgNO_3$, KNO_3, etc.

THALLIUM(I) CARBONATE

Tl_2CO_3

R. MARCHAND, Y. PIFFARD and M. TOURNOUX, 1975. Canad. J. Chem., 53, 2454-2458.

Monoclinic, C2/m, a = 12.486, b = 5.382, c = 7.530 Å, β = 122.35°, D_m = 7.25, Z = 4. Mo radiation, R = 0.033 for 655 reflexions.

The planar carbonate groups form rows along b, which are linked by thallium ions (Fig. 1). The two Tl ions have four and five oxygen neighbours at 2.67-2.82 Å, all on one side, with only weak contacts of >3.2 Å on the other side (Fig. 1); the lone-pairs are thus sterically active.

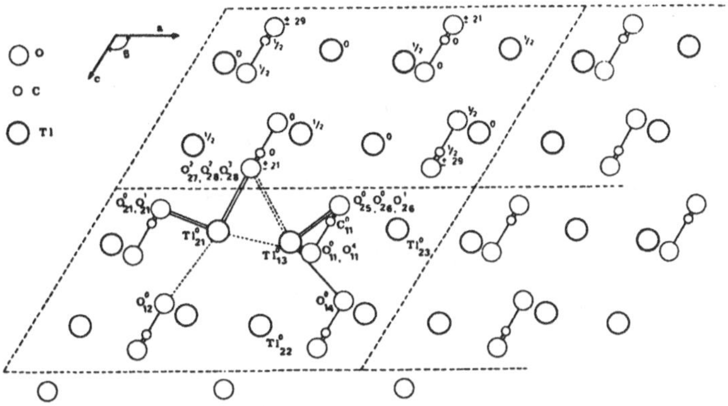

Fig. 1. Structure of Tl_2CO_3.

PHOSGENITE

$Pb_2Cl_2CO_3$

G. GIUSEPPETTI and C. TADINI, 1974. Miner. (Tschermaks) Petrogr. Mitt., 21, 101-109.

Tetragonal, P4/mbm, a = 8.160, c = 8.883 Å, D_m = 6.134, Z = 4. Mo radiation, R = 0.032 for 320 reflexions.

Atomic positions

			x	y	z
Pb	in	8(k)	0.1659	0.6659	0.2594
Cl(1)		4(e)	0	0	0.2428
Cl(2)		4(h)	0.3521	0.8521	1/2
O(1)		4(g)	0.2110	0.7110	0
O(2)		8(k)	0.3726	0.8726	0.1269
C		4(g)	0.3257	0.8257	0

The structure (Fig. 1) is as previously described (1). It contains [Pb,Cl(1)-Cl(2)-Cl(1),Pb] layers parallel to (001), connected by CO_3 groups. Pb has 9 neighbours (4 O and 5 Cl) forming a deformed monocapped square antiprism; Pb-O = 2.36-2.69, Pb-Cl = 3.03-3.34 Å.

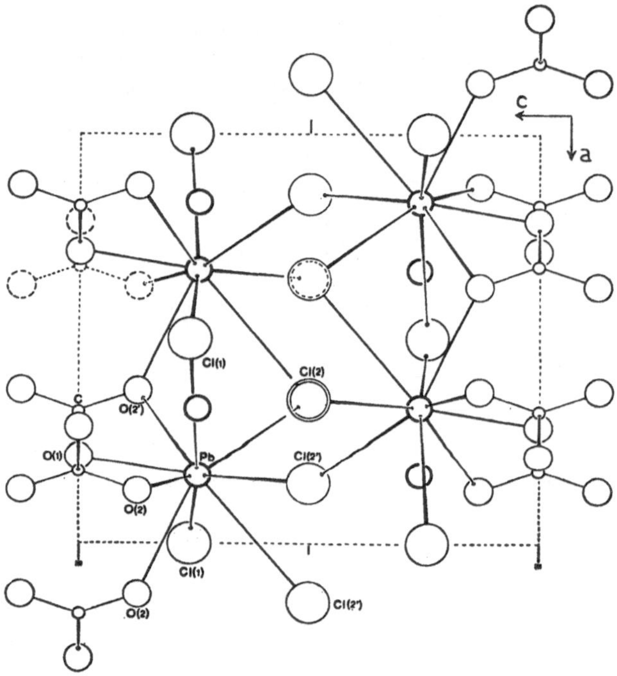

Fig. 1. Structure of phosgenite.

1. Strukturbericht, 3, 89, 416; Structure Reports, 10, 140.

WELOGANITE

$(Sr,Ca)_3Na_2Zr(CO_3)_6.3H_2O$

J.D. GRICE and G. PERRAULT, 1975. Canad. Miner., 13, 209-216.

Triclinic, P1, a = 8.966, b = 8.980, c = 6.730 Å, α = 102.72, β = 116.65, γ = 60.06°, Z = 1. Mo radiation, R = 0.043 for 2354 reflexions.

The structure is pseudo-trigonal, and contains six carbonate ions and three water molecules linked by three Sr^{2+} (10-coordination), one Zr^{4+} (9-coordination), and two Na^{+} ions (octahedral and trigonal-prismatic); Sr-O = 2.59-2.81, Zr-O = 2.16-2.34, Na-O = 2.25-2.54(1), C-O = 1.25-1.34(2) Å. Each water molecule is hydrogen bonded to two oxygen atoms, O-H...O = 2.75-2.86 Å.

SODIUM COPPER(II) CARBONATE TRIHYDRATE

$Na_2Cu(CO_3)_2.3H_2O$

R.L. HARLOW and S.H. SIMONSEN, 1975. Acta Cryst., B31, 1313-1318.

Monoclinic, $P2_1/c$, a = 9.691, b = 6.091, c = 17.111 Å, β = 126.40°, D_m = 2.314, Z = 4. Mo radiation, R = 0.042 for 2373 reflexions.

The structure is as previously reported independently (1). C-O = 1.300 (coordinated), 1.263 Å (not-coordinated).

1. Structure Reports, 39A, 273.

ANCYLITE

$(La,Ce)_{1\cdot4}(Ca,Sr)_{0\cdot6}(CO_3)_2(OH)_{1\cdot4}.0\cdot6H_2O$

A. DAL NEGRO, G. ROSSI and V. TAZZOLI, 1975. Amer. Min., 60, 280-284.

Orthorhombic, Pmcn, a = 5.03, b = 8.53, c = 7.29 Å, D_m = 4.1, Z = 2. Mo radiation, R = 0.059 for 502 reflexions.

Atomic positions

			x	y	z
M	in	4(c)	1/4	0.3399	0.6476
C		4(c)	3/4	0.1905	0.8099
O(1)		4(c)	3/4	0.3181	0.7210
O(2)		8(d)	0.5297	0.1218	0.8520
OH,H_2O(3)		4(c)	1/4	0.4135	0.9749

The structure (Fig. 1) can be derived from that of aragonite (1) by the introduction of OH groups on the mirror plane, bonded to the cations, which are statistically distributed and have 10-coordination, average M-O = 2.61(1) Å.

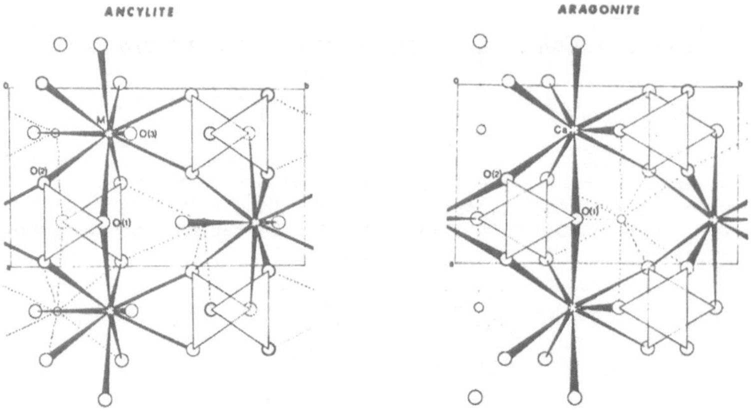

Fig. 1. Structures of ancylite and aragonite.

1. Strukturbericht, 1, 295, 317, 318; Structure Reports, 37A, 280.

SODIUM PENTACARBONATOTHORATE(IV) DODECAHYDRATE

SODIUM PENTACARBONATOCERATE(IV) DODECAHYDRATE

$Na_6[Th(CO_3)_5].12H_2O$
$Na_6[Ce(CO_3)_5].12H_2O$

I. S. VOLIOTIS and A. RIMSKY, 1975. Acta Cryst., B31, 2615-2620.

II. Idem, 1975. Ibid., B31, 2620-2622.

Triclinic, $P\bar{1}$, a = 9.60, 9.53, b = 9.92, 9.84, c = 13.64, 13.58 Å, α = 90.47, 90.46, β = 104.38, 104.50, γ = 95.52, 95.42°, D_m = 2.31, 2.13, for Th and Ce compounds respectively, Z = 2. Mo radiation, R = 0.060 and 0.051 for 4806 and 7157 reflexions.

The compounds are isostructural. All five carbonate groups are bidentate, with Th (Ce) having irregular ten-coordination (Fig. 1); Th-O = 2.45-2.56(1), Ce-O = 2.38-2.50(1) Å. Na ions have distorted octahedral coordination (Fig. 1), Na-O = 2.29-2.83 Å.

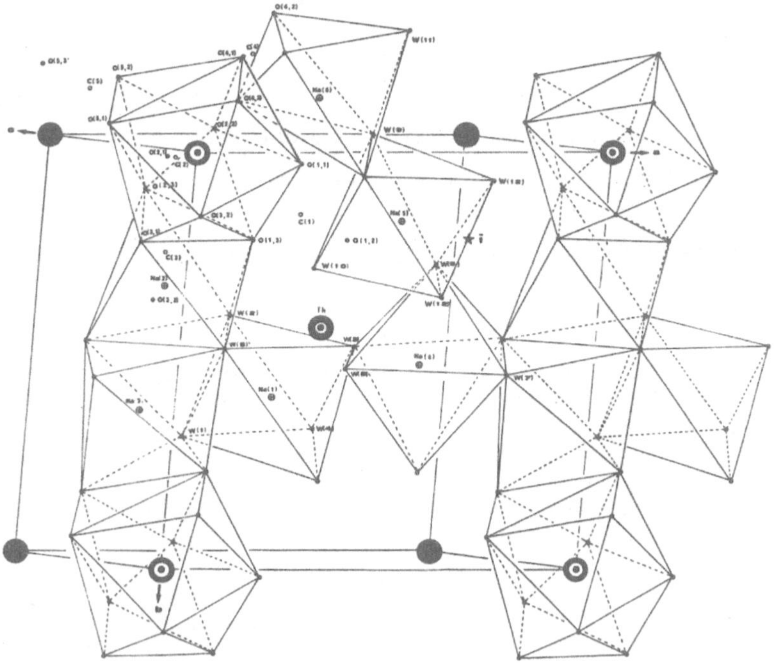

Fig. 1. Structure of $Na_6[Th(CO_3)_5].12H_2O$.

SODIUM NITRITE

$NaNO_2$

M.I. KAY, J.A. GONZALO and R. MAGLIC, 1975. Ferroelectrics, 9, 179-186.

[Orthorhombic, Im2m], a = 3.65, b = 5.65, c = 5.36 Å, at 162°C, [Z = 2]. Neutron diffraction data, $R(F^2)$ = 0.032 for 57 reflexions. Na in 2(a): y = 0.544; N in 2(a): y = 0.077; O in 4(d): y = -0.041, z = 0.193.

Structure as previously described (1).

1. Structure Reports, 26, 473; 39A, 366.

POTASSIUM cis-TETRANITRODIAMMINERHODATE(III)

$K[Rh(NO_2)_4(NH_3)_2]$

Z.G. ALIEV and L.O. ATOVMJAN, 1975. Koordin. Khim., 1, 680-681.

Triclinic, P$\bar{1}$, a = 7.346, b = 17.13, c = 8.945 Å, α = 81.7, β = 107.1, γ = 108.8°, D_m = 2.34, Z = 4 [not 2]. Mo radiation, R = 0.123 for 3766 reflexions (film data).

Rh has cis-octahedral coordination, $Rh-NH_3$ = 2.08, $Rh-NO_2$ = 2.01(2) Å.

POTASSIUM LEAD HEXANITRONICKELATE(II)

$K_2PbNi(NO_2)_6$

I. S. TAKAGI, M.D. JOESTEN and P.G. LENHERT, 1975. Acta Cryst., B31, 1968-1970.

II. Idem, 1976. Ibid., B32, 668.

Cubic, Fm3, a = 10.578 Å, D_m not given, Z = 4. Mo radiation, R = 0.016 for 462 reflexions. Ni in 4(a); Pb in 4(b); K in 8(c); N in 24(e): x = 0.1966; O in 48(h): y = 0.1006, z = 0.2579.

Isostructural with $K_2PbCu(NO_2)_6$ (1). The $Ni(NO_2)_6^{4-}$ anions have m3 symmetry, with neighbouring NO_2 groups mutually perpendicular; Ni-N = 2.080, N-O = 1.245(2) Å, O-N-O = 117.3°.

1. Structure Reports, 20, 361; 34A, 346; 37A, 284.

POTASSIUM BARIUM HEXANITRONICKELATE(II)

$K_2BaNi(NO_2)_6$

I. S. TAKAGI, M.D. JOESTEN and P.G. LENHERT, 1975. Acta Cryst., B31, 1970-1972.

II. Idem, 1976. Ibid., B32, 668.

Cubic, Fm3, a = 10.780 Å, D_m not given, Z = 4. Mo radiation, R = 0.016 for 2913 reflexions.

Isostructural with $K_2PbNi(NO_2)_6$ (1), except that the NO_2 groups are disordered (Fig. 1), either with unequal occupancies (space group Fm3) or equal occupancies (Fm3m).

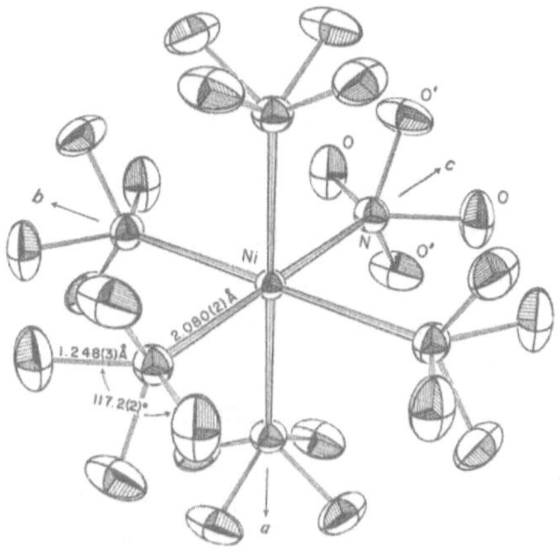

Fig. 1. Disorder of the $Ni(NO_2)_6^{4-}$ ion in $K_2BaNi(NO_2)_6$.

1. Preceding report.

POTASSIUM BARIUM HEXANITROCUPRATE(II)

$K_2BaCu(NO_2)_6$

S. TAKAGI, M.D. JOESTEN and P.G. LENHERT, 1975. Acta Cryst., B31, 596-598.

Orthorhombic, Fmmm, a = 11.219, b = 10.728, c = 10.685 Å, D_m = 2.871, Z = 4. Mo radiation, R = 0.016 for 1470 reflexions (twinned crystal).

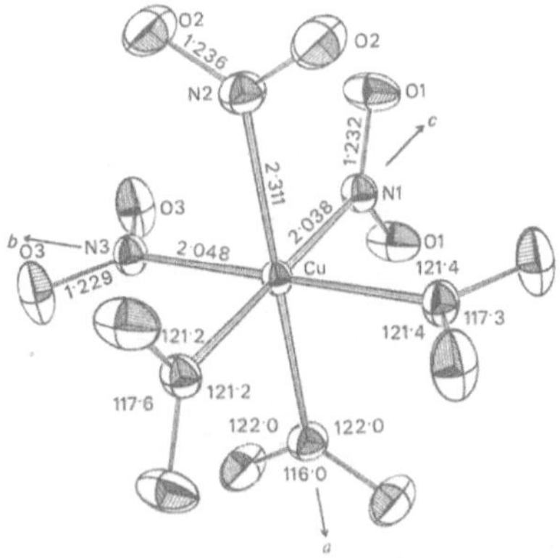

Fig. 1. The $Cu(NO_2)_6^{4-}$ ion in $K_2BaCu(NO_2)_6$.

Isostructural with $K_2CaCu(NO_2)_6$ (1). Cu coordination is tetragonally-elongated octahedral (Fig. 1).

1. Structure Reports, 40A, 230.

NITRIC ACID MONOHYDRATE

$HNO_3.H_2O$

R.G. DELAPLANE, I. TAESLER and I. OLOVSSON, 1975. Acta Cryst., B31, 1486-1489.

Orthorhombic, $P2_1cn$, Z = 4. At 85°K: a = 5.4647, b = 8.6439, c = 6.2308 Å, D_m = 1.816. Cu radiation, R = 0.027 for data from two crystals (310 and 287 reflexions). At 225°K: a = 5.4759, b = 8.7242, c = 6.3275 Å. Cu radiation, R = 0.026 for 320 reflexions.

The structure at both 85 and 225°K is essentially as previously described (1). It contains H_3O^+ ions hydrogen bonded to three different NO_3^- ions (Fig. 1) to form infinite layers (Fig. 2).

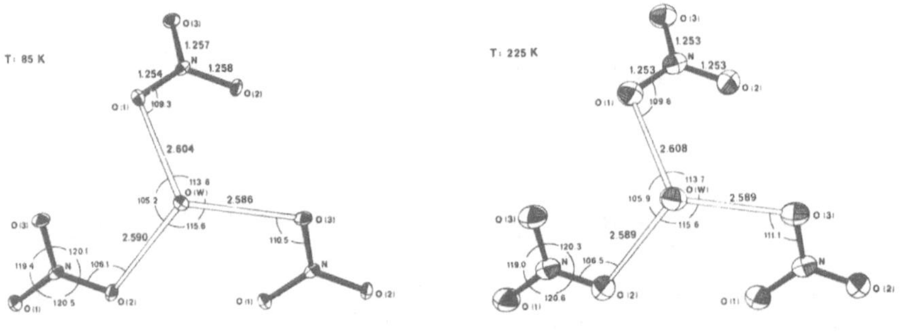

Fig. 1. Bond lengths (Å) and angles (degrees) in nitric acid monohydrate.

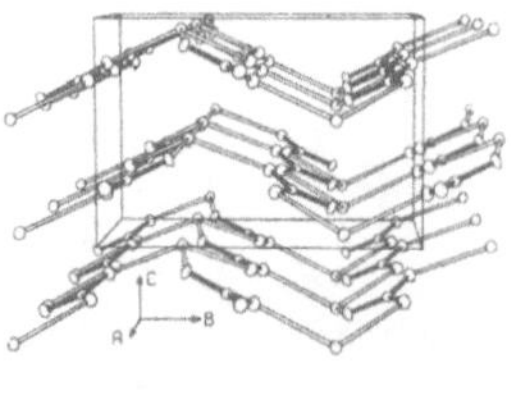

Fig. 2. Structure of nitric acid monohydrate.

1. Structure Reports, 15, 276.

NITRIC ACID TRIHYDRATE

$HNO_3.3H_2O$

I. TAESLER, R.G. DELAPLANE and I. OLOVSSON, 1975. Acta Cryst., B31, 1489-1492.

Orthorhombic, $P2_12_12_1$, a = 9.4845, b = 14.6836, c = 3.4355 Å, at 85°K, D_m = 1.621, Z = 4. Cu radiation, R = 0.026 for 576 reflexions.

The structure is essentially as previously described (1). It contains H_3O^+ ions, each of which is bonded to two water molecules by short hydrogen bonds (2.482 and 2.576 Å, Fig. 1) to form $H_7O_3^+$ ions. A longer hydrogen bond (2.800 Å) connects the $H_7O_3^+$ groups into spirals, which are hydrogen bonded to NO_3^- ions.

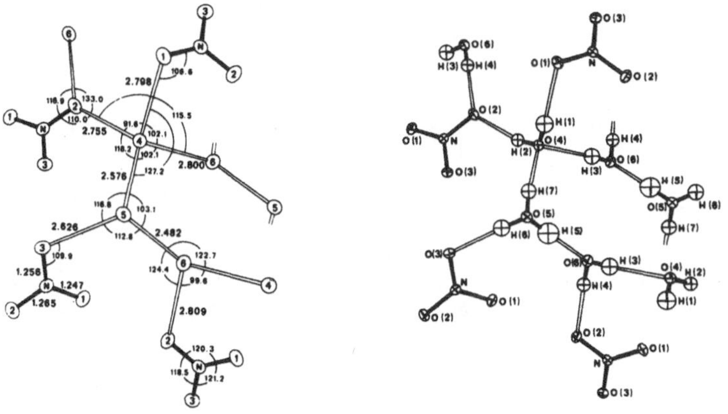

Fig. 1. Bond lengths (Å, σ = 0.002 Å) and angles (degrees) in nitric acid trihydrate.

1. Structure Reports, 15, 276; 17, 506.

AMMONIUM NITRATE

POTASSIUM NITRATE

NH_4NO_3-IV
NH_4NO_3-KNO_3-III
KNO_3-II

J.R. HOLDEN and C.W. DICKINSON, 1975. J. Phys. Chem., 79, 249-256.

NH_4NO_3-IV
Orthorhombic, Pmmn, a = 5.724, b = 5.455, c = 4.945 Å, Z = 2. Mo radiation, no details [given only in microfilm edition]. Structure as in 1.

NH_4NO_3-KNO_3-III
Orthorhombic, Pnma, a = 7.694-7.635, b = 5.827-5.739, c = 7.158-7.026 Å, for 5-37 atomic % K, Z = 4. Structure as in 2.

KNO_3-II
Orthorhombic, Pnma, a = 6.458, b = 5.444, c = 9.211 Å, for a solid solution with 5 atomic % NH_4. Structure as in 3.

1. Strukturbericht, 2, 72, 382; Structure Reports, 24, 418; 27, 567; 38A, 303.
2. Strukturbericht, 2, 71, 382; Structure Reports, 11, 367.
3. Strukturbericht, 2, 381; Structure Reports, 11, 362; 27, 565; 39A, 275.

CAESIUM PENTANITRATOALUMINATE

$Cs_2Al(NO_3)_5$

O.A. D'JAČENKO and L.O. ATOVMJAN, 1975. Ž. Strukt. Khim., 16, 85-91 [J. Struct. Chem., 16, 73-78].

Trigonal, $P3_121$, a = 11.16, c = 10.02 Å, D_m = 2.69, Z = 3. Mo radiation, R = 0.084 for 418 reflexions (films, visual intensities).

The structure contains $Al(NO_3)_5^{2-}$ ions, in which Al has distorted octahedral coordination to one bidentate and four unidentate nitrate groups, Al-O = 1.89-1.98(4) Å. Cs ions have 9- and 10-coordination, Cs-O = 3.13-3.39(4) Å.

THALLIUM(I) NITRATE

$TlNO_3$-III

W.L. FRASER, S.W. KENNEDY and M.R. SNOW, 1975. Acta Cryst., B31, 365-370.

Orthorhombic, Pnma, a = 12.355, b = 8.025, c = 6.298 Å, D_m = 5.556, Z = 8. Mo radiation, R = 0.063 for 659 reflexions.

The structure (Fig. 1) contains a CsCl arrangement of anions and cations. Two independent nitrate ions lie alternately parallel and perpendicular to the mirror planes (but the orientations are different from those previously given, 1). Tl has eight oxygen neighbours at 3.0 Å.

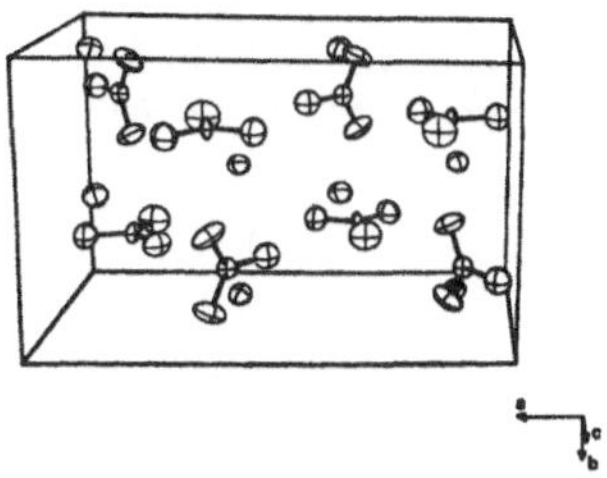

Fig. 1. Structure of $TlNO_3$-III.

1. Structure Reports, 26, 476.

IRON(III) NITRATE DINITROGEN TETROXIDE

(NITRATETRISNITROSONIUM TETRANITRATOFERRATE(III))

$[N_4O_6][Fe(NO_3)_4]_2$ $2[Fe(NO_3)_2.1\cdot5N_2O_4]$

L.J. BLACKWELL, E.K. NUNN and S.C. WALLWORK, 1975. J. Chem. Soc., Dalton, 2068-2072.

Monoclinic, $P2_1/c$, a = 10.419, b = 19.930, c = 14.435 Å, β = 128.89°, Z = 4. Mo radiation, R = 0.039 for 2930 reflexions.

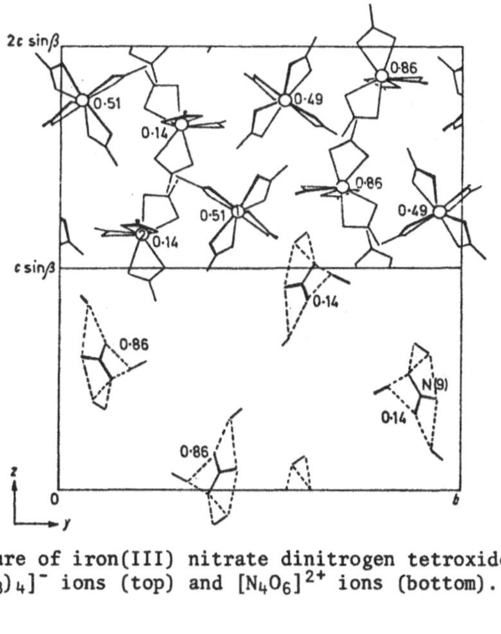

Fig. 1. Structure of iron(III) nitrate dinitrogen tetroxide, showing the $[Fe(NO_3)_4]^-$ ions (top) and $[N_4O_6]^{2+}$ ions (bottom).

The structure (Fig. 1) contains $[N_4O_6]^{2+}$ and $[Fe(NO_3)_4]^-$ ions. The cations consist of an NO_3^- ion surrounded by three NO^+ ions in fairly close contact; N-O (nitrate) = 1.2, N-O(nitrosonium) = 1.0, N...O = 2.4-2.7 Å. In the anion, Fe has dodecahedral coordination to eight oxygen atoms of four bidentate nitrate groups, Fe-O = 2.13 Å.

TETRASILVER TELLURIDE DINITRATE

α-$Ag_4Te(NO_3)_2$

E. SCHULTZE-RHONHOF, 1975. Acta Cryst., B31, 2837-2840.

Cubic, Pa3, a = 8.173 Å, D_m not measured, Z = 4. Cu radiation, R = 0.11 for 28 reflexions (powder data; only four lines with odd indices).

Atomic positions

			x	y	z
4 Te	in	4(a)	0	0	0
4 Ag(1)		4(b)	1/2	1/2	1/2
12 Ag(2)		24(d)	1/4	1/4	0
4 N(1)		8(c)	0.19	0.19	0.19
12 O(1)		24(d)	1/8	1/8	0.31
4 N(2)		8(c)	0.31	0.31	0.31
12 O(2)		24(d)	3/8	3/8	0.19

The structure, which can be compared with that of α-Ag_2Te (1), contains a face-centred arrangement of Te atoms, with Ag atoms and NO_3 groups statistically distributed (Fig. 1).

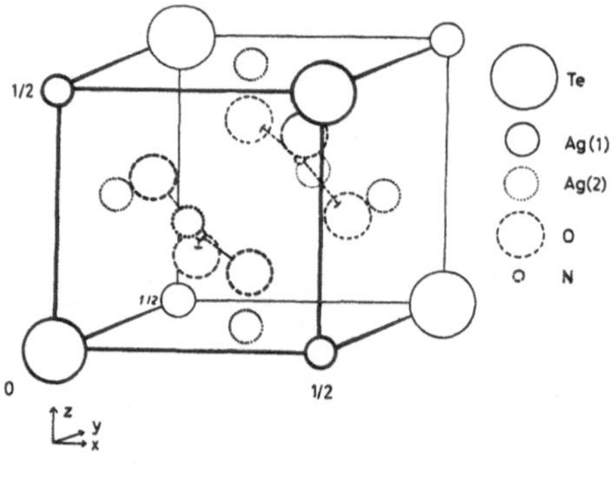

Fig. 1. Structure of α-$Ag_4Te(NO_3)_2$.

1. Strukturbericht, 4, 115.

ZINC NITRATE DIHYDRATE

$Zn(NO_3)_2.2H_2O$

D. PETROVIĆ and B. RIBÁR, 1975. Acta Cryst., B31, 1795-1796.

Monoclinic, $P2_1/c$, a = 5.754, b = 5.978, c = 8.557 Å, β = 91.0°, Z = 2 (all as in 1). Cu radiation, R = 0.076 for 459 reflexions (film data).

Isostructural with $Mg(NO_3)_2.2H_2O$ (2), Zn being in 2(c), and not 2(d) as previously reported (1). Zn-O = 2.04-2.17(1), N-O = 1.22-1.27(2) Å, O-N-O = 118-123°.

1. Structure Reports, 34A, 341.
2. Ibid., 39A, 276 [where $Mg(NO_3)_2.2H_2O$ is given in error as the Mn compound].

MERCURY(I) NITRATE DIHYDRATE

$Hg_2(NO_3)_2.2H_2O$

D. GRDENIĆ, M. SIKIRICA and I. VICKOVIĆ, 1975. Acta Cryst., B31, 2174-2175.

Monoclinic, $P2_1/n$, a = 8.633, b = 7.506, c = 6.256 Å, β = 103.8°, D_m = 4.785, Z = 2. Mo radiation, R = 0.054 for 736 reflexions.

The structure (Fig. 1) is as previously determined (1). It contains $[H_2O\text{-}Hg\text{-}Hg\text{-}OH_2]^{2+}$ and NO_3^- ions. Hg-Hg = 2.508(2), Hg-O = 2.13(2) Å, Hg-Hg-O = 168°, N-O = 1.26(3) Å. The ions are joined by hydrogen bonds, O-H...O = 2.68-2.70 Å, and by further Hg...O interactions, 2.68, 2.75, and 2.81 Å.

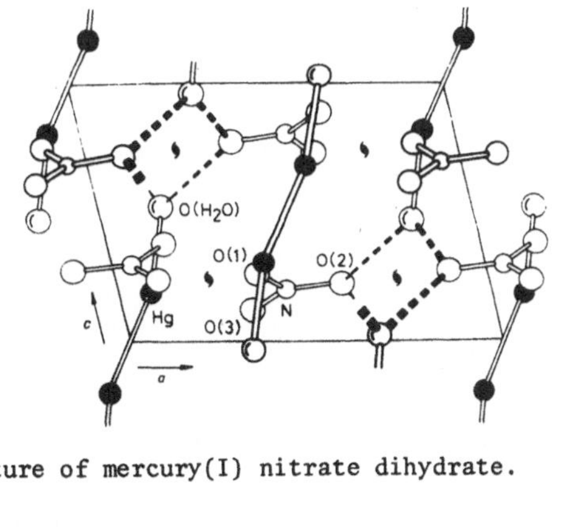

Fig. 1. Structure of mercury(I) nitrate dihydrate.

1. Structure Reports, 20, 368.

NITROSONIUM PENTANITRATOHOLMATE(III)

$(NO)_2[Ho(NO_3)_5]$

G.E. TOOGOOD and C. CHIEH, 1975. Canad. J. Chem., 53, 831-835.

Monoclinic, $P2_1/c$, a = 8.094, b = 11.979, c = 14.170 Å, β = 104.7°, D_m not given, Z = 4. Mo radiation, R = 0.085 for 1514 reflexions.

The structure contains two NO^+ ions, N-O = 1.00(5) Å, and one $Ho(NO_3)_5{}^{2-}$ ion (Fig. 1) in which Ho is 10-coordinated by five essentially symmetric bidentate nitrate groups, which arrange themselves in a trigonal bipyramid around the metal, mean Ho-O = 2.45(2) Å.

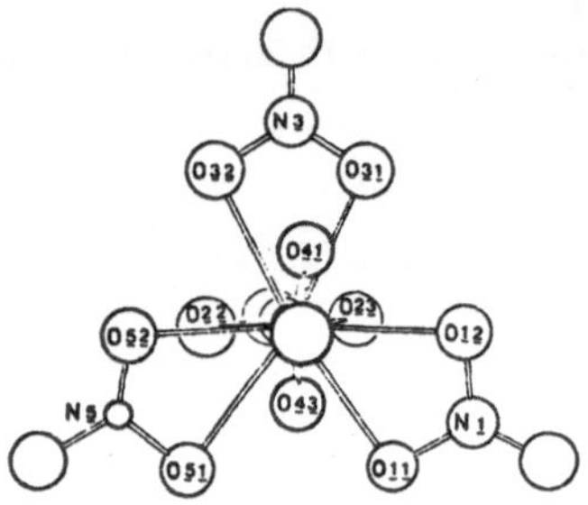

Fig. 1. Structure of the $Ho(NO_3)_5{}^{2-}$ ion.

RUBIDIUM DIFLUOROPHOSPHATE

$RbPO_2F_2$

W. GRANIER, J. DURAND, L. COT and J.L. GALIGNÉ, 1975. Acta Cryst., B31, 2506-2507.

Orthorhombic, Pnma, a = 8.165, b = 6.464, c = 7.786 Å, D_m = 3.00, Z = 4. Cu radiation, R = 0.044 for 373 reflexions.

Atomic positions

	x	y	z
Rb	0.1416	1/4	0.1438
P	0.1037	1/4	0.6856
O	0.1059	0.0524	0.7768
F(1)	0.2422	1/4	0.5464
F(2)	-0.0418	1/4	0.5564

Isostructural with the K, Rb, and NH_4 salts (1). P-O = 1.461(7), P-F = 1.562(9) Å, O-P-O = 121.9, F-P-F = 95.9°. Rb has 6 O and 6 F neighbours (Fig. 1).

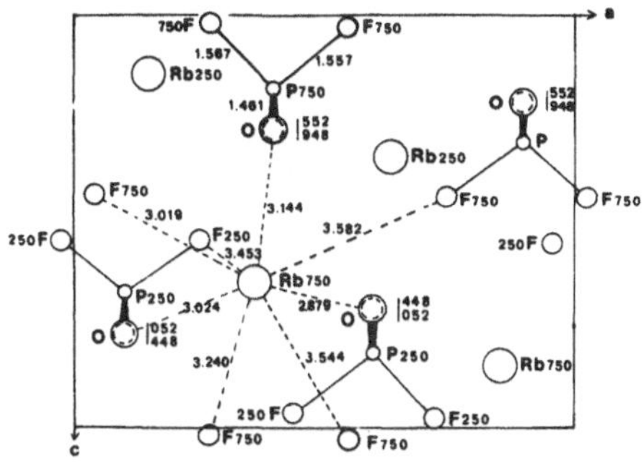

Fig. 1. Structure of $RbPO_2F_2$.

1. Structure Reports, 31A, 192; 32A, 369; 34A, 324.

ERBIUM HYPOPHOSPHITE

$Er(H_2PO_2)_3$

L.A. ASLANOV, V.M. IONOV, M.A. PORAJ-KOŠIC, V.G. LEBEDEV, B.N. KULIKOVSKIJ, O.N. GILJAROV and T.L. NOVODEREŽKINA, 1975. Izv. Akad. Nauk SSSR, Neorg. Mater., 11, 117-119 [Inorganic Materials, 11, 96-98].

Monoclinic, B2/m, a = 14.40, b = 12.10, c = 5.64 Å, γ = 111.3°, D_m not given, Z = 4. Mo radiation, R = 0.12 for 524 reflexions.

The structure contains octahedrally-coordinated Er ions cross-linked by bidentate-bridging H_2PO_2 groups; P-O = 1.57(5) Å, O-P-O = 103-125°, Er-O = 2.08-2.47(7) Å.

POTASSIUM SODIUM FLUOROPHOSPHATE

$K_3Na(PO_3F)_2$

J. DURAND, W. GRANIER, L. COT and J.L. GALIGNÉ, 1975. Acta Cryst., B31, 1533-1535.

Trigonal, $P\bar{3}m1$, a = 5.761, c = 7.374 Å, D_m = 2.596, Z = 1. Cu radiation, R = 0.067 for 258 reflexions. Na in 1(a): 000; K(1) in 1(b): 0,0,1/2; K(2), P, F in 2(d): 2/3,1/3,z, z = 0.1737, 0.7452, 0.5242; O in 6(i): $x\bar{x}z$, x = -0.1876, z = 0.7923.

Isostructural with glaserite (1). The PO_3F ion has C_{3v} symmetry (Fig. 1), P-O = 1.50, P-F = 1.63(1) Å. Na and K(1) coordination octahedra share faces along c; Na-O = 2.42, K-O = 2.86 Å. K(2) has nine O and one F neighbours, K-O = 2.90, 3.17, K-F = 2.58 Å.

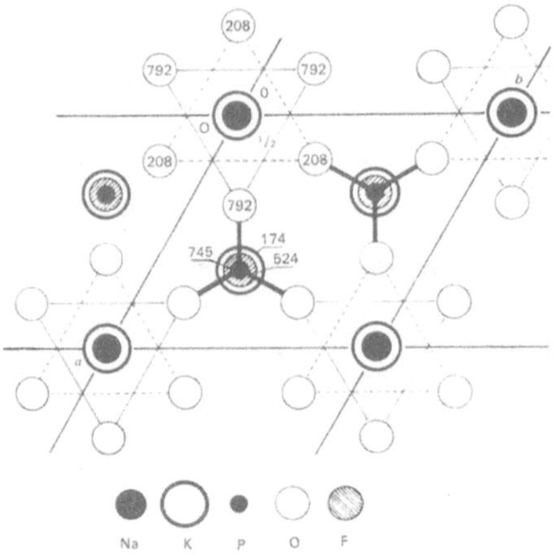

Fig. 1. Structure of $K_3Na(PO_3F)_2$.

1. Strukturbericht, 1, 378.

BARIUM POTASSIUM POLYPHOSPHATE

$Ba_2K(PO_3)_5$

C. MARTIN, I. TORDJMAN and A. DURIF, 1975. Z. Kristallogr., 141, 403-411.

Monoclinic, Pc, a = 8.646, b = 7.329, c = 13.884 Å, β = 129.17°, D_m not given, Z = 2. Mo radiation, R = 0.053 for 3563 reflexions.

The structure contains an infinite $(PO_3)_n$ chain with a period of ten tetrahedra (Fig. 1); P-O = 1.57-1.66 (bridging), 1.46-1.50(1) Å (terminal), O-P-O = 97-122, P-O-P = 130-143°. The K and two Ba ions each have eight-coordination, K-O = 2.84-3.25, Ba-O = 2.64-3.03 Å.

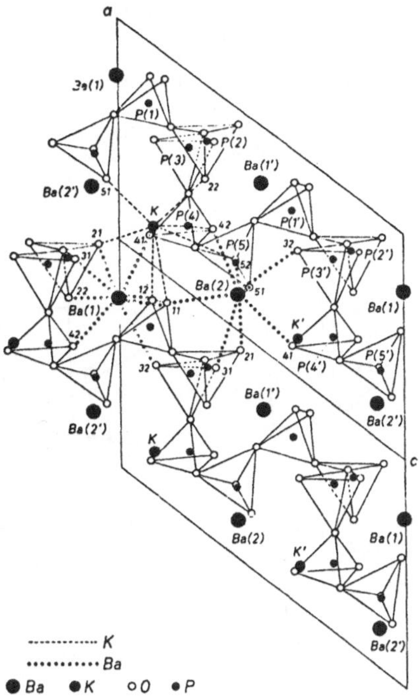

Fig. 1. Structure of barium potassium polyphosphate viewed along b.

BISMUTH POLYPHOSPHATE

BISMUTH HYDROGEN POLYPHOSPHATE

$Bi(PO_3)_3$
$BiH(PO_3)_4$

I. K. PALKINA and K.-H. JOST, 1975. Acta Cryst., B31, 2281-2285.

II. Idem, 1975. Ibid., B31, 2285-2290.

$Bi(PO_3)_3$
Monoclinic, $P2_1/a$, a = 13.732, b = 6.933, c = 7.152 Å, β = 93.35°, D_m = 4.38, Z = 4. Mo radiation, R = 0.046 for 1890 reflexions (twinned crystal).

$BiH(PO_3)_4$
Triclinic, $P\bar{1}$, a = 8.625, b = 8.866, c = 7.062 Å, α = 112.17, β = 108.54, γ = 98.49°, D_m = 3.86, Z = 2. Mo radiation, R = 0.040 for 2390 reflexions.

In $Bi(PO_3)_3$ the anion consists of infinite spiral chains of corner-sharing PO_4 tetrahedra, with six tetrahedra per turn (Fig. 1); P-O = 1.47-1.51 (terminal), 1.57-1.61(1) Å (bridging), P-O-P = 133-139°. Bi has 7-coordination (Fig. 1), the coordination polyhedra forming pairs by edge-sharing.

In $BiH(PO_3)_4$ the anion contains infinite chains of corner sharing PO_4 tetrahedra, with four tetrahedra per motif (Fig. 2). P-O = 1.46-1.51 and 1.55-1.62(1) Å; P(4)-O(12) = 1.54(1) Å suggests that this is the OH group. Bi has 7-coordination (Fig. 2).

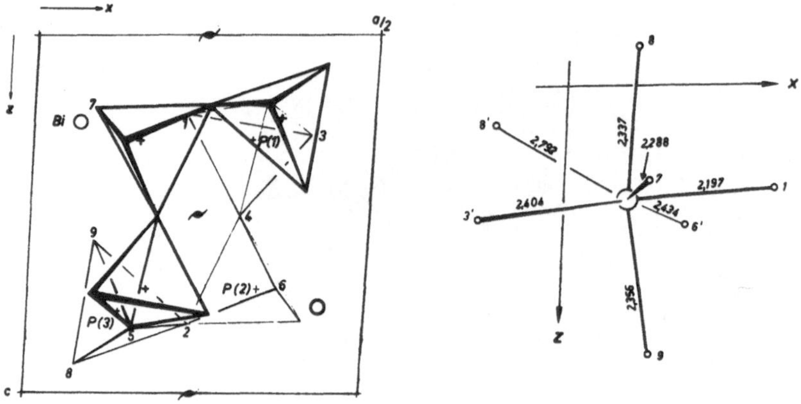

Fig. 1. The polyphosphate chain (left) and Bi coordination (right) in $Bi(PO_3)_3$.

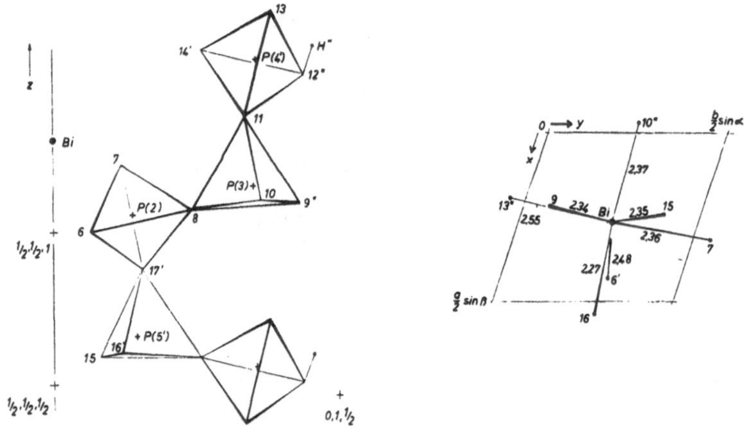

Fig. 2. The polyphosphate chain (left) and Bi coordination (right) in $BiH(PO_3)_4$.

BARIUM COPPER HEXAMETAPHOSPHATE

$Ba_2Cu(PO_3)_6$

M. LAÜGT and J.-C. GUITEL, 1975. Acta Cryst., B31, 1148-1153.

Monoclinic, $P2_1/a$, a = 21.382, b = 7.286, c = 9.520 Å, β = 97.96°, D_m not given, [Z = 4]. Mo radiation, R = 0.050 for 2507 reflexions.

The structure (Fig. 1) contains $(PO_3)_\infty$ chains along b, joined by the cations; P-O = 1.57-1.61 (bridging), 1.46-1.50 Å (terminal). Cu has very distorted octahedral coordination, Cu-O = 1.92-2.07, 2.23, 3.07 Å, and Ba ions have 8- and 9-coordination, Ba-O = 2.60 - 3.12 Å.

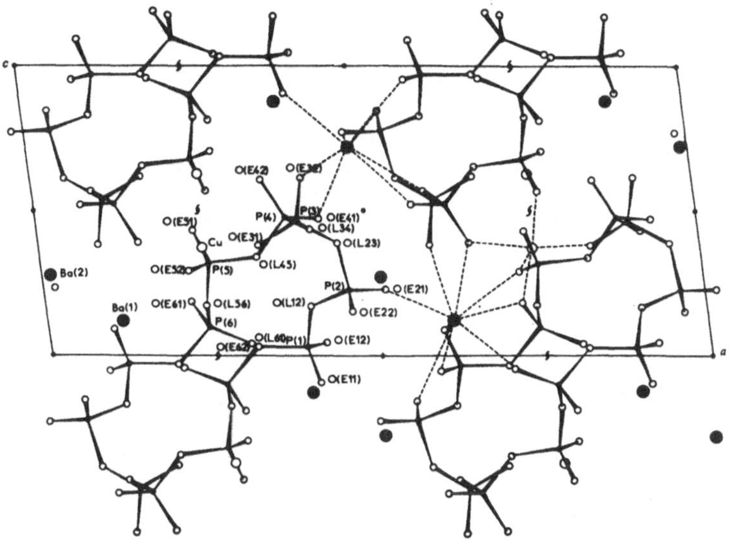

Fig. 1. Structure of $Ba_2Cu(PO_3)_6$.

CADMIUM POLYPHOSPHATE

$Cd(PO_3)_2$

M. BAGIEU-BEUCHER, J.-C. GUITEL, I. TORDJMAN and A. DURIF, 1974. Bull. Soc. Fr. Minér. Crist., 97, 481-484.

Orthorhombic, Pbca, a = 9.607, b = 13.70, c = 7.037 Å, Z = 8. Mo radiation, R = 0.035 for 1714 reflexions.

The structure is as previously described (1).

1. Structure Reports, 33A, 417.

BARIUM CADMIUM POLYPHOSPHATE

$BaCd(PO_3)_4$

I. M.T. AVERBUCH-POUCHOT, 1975. J. Appl. Cryst., 8, 389-390.

II. M.T. AVERBUCH-POUCHOT, A. DURIF and J.C. GUITEL, 1975. Acta Cryst., B31, 2453-2456.

Monoclinic, $P2_1/n$, a = 14.94, b = 9.192, c = 7.219 Å, β = 90.79°, D_m not given, Z = 4. Mo radiation, R = 0.033 for 2822 reflexions.

The structure (Fig. 1) contains infinite $(PO_3^-)_n$ chains along c, with a period of four tetrahedra; P-O(bridging) = 1.58-1.60, P-O(terminal) = 1.46-1.49 Å, P-O-P = 132-141°. Cd has octahedral coordination, Cd-O = 2.25-2.31 Å, and Ba has 9-coordination, Ba-O = 2.67-3.06 Å; none of bridging oxygen atoms participates in the metal coordination.

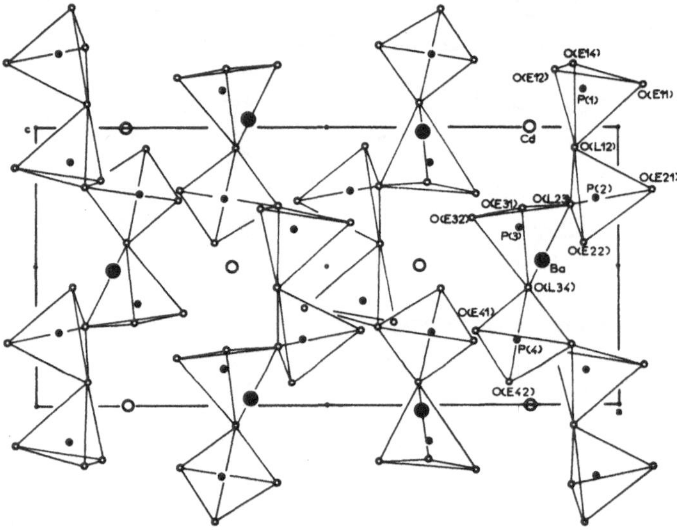

Fig. 1. Structure of $BaCd(PO_3)_4$.

LITHIUM NEODYMIUM METAPHOSPHATE

$LiNd(PO_3)_4$

H.Y.-P. HONG, 1975. Mater. Res. Bull., 10, 635-640.

Monoclinic, C2/c, a = 16.408, b = 7.035, c = 9.729 Å, β = 126.38°, D_m not given, Z = 4. Mo radiation, R = 0.072 for 590 reflexions.

The structure contains helical $(PO_3)_n$ chains along b of corner sharing PO_4 tetrahedra; the chains are linked by isolated NdO_8 dodecahedra. The Li position is uncertain, but was assumed to be 4(a): (0,0,0). P-O = 1.45-1.61, Nd-O = 2.38-2.56(1), Li-O (4-coordination, rectangular) = 1.96, 2.14 Å.

POTASSIUM NEODYMIUM METAPHOSPHATE

$KNd(PO_3)_4$

H.Y.-P. HONG, 1975. Mater. Res. Bull., 10, 1105-1110.

Monoclinic, $P2_1$, a = 7.266, b = 8.436, c = 8.007 Å, β = 91.97°, D_m not given, Z = 2. Mo radiation, R = 0.060 for 724 reflexions.

The structure contains helical $(PO_3)_n$ chains along a of corner-sharing PO_4 tetrahedra; the chains are linked by isolated NdO_8 dodecahedra and by irregularly coordinated K^+ ions. P-O = 1.38-1.60, Nd-O = 2.35-2.55, K-O = 2.77-3.03(2) Å.

POTASSIUM BERYLLIUM PHOSPHATE

$KBePO_4$

G. NITSCH and H. SCHÄFER, 1975. Z. anorg. Chem., 417, 11-18.

Orthorhombic, $Pna2_1$, a = 8.30, b = 8.47, c = 5.00 Å, D_m = 2.70, Z = 4. R = 0.11 for 224 reflexions.

Isostructural with NH_4LiSO_4 (1), with Be in the Li position. In the PO_4 tetrahedron, P-O = 1.45-1.57 Å, O-P-O = 94-115°. Be has tetrahedral coordination, Be-O = 1.46-1.96 Å, O-Be-O = 90-140°, and K has eight oxygen neighbours at 2.71-3.61 Å. The Rb compound is isostructural, and the Na compound is identical with natural beryllonite (2).

1. Structure Reports, 34A, 305.
2. Ibid., 26, 456; 27, 580; 39A, 283.

HURLBUTITE

$CaBe_2P_2O_8$

V.V. BAKAKIN, G.M. RYLOV and V.I. ALEKSEEV, 1974. Kristallografija, 19, 1283-1285 [Soviet Physics - Crystallography, 19, 798-799].

Monoclinic, $P2_1/a$, a = 8.306, b = 8.790, c = 7.804 Å, β = 89.48°, D_m = 2.88, Z = 2. Mo radiation, R = 0.071 for 960 reflexions.

The structure is as previously described (1).

1. Structure Reports, 24, 406; 40A, 238.

ROSCHERITE

$Ca(Mg,Fe)_2AlBe_2(PO_4)_3(OH)_3.2H_2O$

L. FANFANI, A. NUNZI, P.F. ZANAZZI and A.R. ZANZARI, 1975. Miner. (Tschermaks) Petrogr. Mitt., 22, 266-277.

Monoclinic, C2/c, a = 15.874, b = 11.854, c = 6.605 Å, β = 95.34°, D_m = 2.77, Z = 4. Mo radiation, R = 0.073 for 2129 reflexions.

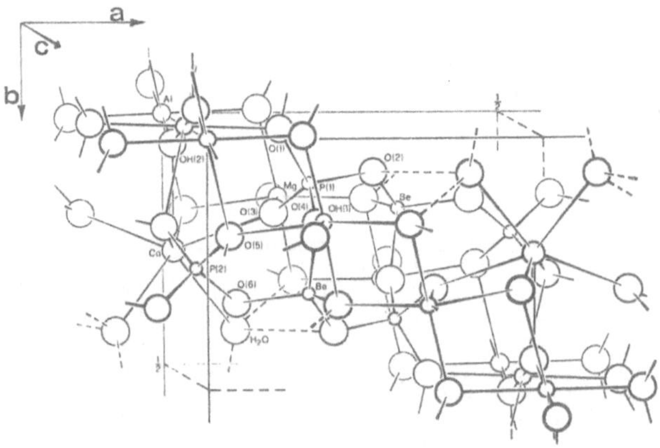

Fig. 1. Structure of roscherite.

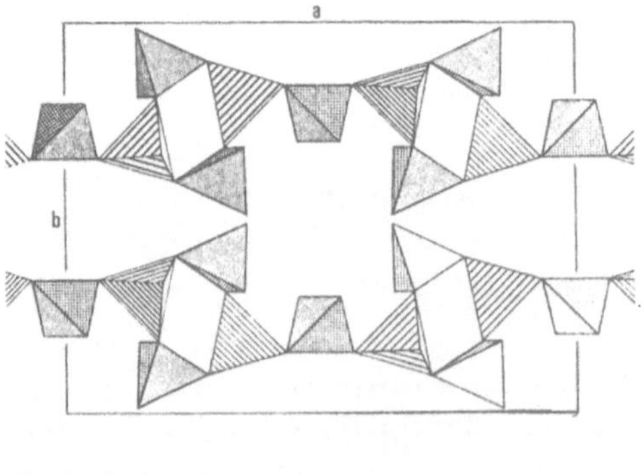

Fig. 2. Tetrahedral chains in roscherite.

The structure (Fig. 1) contains rings of four tetrahedra (two PO_4 and two $BeO_3(OH)$) alternating with single PO_4 tetrahedra to form chains along [101] (Fig. 2). (Mg,Fe) has octahedral coordination, M-O = 2.047-2.150(3) Å, and Ca has sevenfold distorted pentagonal bipyramidal coordination, Ca-O = 2.415-2.573 Å. The Al site at (0,0,0) is only 2/3-occupied, with highly distorted octahedral coordination, Al-O = 1.727-2.529 Å. P-O = 1.524-1.549(3), Be-O = 1.615-1.655(6) Å.

COLLINSITE

$MgCa_2(PO_4)_2.2H_2O$

P.D. BROTHERTON, E.N. MASLEN, M.W. PRYCE and A.H. WHITE, 1974. Austral. J. Chem., 27, 653-656.

Triclinic, $P\bar{1}$, a = 5.734, b = 6.780, c = 5.441 Å, α = 97.29, β = 108.56, γ = 107.28°, Z = 1. Cu radiation, R = 0.033 for 379 reflexions.

The structure contains PO_4 tetrahedra, $MgO_4(H_2O)_2$ distorted octahedra, and Ca with irregular eight-coordination. P-O = 1.53-1.56, Mg-O = 2.11, 2.14, Mg-OH_2 = 2.00, Ca-O = 2.38-2.70(1) Å.

WHITLOCKITE

$Ca_{18}(Mg,Fe)_2H_2(PO_4)_{14}$

C. CALVO and R. GOPAL, 1975. Amer. Min., 60, 120-133.

Rhombohedral, R3c, a = 13.729 Å, α = 44.20° (hexagonal cell has a = 10.330, c = 37.103 Å), D_m = 3.12, Z = 1. Mo radiation, R = 0.048 for 1857 reflexions.

The structure is as previously described for magnesium whitlockite (1), with random distribution of Mg and Fe in octahedral sites.

1. Structure Reports, 40A, 239.

CALCIUM DIHYDROGEN PHOSPHATE MONOHYDRATE

$Ca(H_2PO_4)_2.H_2O$

L.W. SCHROEDER, E. PRINCE and B. DICKENS, 1975. Acta Cryst., B31, 9-12.

Triclinic, $P\bar{1}$, a = 5.6261, b = 11.889, c = 6.473 Å, α = 98.633, β = 118.2, γ = 83.344°, Z = 2 (1). Neutron diffraction data, R = 0.055 for 1045 reflexions.

The structure is as previously described (1, 2), except that the previous description of one of the hydrogen bonds as bifurcated is considered to be inappropriate, since the H...O distances are unequal, 2.10 and 2.32 Å.

1. Structure Reports, 37A, 294.
2. Ibid., 20, 308; 26, 458.

BARIUM SODIUM PHOSPHATE

$BaNaPO_4$

C. CALVO and R. FAGGIANI, 1975. Canad. J. Chem., 53, 1849-1853.

Trigonal, $P\bar{3}m1$, a = 5.622, c = 7.259 Å, D_m = 4.17, Z = 2. Mo radiation, R = 0.042 for 218 reflexions.

Glaserite-type structure (1), with Na in 1(a), Ba in 1(b), and (Na+Ba) in 2(d), with 6-, 12-, and 10-coordination, respectively; Na-O = 2.35, Ba-O = 2.79 and 3.25, (Na+Ba)-O = 2.55-3.02(1) Å. Mean P-O = 1.54 Å.

1. Strukturbericht, 1, 378.

FOGGITE

$CaAl(OH)_2PO_4.H_2O$

P.B. MOORE, A.R. KAMPF and T. ARAKI, 1975. Amer. Min., 60, 965-971.

Orthorhombic, $A2_122$, a = 9.270, b = 21.324, c = 5.190 Å, Z = 8. Mo radiation, R = 0.059 for 888 reflexions.

The formula may be written $Ca(H_2O)_2[CaAl_2(OH)_4(PO_4)_2]$, and the structure (Fig. 1) is related to that of calcium Tschermak's pyroxene (1). A feature is a zigzag chain of edge-sharing Al coordination octahedra, linked into a sheet by phosphate tetrahedra and $Ca(1)O_8$ polyhedra; equivalent sheets are weakly linked by $Ca(2)O_4(H_2O)_4$ polyhedra, the water molecules being disordered. Average distances are Ca-O = 2.43, Al-O = 1.90, P-O = 1.54 Å.

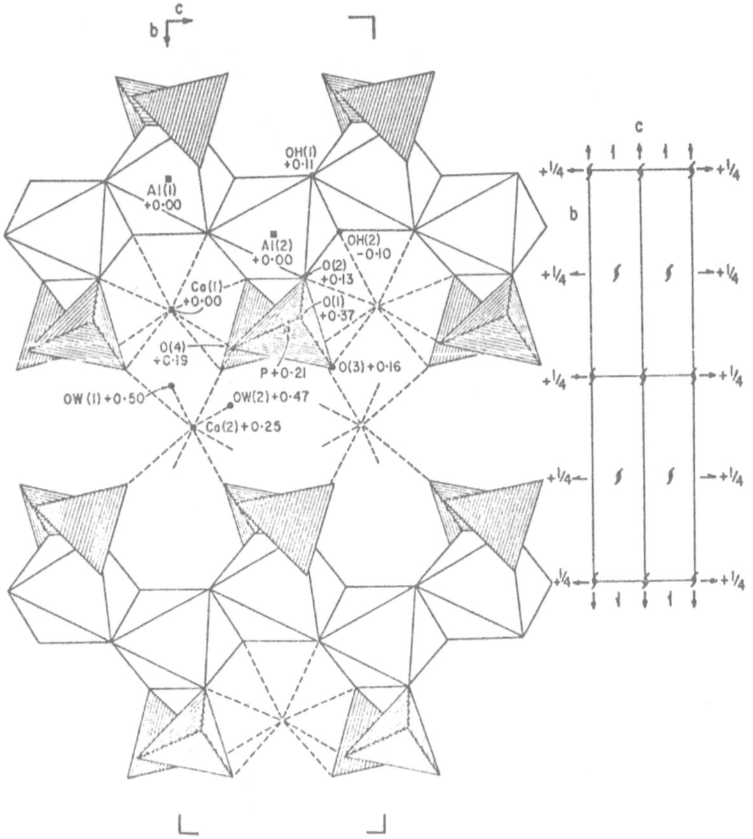

Fig. 1. Structure of foggite.

1. Structure Reports, 40A, 288.

PALERMOITE

$SrLi_2[Al_4(OH)_4(PO_4)_4]$

P.B. MOORE and T. ARAKI, 1975. Amer. Min., 60, 460-465.

Orthorhombic, Imcb, a = 11.556, b = 15.847, c = 7.315 Å, Z = 4. Mo radiation, R = 0.090 for 1471 reflexions.

The structure (Fig. 1) is related to that of carminite, $Pb_2[Fe_4(OH)_4(AsO_4)_4]$ (1). Both contain the same $[M_4(OH)_4(XO_4)_4]^{4-}$ octahedral and tetrahedral slabs parallel to (010), but the slabs are linked differently across the glide planes at y = 1/4. The octahedra form chains of edge-linked dimers which are corner-linked to symmetry-equivalent dimers; one-eight of the tetrahedral oxygens are not bonded to the octahedra. Al-O = 1.82-2.09, P-O = 1.51-1.57, Sr-O = 2.59, 2.65 (8-fold cubic coordination), Li-O = 1.94-2.29 Å (tetrahedral, with a fifth oxygen at 2.46 Å).

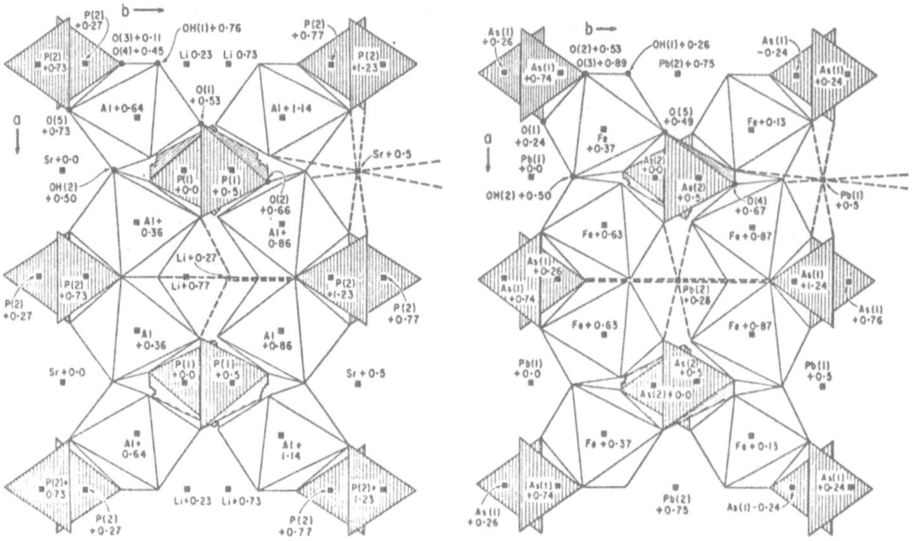

Fig. 1. Structures of palermoite (left) and carminite (right).

1. Structure Reports, 27, 602; 28, 197.

LEAD(II) PHOSPHATE

$Pb_3P_2O_8$

H.N. NG and C. CALVO, 1975. Canad. J. Phys., 53, 42-51.

α-$Pb_3P_2O_8$ (below 180°C)
Monoclinic, C2/c, a = 13.81, b = 5.71, c = 9.31 Å, β = 102.4°, at 25°C, Z = 4. Mo radiation, R = 0.072, 0.087, 0.130 for 874, 317, 336 reflexions at 25, 100, 150°C, respectively.

β-$Pb_3P_2O_8$ (above 180°C)
Rhombohedral, $R\bar{3}m$, a = 5.56, c = 20.39 Å, at 200°C, Z = 3. Mo radiation, R = 0.096 for 162 reflexions. Pb(1) in 3(a): Pb(2), P, O(4) in 6(c): z = 0.2126, 0.4021, 0.329; O(1) in 18(h): x = 0.181, z = 0.096.

The structure of the low-temperature α-form is as previously described (1), and the high-temperature β-form is isostructural with $Ba_3V_2O_8$ (2). The transition occurs via an F2/m form.

1. Structure Reports, 35A, 330.
2. Ibid., 35A, 363.

DITELLURIUM(IV) TRIOXIDE HYDROGENPHOSPHATE

$Te_2O_3(HPO_4)$

H. MAYER, 1975. Z. Kristallogr., 141, 354-362.

Orthorhombic, $Pca2_1$, a = 10.232, b = 7.012, c = 7.928 Å, D_m = 4.59, Z = 4. Cu radiation, R = 0.10 for 513 reflexions (films, densitometer intensities).

The structure (Fig. 1) contains puckered Te-O layers, in which Te has trigonal bipyramidal coordination with one equatorial site occupied by the lone electron-pair; Te-O((eq) = 1.85 and 1.89, Te-O(axial) = 2.06 and 2.07 Å. The HPO_4 groups are between the layers, bound strongly to two Te atoms of one layer and weakly to one Te atom of the adjacent layer; P-O = 1.48-1.54 Å, O-P-O = 102-115°.

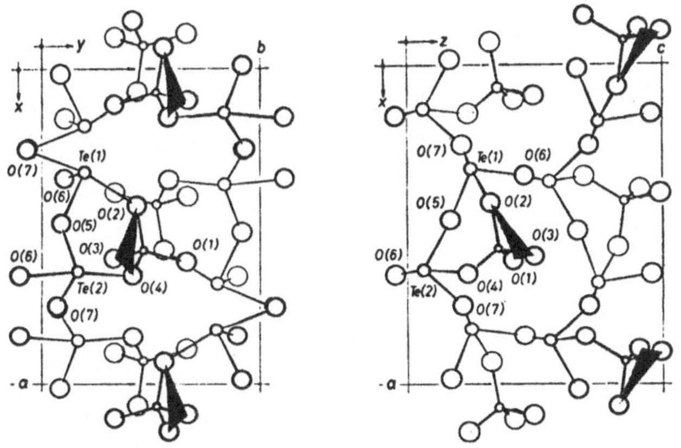

Fig. 1. Structure of $Te_2O_3(HPO_4)$, viewed along c and b.

DIZIRCONIUM DIORTHOPHOSPHATE

$Zr_2O(PO_4)_2$

W. GEBERT and E. TILLMANNS, 1975. Acta Cryst., B31, 1768-1770.

Orthorhombic, Cmca, a = 6.624, b = 8.637, c = 11.872 Å, D_m not given, Z = 4. Mo radiation, R = 0.013 for 724 reflexions.

ZrO_7 distorted pentagonal bipyramids and PO_4 tetrahedra share edges to form infinite chains along c of composition $[ZrO_3PO_4]^{5-}$ (Fig. 1). The chains form a three-dimensional structure by sharing corners. Zr-O = 1.927-2.287(1), P-O = 1.500-1.558(1) Å.

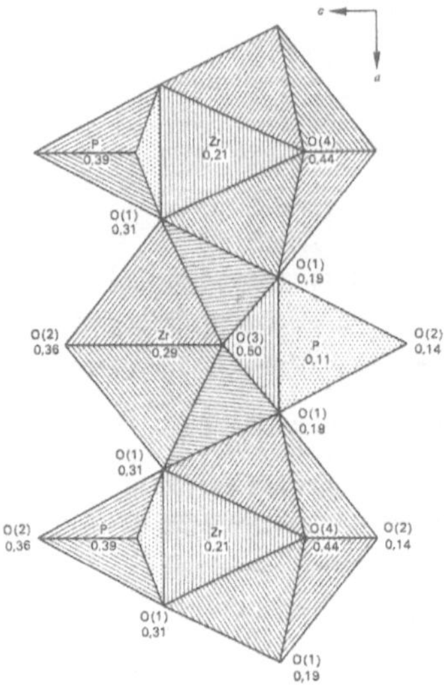

Fig. 1. Chains in $Zr_2O(PO_4)_2$ structure.

ZINC VANADATE PHOSPHATE

$Zn_3(V_{0.5}P_{1.5})O_8$

K.L. IDLER and C. CALVO, 1975. Canad. J. Chem., 53, 3665-3668.

Monoclinic, C2/c, a = 15.941, b = 5.314, c = 8.265 Å, β = 106.96°, D_m not measured, Z = 4. Mo radiation, R = 0.073 for 910 reflexions.

Atomic positions

Zn(1) in 4(e), other atoms in 8(f); X = (0.25 V + 0.75 P)

	x	y	z
Zn(1)	0	0.0587	3/4
Zn(2)	0.18558	0.7929	0.7137
X	0.1209	0.292	0.5294
O(1)	0.6021	0.672	0.3456
O(2)	0.6491	0.931	0.0253
O(3)	0.4601	0.747	0.9089
O(4)	0.2964	0.638	0.8495

The structure (Fig. 1) contains sheets parallel to (100) of corner sharing Zn(1)O_4 and (V,P)O_4 tetrahedra; the Zn(2)O_4 tetrahedra corner share to form chains parallel to b. Zn-O = 1.90-2.01(1), (V,P)-O = 1.54-1.63(1) Å, O-Zn-O = 101-117, O-(V,P)-O = 106-116°.

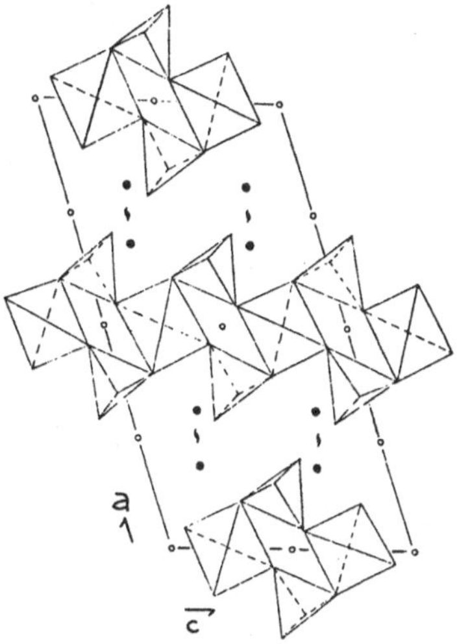

Fig. 1. Structure of $Zn_3(V,P)_2O_8$; Zn(1)O_4 and (V,P)O_4 tetrahedra are outlined and Zn(2) ions are indicated by filled circles.

MANGANESE CHLORAPATITE

$Mn_5(PO_4)_3Cl_{0.9}(OH)_{0.1}$

G. ENGEL, J. PRETZSCH, V. GRAMLICH and W.H. BAUR, 1975. Acta Cryst., B31, 1854-1860.

Hexagonal, $P6_3/m$, a = 9.532, c = 6.199 Å, D_m not given, Z = 2. Mo radiation, R = 0.054 for 2107 reflexions.

Fluorapatite structure (1), with (Cl,OH) at (0,0,1/4).

1. Strukturbericht, 2, 99; Structure Reports, 38A, 308.

IRON(III) PHOSPHATE

$FePO_4$ (room-temperature form)

HOK NAM NG and C. CALVO, 1975. Canad. J. Chem., 53, 2064-2067.

Trigonal, $P3_121$, a = 5.036, c = 11.255 Å, D_m not given, Z = 3. Mo radiation, R = 0.078 for 487 reflexions (twinned crystal).

Isostructural with berlinite, $AlPO_4$ (1). The structure is related to that of α-quartz, but with a doubled c-axis because of ordering of Fe and P, both of which have tetrahedral coordination, Fe-O = 1.853, P-O = 1.526(7) Å.

1. Strukturbericht, 3, 423; Structure Reports, 19, 359; 29, 367; 31A, 183.

IRON(II) PHOSPHATE MONOHYDRATE

$Fe_3(PO_4)_2.H_2O$

P.B. MOORE and T. ARAKI, 1975. Amer. Min., 60, 454-459.

Monoclinic, $P2_1/a$, a = 9.431, b = 10.066, c = 8.040 Å, β = 117.63°, D_m not given, Z = 4. Mo radiation, R = 0.068 for 2457 reflexions.

The structure (Fig. 1) consists of a complex open framework of edge- and corner-linked distorted octahedra (Fe(1) and Fe(3)), five-coordinate polyhedra (Fe(2)), and PO_4 tetrahedra. Fe-O = 2.07-2.38 (octahedra), 1.99-2.24 (five-coordinate), P-O = 1.53-1.56 Å. The five-coordinate polyhedron is between a trigonal bipyramid and square pyramid.

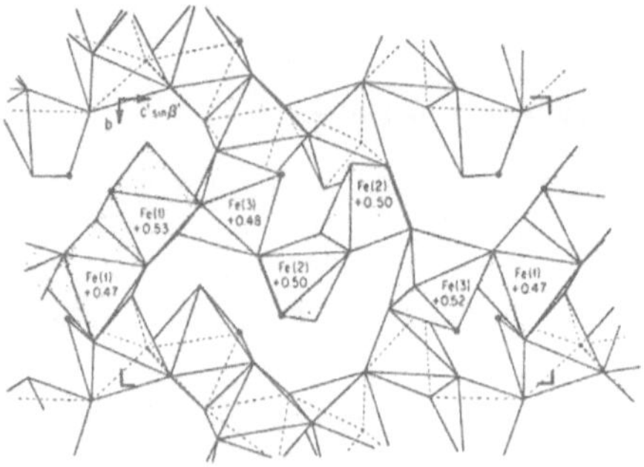

Fig. 1. Fe-O framework in $Fe_3(PO_4)_2.H_2O$.

COBALT(II) PHOSPHATE

$Co_3(PO_4)_2$

J.B. ANDERSON, E. KOSTINER, M.C. MILLER and J.R. REA, 1975. J. Solid State Chem., 14, 372-377.

Monoclinic, $P2_1/c$, a = 5.063, b = 8.361, c = 8.788 Å, β = 121.00°, D_m not given, Z = 2. Mo radiation, R = 0.039 for 1032 reflexions.

Isostructural with α-$Zn_3(PO_4)_2$ (1) and $Mg_3(PO_4)_2$ (2), as previously reported (3). Co ions have five- and six-coordination.

1. Structure Reports, 28, 189.
2. Ibid., 33A, 394.
3. Ibid., 40A, 249.

NICKEL ORTHOPHOSPHATE

$Ni_3P_2O_8$

C. CALVO and R. FAGGIANI, 1975. Canad. J. Chem., 53, 1516-1520.

Monoclinic, $P2_1/c$, a = 5.830, b = 4.700, c = 10.107 Å, β = 91.22°, D_m = 4.38, Z = 2. Mo radiation, R = 0.035 for 986 reflexions.

Isostructural with sarcopside (1). Ni ions have octahedral coordination, Ni-O = 2.00-2.19 Å; in the PO_4 tetrahedra, P-O = 1.52-1.60 Å, O-P-O = 102-113°.

1. Structure Reports, 38A, 313.

HOPEITE

$Zn_3(PO_4)_2.4H_2O$

A. WHITAKER, 1975. Acta Cryst., B31, 2026-2035.

Orthorhombic, Pnma, a = 10.629, b = 18.339, c = 5.040 Å, D_m not given, Z = 4. Mo radiation, R = 0.068 for 1467 reflexions.

The structure is different from two previous descriptions based on projection data (1). It contains edge- and corner-sharing $ZnO_2(H_2O)_4$ octahedra, ZnO_4 tetrahedra, and PO_4 tetrahedra (Fig. 1), none of which are regular; Zn-O = 2.05-2.16 (octahedral), 1.91-2.00 (tetrahedral), P-O = 1.52-1.56(1) Å. The crystal also contains a system of O-H...O hydrogen bonds.

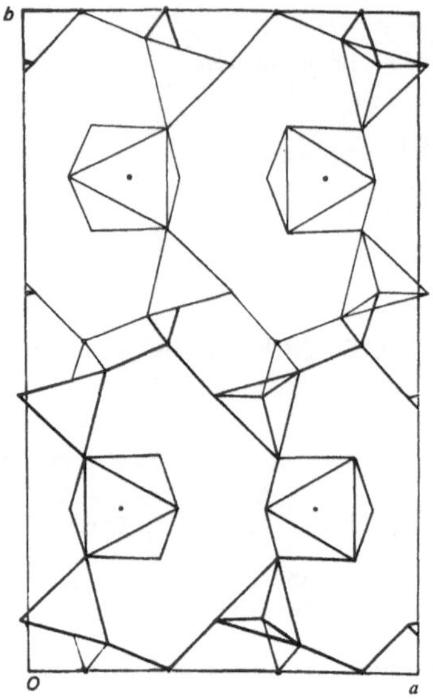

Fig. 1. Structure of hopeite; small circles are Zn atoms, and tetrahedra are PO_4 groups.

1. Structure Reports, 26, 462; 27, 586; 28, 190.

POTASSIUM ZINC HYDROGEN PHOSPHATE HYDRATE

$KZn_2H(PO_4)_2.2 \cdot 5H_2O$

I. TORDJMAN, A. DURIF, M.T. AVERBUCH-POUCHOT and J.C. GUITEL, 1975. Acta Cryst., B31, 1143-1148.

Triclinic, $P\bar{1}$, a = 9.109, b = 13.543, c = 8.814 Å, α = 102.21, β = 113.35, γ = 95.92°, D_m not given, Z = 4. Mo radiation, R = 0.06 for 3034 reflexions.

The structure (Fig. 1) contains ZnO_6 octahedra, ZnO_4 tetrahedra, and PO_4 tetrahedra; K ions have 8- and 9-coordination. Zn-O = 2.02-2.19 (octahedra), 1.91-1.96 (tetrahedra), P-O = 1.49-1.58, K-O = 2.74-3.26 Å.

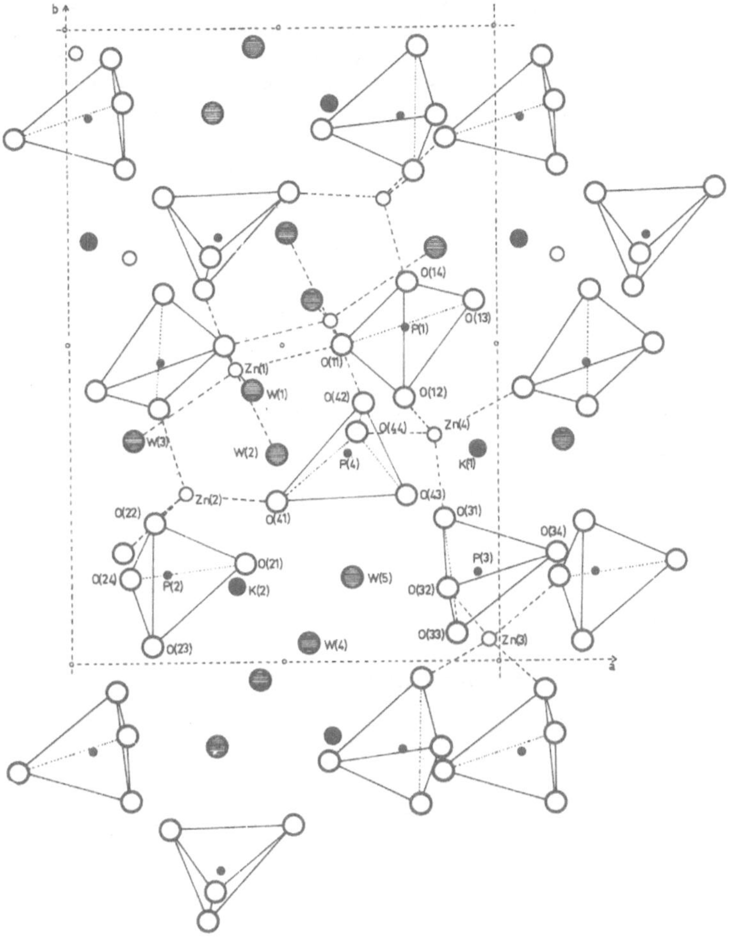

Fig. 1. Structure of $KZn_2H(PO_4)_2.2{\cdot}5H_2O$.

SCHOLZITE

$CaZn_2(PO_4)_2.2H_2O$

K. TAXER, 1975. Amer. Min., 60, 1019-1022.

Orthorhombic, $Pbc2_1$ [compare 1], a = 17.149, b = 22.236, c = 6.667 Å, Z = 12. Cu radiation, R = 0.081 for 1924 reflexions.

Scholzite has a substructure, Pbcn, b' = b/3, and the actual structure has small shifts (mainly parallel to a) of the atomic positions in three adjacent unit cells. The structure (Fig. 1) contains chains of ZnO_4 tetrahedra along c, linked by isolated PO_4 tetrahedra. Ca ions have distorted octahedral coordination, and the water molecules, in groups of four, occupy cages surrounded by two CaO_6 octahedra and four PO_4 tetrahedra.

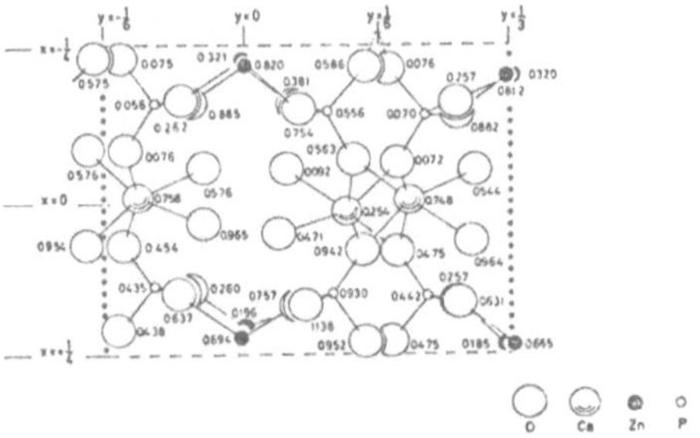

Fig. 1. Structure of scholzite.

1. Structure Reports, 35A, 333.

AMMONIUM CADMIUM PHOSPHATE

$NH_4Cd(PO_3OH)(OH)$

Ju.A. IVANOV, Ju.K. EGOROV-TISMENKO, M.A. SIMONOV and N.V. BELOV, 1974. Kristallografija, 19, 1076-1077 [Soviet Physics - Crystallography, 19, 665-666].

Orthorhombic, Pnma, a = 17.21, b = 5.887, c = 5.111 Å, D_m = 4.6, Z = 4. Mo radiation, R = 0.10 for 1570 reflexions.

Atomic positions

	x	y	z
Cd	0.2638	1/4	0.1152
P	0.144	1/4	0.681
O(1)	0.161	1/4	0.390
O(2)	0.181	0.042	0.820
OH(3)	0.056	1/4	0.722
OH(4)	0.362	1/4	0.792
NH_4	0.486	1/4	0.292

The structure (Fig. 1) contains layers of corner-sharing Cd octahedra, with phosphate tetrahedra (which share an edge and a corner with octahedra) above and below. The fourth corners of the tetrahedra are directed towards interlayer NH_4^+ ions. P-O = 1.54, Cd-O = 2.23-2.41, N-O = 2.76-3.16 Å.

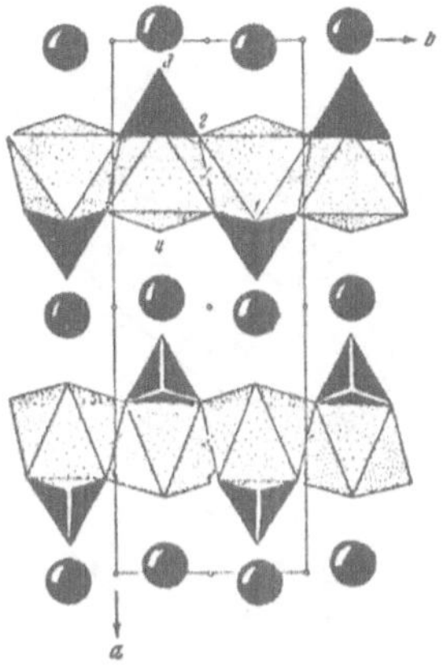

Fig. 1. Structure of $NH_4Cd(PO_3OH)(OH)$; black spheres are NH_4^+ ions.

MERCURY(I) DIHYDROGENPHOSPHATE

$Hg_2(H_2PO_4)_2$

B.O. NILSSON, 1975. Z. Kristallogr., 141, 321-329.

Monoclinic, $P2_1/n$, a = 6.075, b = 14.503, c = 4.728 Å, β = 92.17°, D_m = 4.7, Z = 2. Mo radiation, R = 0.059 for 996 reflexions.

Atomic positions

Atoms in 4(e); (x, y, z; 1/2 + x, 1/2 - y, 1/2 + z)

	x	y	z
Hg	0.30421	0.01696	0.05308
P	-0.0114	0.1523	0.4050
O(1)	0.001	0.0677	0.210
O(2)	0.188	0.1442	0.624
O(3)	0.771	0.1379	0.567
O(4)	-0.008	0.2450	0.263

The structure (Fig. 1) contains mercury atoms linked in pairs across centres of symmetry, Hg-Hg = 2.499(1) Å, with each Hg atom also bonded to an oxygen atom, Hg-O = 2.14(1) Å, Hg-Hg-O = 167.2(4)°; $Hg_2(H_2PO_4)_2$ units are thus formed. These are joined by a weaker Hg-O bond, 2.51 Å, and by hydrogen bonds.

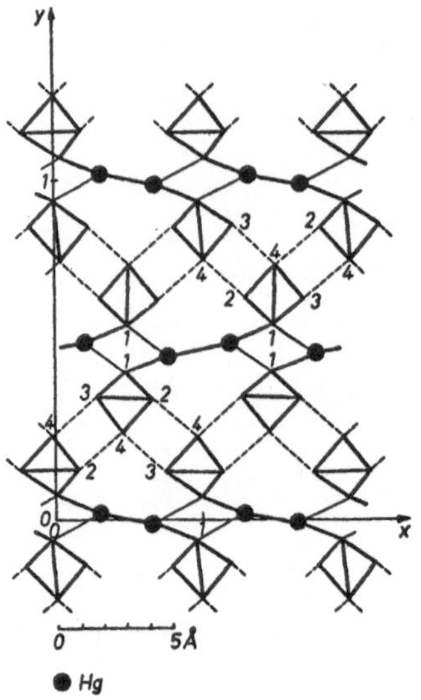

Fig. 1. Structure of $Hg_2(H_2PO_4)_2$; thin lines are weak Hg-O interactions, and broken lines are possible hydrogen bonds.

MERCURY(II) PHOSPHATE

$Hg_3(PO_4)_2$

K. AURIVILLIUS and B.O. NILSSON, 1975. Z. Kristallogr., 141, 1-10.

Monoclinic, $P2_1/c$, a = 9.737, b = 11.466, c = 6.406 Å, β = 99.51°, D_m = 7.32, Z = 4. Mo radiation, R = 0.043 for 1491 reflexions.

Hg atoms have nearly-linear coordination to two oxygen atoms of two PO_4 tetrahedra; Hg-O = 2.06-2.13(1) Å, O-Hg-O = 163-170°, P-O = 1.53-1.56(1) Å, O-P-O = 103-114°. This coordination results in puckered nets; the shortest Hg...O distance between the nets is 2.42 Å.

CALCIUM PYROPHOSPHATE DIHYDRATE

$Ca_2P_2O_7.2H_2O$

N.S. MANDEL, 1975. Acta Cryst., B31, 1730-1734.

Triclinic, $P\bar{1}$, a = 7.365, b = 8.287, c = 6.691 Å, α = 102.96, β = 72.73, γ = 95.01°, D_m = 2.55, Z = 2. Mo radiation, R = 0.037 for 4685 reflexions.

The pyrophosphate anion is about 20° from an eclipsed conformation; P-O (bridging) = 1.623, P-O(terminal) = 1.499-1.538(1) Å (the variation being related to oxygen environment), P-O-P = 123.1°. Both Ca ions have seven-coordination (Fig. 1), Ca-O = 2.29-2.67 Å, and the water molecules are involved in hydrogen bonding, O-H...O = 2.76-2.89 Å.

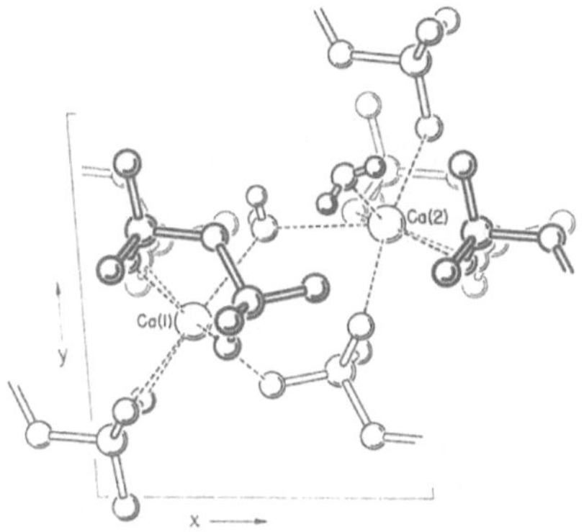

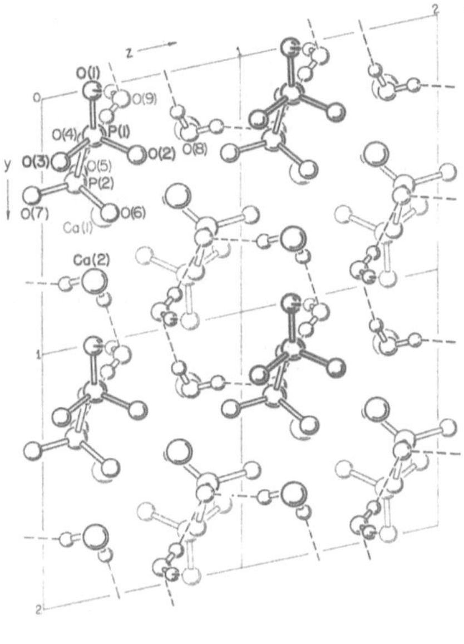

(a) (b)

Fig. 1. (a) Ca coordination, and (b) hydrogen bonding in calcium pyrophosphate dihydrate.

ZIRCONIUM(IV) PYROPHOSPHATE

TIN(IV) PYROPHOSPHATE

ZrP_2O_7
SnP_2O_7

C.-S. HUANG, O. KNOP, D.A. OTHEN, F.W.D. WOODHAMS and R.A. HOWIE, 1975. Canad. J. Chem., 53, 79-91.

Refinements in a Pa3 subcell give parameters which are not very accurate but are in agreement with previous work (1, 2, 3).

1. Strukturbericht, 3, 140, 503.
2. L.-O. HAGMAN and P. KIERKEGAARD, 1969. Acta Chem. Scand., 23, 327.
3. Structure Reports, 37A, 299.

AMMONIUM CALCIUM TRIMETAPHOSPHATE

$NH_4CaP_3O_9$

R. MASSE, A. DURIF and J.C. GUITEL, 1975. Z. Kristallogr., 141, 113-125.

Monoclinic, $P2_1/n$, a = 7.446, b = 12.461, c = 10.050 Å, β = 90.11°, D_m not given, Z = 4. Mo radiation, R = 0.05 for 1000 reflexions.

The structure (Fig. 1) is a distortion of the orthorhombic $NH_4MgP_3O_9$ structure (1). The six-membered P_3O_3 ring is a boat, as in the Mg compound, P-O = 1.61-1.62 (ring), 1.45-1.49(1) Å (terminal), P-O-P = 124-137°. Ca has octahedral coordination, Ca-O = 2.28-2.37 Å, and N has eight oxygen neighbours at 2.97-3.38 Å.

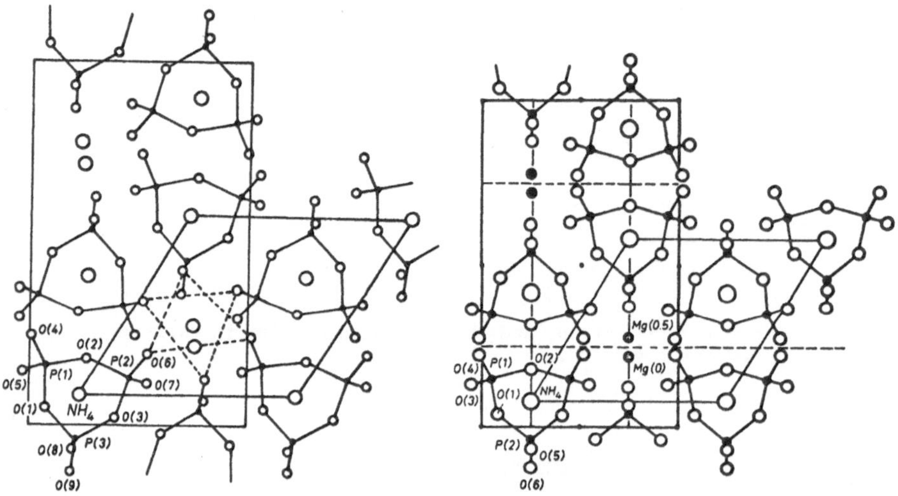

Fig. 1. Structures of $NH_4CaP_3O_9$ (left) and $NH_4MgP_3O_9$ (right) viewed along c; the benitoite pseudo-cells are shown.

1. Structure Reports, 33A, 416.

STRONTIUM SODIUM TRIMETAPHOSPHATE TRIHYDRATE

$SrNaP_3O_9.3H_2O$

R. ZILBER, I. TORDJMAN, A. DURIF and J.-C. GUITEL, 1974. Z. Kristallogr., 140, 350-359.

Orthorhombic, Pnma, a = 16.167, b = 12.013, c = 10.651 Å, Z = 8. Mo radiation, R = 0.053 for 796 reflexions.

The six-membered P_3O_3 ring (Fig. 1) is nearly planar, P-O = 1.60-1.61 (ring), 1.46-1.50(1) Å (terminal), P-O-P = 127°. Sr ions have 8- and 9-coordination, Sr-O = 2.54-2.75 Å, and Na ions have 6- and 7-coordination, Na-O = 2.38-2.70 Å.

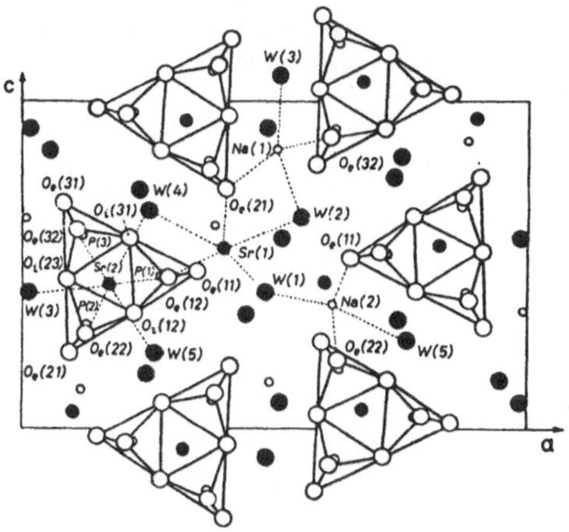

Fig. 1. Structure of $SrNaP_3O_9.3H_2O$.

AMMONIUM BARIUM TRIMETAPHOSPHATE MONOHYDRATE

$NH_4BaP_3O_9.H_2O$

A. DURIF, C. MARTIN and G. BASSI, 1975. Bull. Soc. Fr. Minér. Crist., 98, 19-24.

Monoclinic, $P2_1/n$, a = 11.70, b = 12.12, c = 7.559 Å, β = 101.05°, Z = 4. Cu radiation, R = 0.058 for 2150 reflexions.

The structure contains $P_3O_9{}^{3-}$ anions which have six-membered rings, P-O = 1.60 (ring), 1.47(1) Å (terminal). Ba and NH_4 ions are each coordinated to one water molecule and seven phosphate oxygen atoms, Ba-O = 2.73-3.39, N-O = 2.83-3.28 Å.

SODIUM GALLIUM TRIMETAPHOSPHIMATE HYDRATE

$Na_3Ga[(PO_2NH)_3]_2.12H_2O$

V.I. SOKOL, M.A. PORAJ-KOŠIC, V.R. BERDNIKOV, I.A. ROZANOV and L.A. BUTMAN, 1975. Koordin. Khim., 1, 429-434.

Triclinic, $P\bar{1}$, a = 8.729, b = 9.902, c = 8.716 Å, α = 97.84, β = 87.88, γ = 93.45°, Z = 2. Mo radiation, R = 0.10 for 2500 reflexions.

The structure contains complex $Ga[(PO_2NH)_3]_2^{3-}$ anions, in which Ga is coordinated octahedrally to three oxygen of each of the two trimetaphosphimate groups (Fig. 1), Ga-O = 2.01-2.04(1), P-N = 1.70-1.75(1), P-O = 1.50-1.55(1) Å, N-P-N = 103-106, P-N-P = 119-120°. Each Na ion has octahedral coordination, one at a centre of symmetry to six water molecules, and one in a general position to five water molecules and one anionic oxygen atom, Na-O = 2.35-2.66 Å; the octahedra share edges. The structure contains an extensive system of hydrogen bonds, N-H...O = 2.91-3.28, O-H...O = 2.85-3.22 Å.

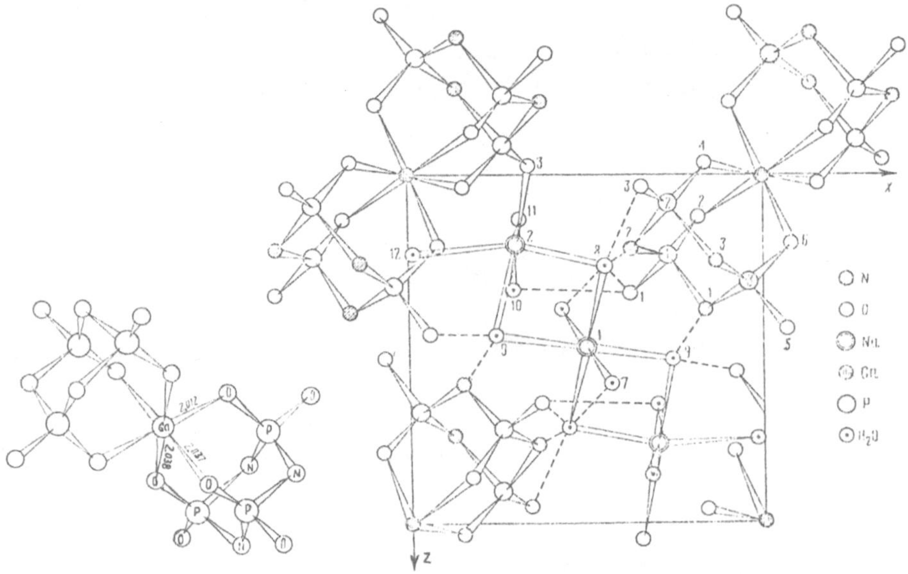

Fig. 1. The $Ga[(PO_2NH)_3]_2^{3-}$ anion (left) and the structure of $Na_3Ga[(PO_2NH)_3]_2.$-$12H_2O$ (right).

SILVER TRIMETAPHOSPHATE MONOHYDRATE

$Ag_3P_3O_9.H_2O$

M. BAGIEU-BEUCHER, A. DURIF and J.C. GUITEL, 1975. Acta Cryst., B31, 2264-2267.

Triclinic, $P\bar{1}$, a = 7.800, b = 7.796, c = 9.276 Å, α = 115.15, β = 115.15, γ = 88.93°, D_m not given, Z = 2. Mo radiation, R = 0.048 for 1874 reflexions.

The six-membered P_3O_3 ring of the cyclic $P_3O_9{}^{3-}$ anion has a chair conformation; P-O(ring) = 1.61(1), P-O(terminal) = 1.49(1) Å, P-O-P = 127°. The three Ag ions have very distorted tetrahedral coordination to terminal oxygen atoms, with the water molecule completing five-coordination for two of the Ag ions; Ag-O = 2.33-2.65 Å.

ZINC TRIPOLYPHOSPHATE HEPTADECAHYDRATE

$Zn_5(P_3O_{10})_2.17H_2O$

I. M.T. AVERBUCH-POUCHOT and A. DURIF, 1975. J. Appl. Cryst., 8, 564.

II. M.T. AVERBUCH-POUCHOT, A. DURIF and J.C. GUITEL, 1975. Acta Cryst., B31, 2482-2486.

Triclinic, $P\bar{1}$, a = 10.766, b = 10.316, c = 8.525 Å, α = 111.39, β = 115.08, γ = 70.19°, D_m not given, Z = 1. Mo radiation, R = 0.04 for 3558 reflexions.

The $P_3O_{10}{}^{5-}$ anion contains a chain of three corner-sharing tetrahedra (Fig. 1); P-O(bridging) = 1.59-1.63, P-O(terminal) = 1.49-1.52 Å, P-O-P = 128, 131°. Zn ions have tetrahedral and octahedral coordinations (Fig. 1); Zn-O = 1.93-1.96 (tetrahedral), 2.06-2.18 Å (octahedral). Four of the water molecules are not bonded to Zn, and W(8), which occupies a centre of symmetry, may be disordered.

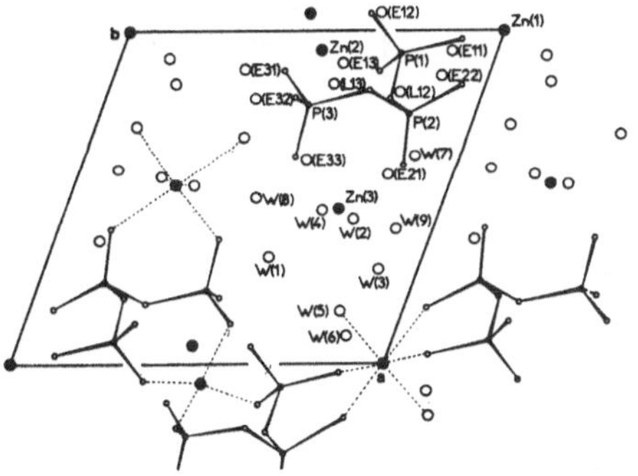

Fig. 1. Structure of $Zn_5(P_3O_{10})_2.17H_2O$.

BARIUM ZINC TRIMETAPHOSPHATE DECAHYDRATE

$Ba_2Zn(P_3O_9)_2.10H_2O$

A. DURIF, M.T. AVERBUCH-POUCHOT and J.C. GUITEL, 1975. Acta Cryst., B31, 2680-2682.

Monoclinic, C2/c, a = 26.52, b = 7.625, c = 12.92 Å, β = 100.93°, D_m not given, Z = 4. Mo radiation, R = 0.059 for 2898 reflexions.

The $P_3O_9^{3-}$ ion has a chair conformation (Fig. 1), P-O(ring) = 1.57-1.65, P-O(terminal) = 1.44-1.52(1) Å, P-O-P = 124-130°. Zn has octahedral coordination (Fig. 2), Zn-O = 2.03-2.09 Å, and Ba has 9-coordination, Ba-O = 2.71-3.20 Å; two water molecules are not coordinated to the cations, but participate in hydrogen bonding.

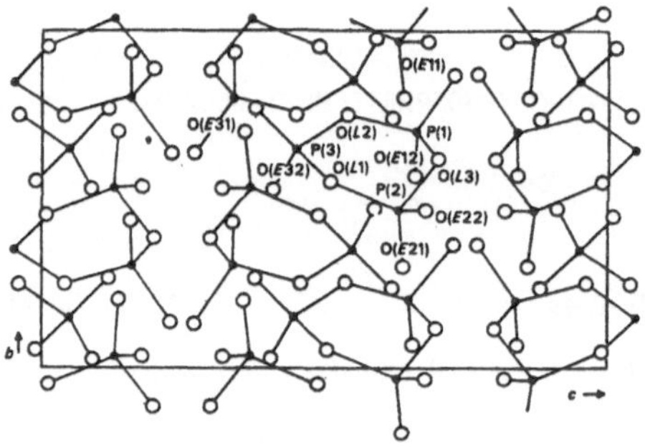

Fig. 1. The $P_3O_9^{3-}$ ions in $Ba_2Zn(P_3O_9)_2.10H_2O$.

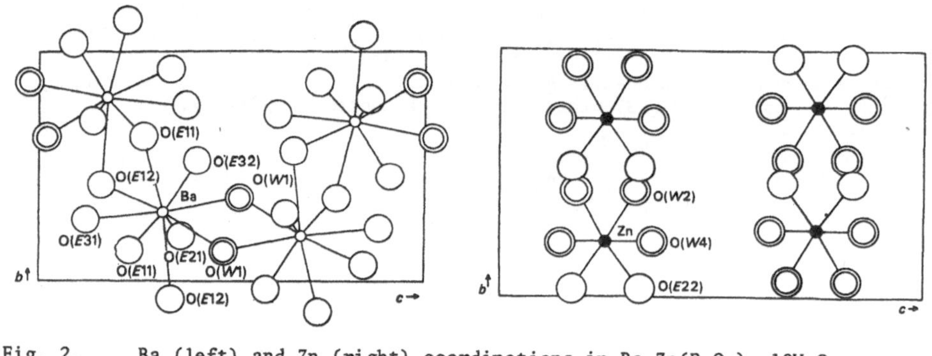

Fig. 2. Ba (left) and Zn (right) coordinations in $Ba_2Zn(P_3O_9)_2.10H_2O$.

MAGNESIUM TETRAMETAPHOSPHATE

$Mg_2P_4O_{12}$

A.G. NORD and K.B. LINDBERG, 1975. Acta Chem. Scand., A29, 1-6.

Monoclinic, C2/c, a = 11.756, b = 8.285, c = 9.917 Å, β = 118.96°, D_m = 2.85 (1), Z = 4. Cu radiation, R = 0.060 for 781 reflexions.

Four PO_4 units, linked by shared corners, yield the centrosymmetric rings shown in Fig. 1. The four P-O distances for shared oxygen atoms average 1.595 Å, for unshared oxygen atoms, 1.537 Å, though for the latter there are significant variations. The two independent Mg ions have octahedral coordinations with C_i and C_2 symmetries, Mg-O = 1.990-2.161 Å.

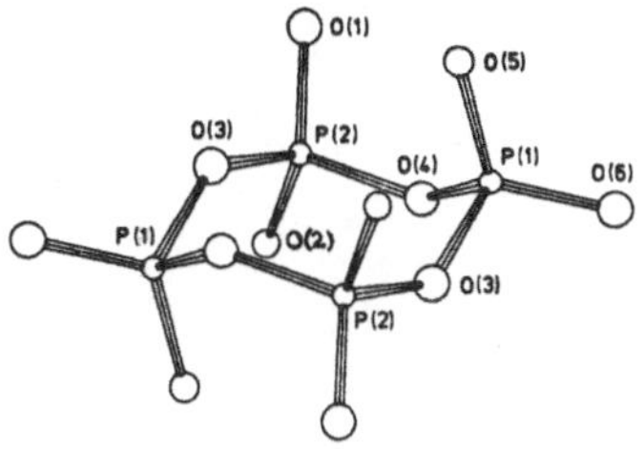

Fig. 1. The tetrametaphosphate anion in $Mg_2P_4O_{12}$.

1. M. BEUCHER and J.-C. GRENIER, 1968. Mater. Res. Bull., 3, 643.

MANGANESE(II) ULTRAPHOSPHATE

MnP_4O_{11}

L.K. MINAČEVA, M.A. PORAJ-KOŠIC, A.S. ANCYŠKINA, V.G. IVANOVA and A.V. LAVROV, 1975. Koordin. Khim., 1, 421-428.

Monoclinic, $P2_1/n$, a = 8.608, b = 8.597, c = 12.464 Å, β = 97.30°, Z = 4. Cu radiation, R = 0.13 for 1140 reflexions.

The structure contains polymeric $(P_4O_{11}{}^{2-})_\infty$ anions parallel to (110), consisting of corner-sharing PO_4 tetrahedra (Fig. 1); P-O = 1.42-1.52 (terminal), 1.56-1.68(2) Å (shared), P-O-P = 125-144°. Mn has octahedral coordination, Mn-O = 2.14-2.27(2) Å.

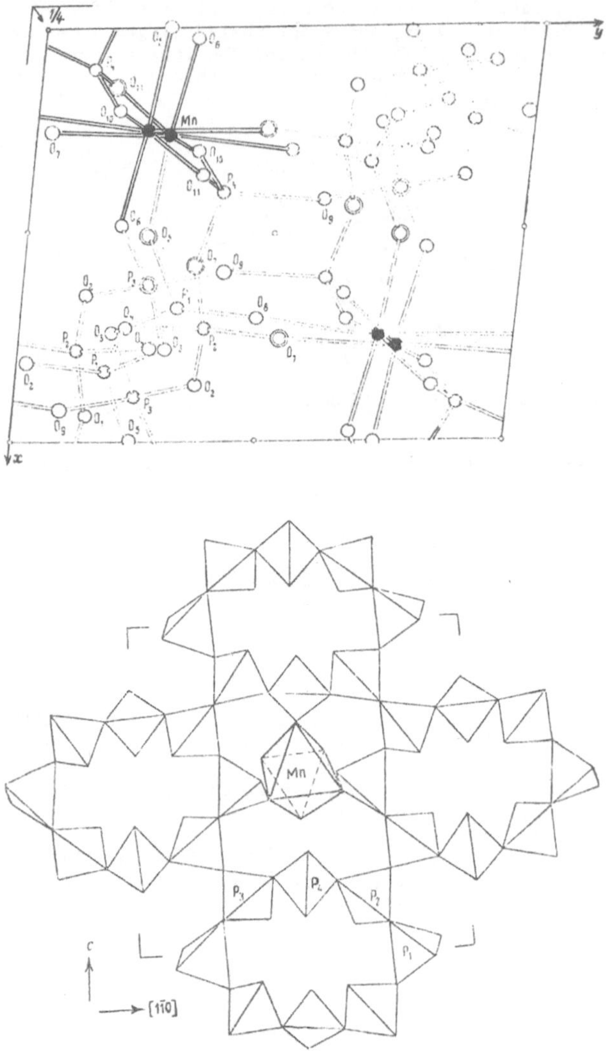

Fig. 1. Structure of MnP_4O_{11} projected on (001) (top) and (110) (bottom).

AMMONIUM COPPER OCTAMETAPHOSPHATE

$(NH_4)_2Cu_3P_8O_{24}$

M. LAÜGT and J.C. GUITEL, 1975. Z. Kristallogr., 141, 203-216.

Triclinic, $P\bar{1}$, a = 9.846, b = 7.962, c = 7.261 Å, α = 80.98, β = 110.79, γ = 110.61°, D_m = 2.858, Z = 1. Mo radiation, R = 0.031 for 1986 reflexions.

The structure contains the $P_8O_{24}^{8-}$ anion, with a centrosymmetrical sixteen-membered phosphorus-oxygen ring, P-O = 1.571-1.602 (ring), 1.477-1.495(4) Å (exocyclic), P-O-P = 129-146°. One Cu ion lies on a centre of symmetry and has tetragonally-distorted octahedral coordination, Cu-O = 1.96 (4 distances), 2.34 (x 2) Å; the other Cu ion is in a general position and has a more-distorted octahedral coordination, Cu-O = 1.93-2.01, 2.19, 2.87 Å. The ammonium ion has nine oxygen neighbours at 2.90-3.37 Å.

SODIUM DIHYDROGEN ARSENATE MONOHYDRATE

$NaH_2AsO_4.H_2O$

G. FERRARIS, D.W. JONES and J.M. SOWDEN, 1974. Atti Accad. Sci. Torino, Cl. Sci. Fiz. Mat. Nat., 108, 507-527.

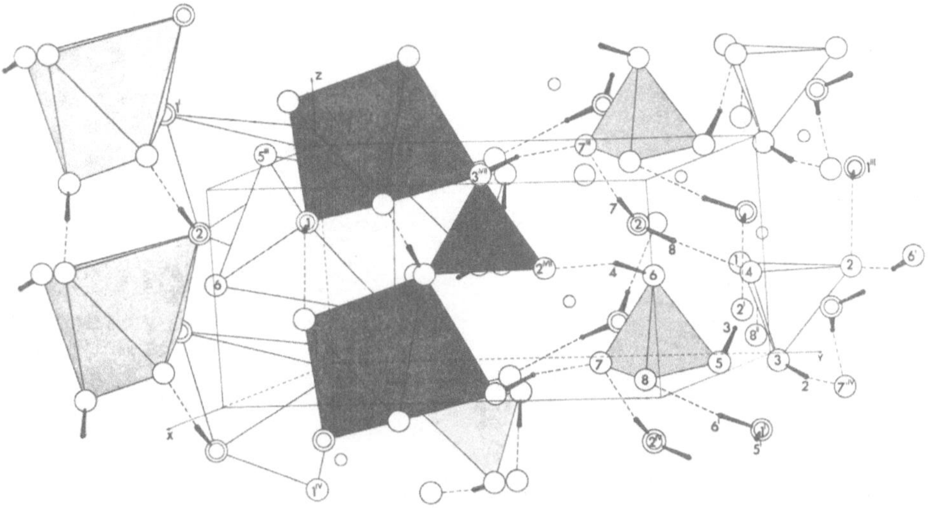

Fig. 1. Structure of $NaH_2AsO_4.H_2O$, showing the Na octahedra (on the left) and As tetrahedra (on the right).

Monoclinic, $P2_1$ [pseudo $P2_12_12_1$], a = 8.452, b = 10.360, c = 5.166 Å, β = 90.0°, D_m = 2.67, Z = 4. Cu radiation, 610 reflexions, and neutron diffraction, 537 reflexions (one octant only i.e. only half of the unique reflexions for $P2_1$), R = 0.041 for the neutron data.

The structure is described in $P2_1$ but is very close to $P2_12_12_1$ [the experimental data correspond to $P2_12_12_1$ except for weak 900 and 005 reflexions in the neutron measurements]. The structure (Fig. 1) contains bent chains along c of $NaO_4(H_2O)_2$ octahedra which share their water molecules; the chains are linked into a three-dimensional framework by $AsO_2(OH)_2$ tetrahedra and by hydrogen bonds. The measured values of the interatomic distances are unreliable as a result of the possible pseudo-symmetry.

SARKINITE

$Mn_2AsO_4(OH)$

A. DAL NEGRO, G. GIUSEPPETTI and J.M. MARTIN POZAS, 1974. Miner. (Tschermaks) Petrogr. Mitt., 21, 246-260.

Monoclinic, $P2_1/a$, a = 12.779, b = 13.596, c = 10.208 Å, β = 108°53', D_m = 4.15, Z = 16. Mo radiation, R = 0.052 for 3519 reflexions.

Isostructural with triploidite (1). The structure contains a three-dimensional framework of edge- and corner-sharing $MnO_4(OH)_2$ octahedra, $MnO_4(OH)$ trigonal bipyramids, and AsO_4 tetrahedra; Mn-O = 2.15-2.34 (octahedra), 2.17-2.22 (bipyramids), As-O = 1.66-1.70(1) Å (Fig. 1).

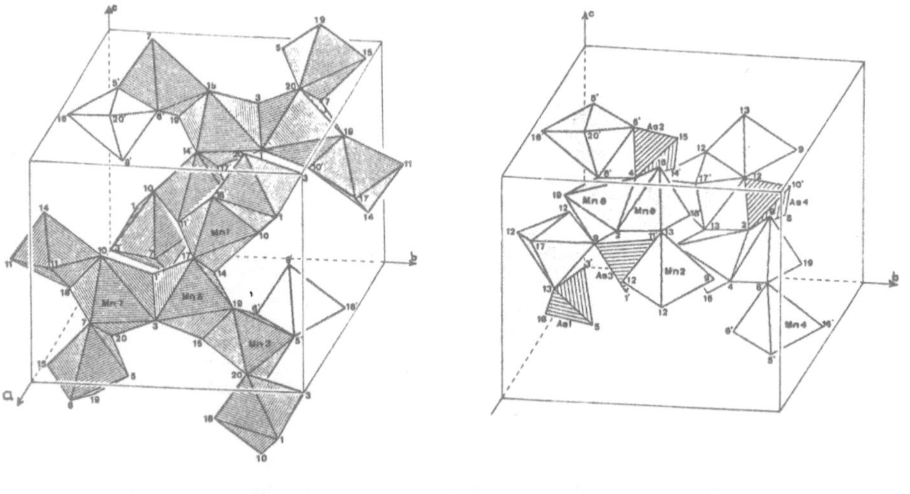

Fig. 1. Structure of sarkinite, showing linkage of the octahedra (left) and trigonal bipyramids (right). Primed atoms are common to both diagrams.

1. Structure Reports, 35A, 432.

SCORODITE

$FeAsO_4.2H_2O$

K. KITAHAMA, R. KIRIYAMA and Y. BABA, 1975. Acta Cryst., B31, 322-324.

Orthorhombic, Pbca, a = 10.325, b = 8.953, c = 10.038 Å, D_m = 3.27, Z = 8. Mo radiation, R = 0.074 for 1695 reflexions.

The structure is essentially as previously described (1), with AsO_4 tetrahedra and $FeO_4(OH_2)_2$ octahedra sharing corners, and O-H...O hydrogen bonds (one H does not participate). As-O = 1.670-1.685(5), Fe-O = 1.959-1.990(5), Fe-OH_2 = 2.061 and 2.125(5) Å.

1. Structure Reports, 12, 251.

COBALT(II) HYDROXIDE ARSENATE

$Co_2(OH)AsO_4$

H. RIFFEL, F. ZETTLER and H. HESS, 1975. Neues Jb. Miner., Mh., 514-517.

Orthorhombic, Pnnm, a = 8.248, b = 8.551, c = 6.036 Å, Z = 4. Mo radiation, R = 0.079 for 252 reflexions.

Atomic positions

	x	y	z
Co(1)	0	0	0.2464
Co(2)	0.3644	0.1352	1/2
As	0.2468	0.2554	0
OH(1)	0.3890	0.3744	1/2
O(2)	0.4223	0.3530	0
O(3)	0.1031	0.3932	0
O(4)	0.2304	0.1357	0.2227

Isostructural with adamite (1). Co ions have distorted octahedral (Co-O = 2.05, 2.08, 2.23 Å) and trigonal bipyramidal (Co-O = 1.98-2.00 (equatorial), 2.05, 2.09 Å (axial)) coordinations. In the AsO_4 tetrahedron, As-O = 1.67-1.69(3) Å, O-As-O = 105-112°.

1. Strukturbericht, 5, 17, 95; 6, 22.

TRIPPKEITE

$CuAs_2O_4$

F. PERTLIK, 1975. Miner. (Tschermaks) Petrogr. Mitt., 22, 211-217.

Tetragonal, $P4_2/mbc$, a = 8.59, c = 5.57 Å, Z = 4. Mo radiation, R = 0.059 for 270 reflexions. Cu in 4(d); As in 8(h): x = 0.2008, y = 0.1613; O(1) in 8(g): x = 0.7034; O(2) in 8(h): x = 0.0971, y = 0.6246.

The material is synthetic, with a structure similar to that of the natural material (1). AsO_3 trigonal pyramids share corners to form chains along c, As-O = 1.814 (shared), 1.765(9) Å (terminal). Cu has distorted octahedral coordination, Cu-O = 1.945 (x 4), 2.472(8) (x 2) Å; the octahedra share edges.

1. Structure Reports, 15, 271.

SODIUM NITROSYL-N,N-DISULPHONATE TRIHYDRATE

$Na_3[ON(SO_3)_2].3H_2O$

J.S. RUTHERFORD and B.E. ROBERTSON, 1975. Inorg. Chem., 14, 2537-2540.

Orthorhombic, $Cmc2_1$, a = 14.234, b = 11.052, c = 5.861 Å, D_m = 2.25, Z = 4. Mo radiation, R = 0.033 for 928 reflexions.

The $[ON(SO_3)_2]^{3-}$ ion has C_s symmetry, the configuration at N being pyramidal, N-S = 1.727(2), N-O = 1.415 Å (compare the $[ON(SO_3)_2]^{2-}$ ion (1), which is planar at N, with N-O = 1.28 Å). The N-O bond is directed at the centroid of a triangle of three Na^+ ions, which have 6-, 6-, and 7-coordination, Na-O = 2.40-2.62 Å.

1. Structure Reports, 33A, 386.

HYDRAZINIUM SULPHATE

$N_2H_6SO_4$

L.F. POWER, K.E. TURNER, J.A. KING and F.H. MOORE, 1975. Acta Cryst., B31, 2470-2474.

Orthorhombic, $P2_12_12_1$, a = 8.232, b = 9.145, c = 5.535 Å, [D_m = 2.02], Z = 4. Neutron diffraction data, R = 0.049 for 900 reflexions.

The results are in agreement with previous X-ray and neutron studies (1).

1. Structure Reports, 11, 387; 15, 247; 35A, 366.

SCHAIRERITE

$Na_{21}(SO_4)_7F_6Cl$

L. FANFANI, A. NUNZI, P.F. ZANAZZI, A.R. ZANZARI and C. SABELLI, 1975. Miner. Mag., 40, 131-139.

Trigonal, P31m, a = 12.197, c = 19.259 Å, D_m = 2.616, Z = 3. Mo radiation, R = 0.070 for 2536 reflexions.

There is a marked sub-cell (a = 7.042 Å, P3ml), which is related to the cubic unit cell of sulphohalite (Fig. 1). The structure of schairerite contains seven sheets of Na^+ ions perpendicular to c, connected to form a three-dimensional framework. Na^+ ions in the sheets are arranged in an array of hexagons and triangles; S atoms lie in the sheets at the centres of each hexagon, and halogen atoms are between the sheets midway between the centres of two triangles. Sulphohalite (1) has six Na^+ sheets and different orientation of SO_4 tetrahedra.

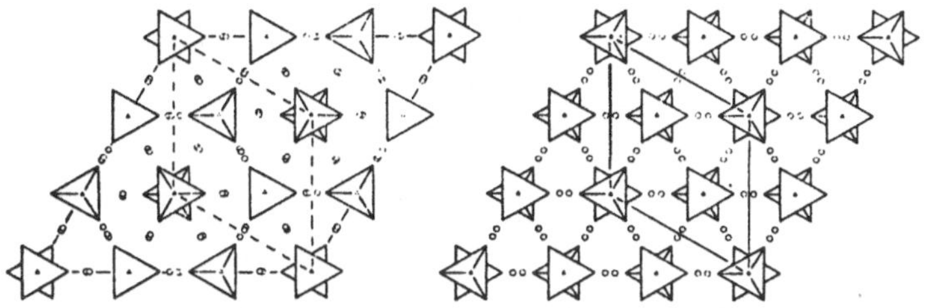

Fig. 1. SO_4 tetrahedra and Na^+ ions in schairerite (left) and sulphohalite (right); F and Cl atoms lie on 3-fold and pseudo 3-fold axes.

1. Strukturbericht, 3, 118, 470; Structure Reports, 33A, 377.

GALEITE

$Na_{15}(SO_4)_5F_4Cl$

L. FANFANI, A. NUNZI, P.F. ZANAZZI and A.R. ZANZARI, 1975. Miner. Mag., 40, 357-361.

Trigonal, P31m, a = 12.197, c = 13.955 Å, D_m = 2.605, Z = 3. Mo radiation, R = 0.09 for 2078 reflexions.

There is a marked sub-cell, P3ml, a = 7.042, c = 13.955 Å. The structure consists of a three-dimensional framework of five sheets of face-, edge-, and corner-sharing Na coordination octahedra (NaO_4X_2, X_2 = trans-FCl or cis- or trans-F_2) and SO_4 tetrahedra. S-O = 1.40-1.57, Na-Cl = 2.65-2.90, Na-F = 2.22-2.46, Na-O = 2.33-2.55 Å. The framework of schairerite (1) contains seven octahedral sheets.

1. Preceding report.

POTASSIUM HYDROGEN SULPHATE

(MERCALLITE)

$KHSO_4$

F.A. COTTON, B.A. FRENZ and D.L. HUNTER, 1975. Acta Cryst., B31, 302-304, 2747.

Orthorhombic, Pbca, a = 8.421 [in text, 8.412 in abstract], b = 9.800, c = 18.957 Å, D_m = 2.329, Z = 16. Mo radiation, R = 0.040 for 1151 reflexions.

The structure is as previously described (1), and H atoms have been located. S-O = 1.440, 1.441, and 1.465(3) Å (the last accepts a hydrogen bond), S-OH = 1.564(4) Å.

1. Structure Reports, 22, 458; 29, 346.

ANHYDRITE

$CaSO_4$

I. H. MORIKAWA, I. MINATO, T. TOMITA and S. IWAI, 1975. Acta Cryst., B31, 2164-2165.

II. F.C. HAWTHORNE and R.B. FERGUSON, 1975. Canad. Miner., 13, 289-292.

Orthorhombic, Amma, a = 6.999, 6.993, b = 6.992, 6.995, c = 6.240, 6.245 Å, in I, II, respectively, Z = 4. Mo radiation, R = 0.021 and 0.016 for 266 and 284 reflexions.

Atomic positions [mean of I and II]

	x	y	z
Ca	3/4	0	0.3477
S	1/4	0	0.1556
O(1)	1/4	0.1700	0.0160
O(2)	0.0817	0	0.2974

The structure is as previously described (1). S-O = 1.474(2) Å, O-S-O = 106.1-110.8°, Ca-O = 2.34-2.56 Å (8-coordination).

1. Strukturbericht, 1, 340, 380; Structure Reports, 16, 286; 27, 607; 28, 207.

CELESTITE

THENARDITE

$SrSO_4$
Na_2SO_4

F.C. HAWTHORNE and R.B. FERGUSON, 1975. Canad. Miner., 13, 181-187.

Celestite, $SrSO_4$
Orthorhombic, Pnma, a = 8.360, b = 5.352, c = 6.858 Å, Z = 4. Mo radiation, R = 0.041 for 355 reflexions.

Thenardite, Na_2SO_4
Orthorhombic, Fddd, a = 9.829, b = 12.302, c = 5.868 Å, Z = 8. Mo radiation, R = 0.024 for 283 reflexions.

The structures are as previously described (1, 2). S-O = 1.46-1.48(1) Å in celestite, 1.479(1) Å in thenardite.

1. Strukturbericht, 1, 344, 382; Structure Reports, 30A, 365.
2. Strukturbericht, 2, 88, 421; Structure Reports, 8, 189; 18, 474; 39A, 306.

ALUNOGEN

$Al_2(SO_4)_3.16\cdot 4H_2O$

S. MENCHETTI and C. SABELLI, 1974. Miner. (Tschermaks) Petrogr. Mitt., 21, 164-178.

Triclinic, $P\bar{1}$, a = 7.425, b = 26.975, c = 6.061 Å, α = 90.03, β = 97.66, γ = 91.94°, D_m = 1.77, Z = 2. Cu radiation, R = 0.064 for 1631 reflexions (film data from a twinned crystal).

The structure (Fig. 1) contains two discrete $Al(H_2O)_6$ octahedra and three SO_4 tetrahedra, joined by hydrogen bonding. Four other water molecules and a fifth with partial occupancy lie in a channel parallel to c. Al-O = 1.85-1.90(1), S-O = 1.45-1.48(1), O-H...O = 2.55-3.09 Å.

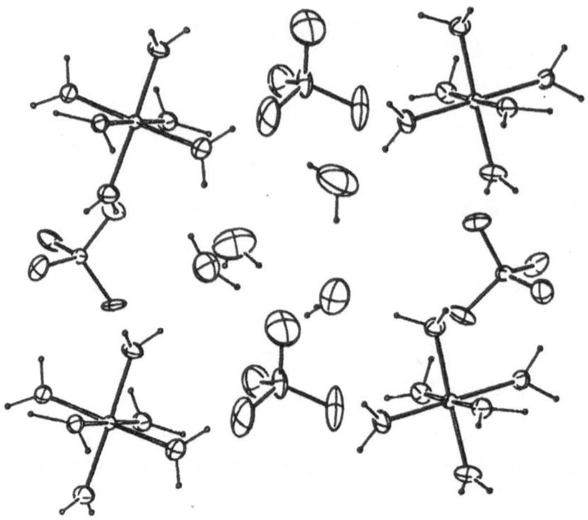

Fig. 1. Structure of alunogen viewed along c.

TRITIN(II) DIHYDROXIDE OXIDE SULPHATE

$[Sn_3O(OH)_2]SO_4$

I. C.G. DAVIES, J.D. DONALDSON, D.R. LAUGHLIN, R.A. HOWIE and R. BEDDOES, 1975. J. Chem. Soc., Dalton, 2241-2244.

II. S. GRIMVALL, 1975. Acta Chem. Scand., A29, 590-598.

Orthorhombic, $Pbc2_1$, a = 4.983, 4.938, b = 13.128, 13.045, c = 12.214, 12.140 Å, in I, II, respectively, D_m = 4.12, Z = 4. I, Mo radiation, R = 0.028 for 1070 reflexions; II, Cu radiation, R = 0.069 for 452 reflexions.

The structure (Fig. 1) contains $[Sn_3O(OH)_2]^{2+}$ ions, which consist of two four-membered rings sharing an edge. The central Sn has trigonal pyramidal coordination, Sn-O = 2.06-2.18 Å (next shortest, 2.95 Å), O-Sn-O = 87-94°; the two terminal Sn atoms are coordinated to two oxygen atoms of the cation, Sn-O = 2.06-2.16 Å, with sulphate oxygens at 2.38-2.48(1) Å completing square-pyramidal four-coordination. In the sulphate ion, S-O = 1.47-1.50(1) Å, O-S-O = 106-111°.

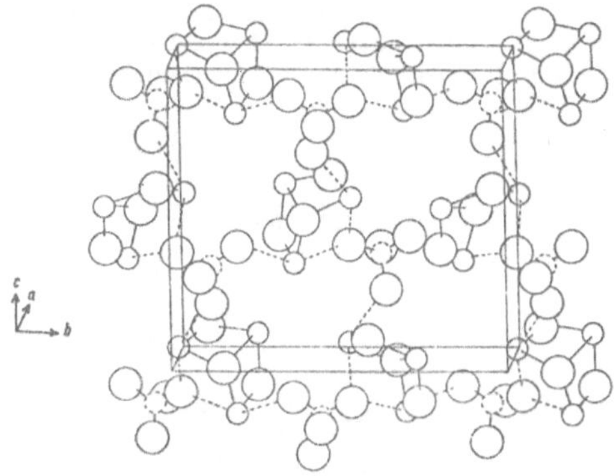

Fig. 1. Structure of $[Sn_3O(OH)_2]SO_4$; small open circles are Sn, large open circles O, dashed circles S.

FLEISCHERITE

$Pb_3Ge(OH)_6(SO_4)_2.3H_2O$

H.H. OTTO, 1975. Neues Jb. Miner., Abh., 123, 160-190.

Hexagonal, $P\bar{6}2c$, a = 8.867, c = 10.875 Å, D_m = 4.55, Z = 2. Cu radiation, R = 0.039 for 320 reflexions (film data).

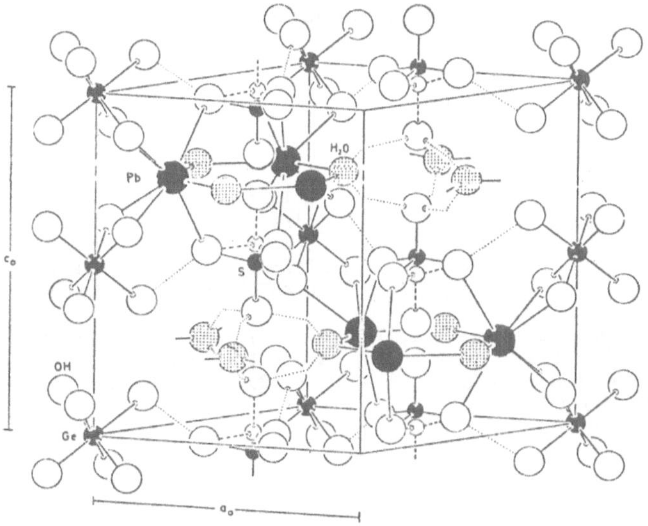

Fig. 1. Structure of fleischerite.

The structure (Fig. 1) contains $Ge(OH)_6$ octahedra (Ge-O = 1.90(1) Å), joined by Pb coordination polyhedra (9-coordinate, Pb-O = 2.62-2.79 Å). Sulphate ions are disordered along c (or the crystal is twinned), and hydrogen bonds are present.

LEAD OXIDE SULPHATE

$4PbO.PbSO_4$

K. SAHL, 1975. Z. Kristallogr., 141, 145-150.

Monoclinic, $P2_1/c$, a = 14.610, b = 11.703, c = 11.526 Å, β = 91°, D_m not given, Z = 8. Mo radiation, R = 0.23 for 1677 reflexions for a disordered arrangement of Pb and SO_4 groups in a cell with a' = a/2.

[The structure has not been properly determined.]

ARSENIC OXIDE SULPHATE

(ARSENIC(III) OXIDE - SULPHUR TRIOXIDE)

$As_2O_3.SO_3$

R. MERCIER, 1975. Rev. Chim. Miner., 12, 508-517.

Orthorhombic, $P2_12_12$, a = 4.655, b = 11.54, c = 4.795 Å, D_m = 3.52, Z = 2. Mo radiation, R = 0.13 for 207 reflexions (film data).

Atomic positions

S in 2(a), other atoms in 4(c)

	x	y	z
As	0.2660	0.1914	0.1780
S	1/2	1/2	0.2117
O(1)	0.7170	0.4526	0.0210
O(2)	0.9052	0.2395	0.1140
O(3)	0.3830	0.4070	0.3825

The structure contains layers parallel to (001), which consist of $(AsO^+)_n$ chains along a linked by SO_4^{2-} tetrahedra. As has trigonal pyramidal coordination, As-O = 1.74-1.93(2) Å, O-As-O = 92-94, As-O-As = 129°. S-O = 1.45, 1.47(2) Å, O-S-O = 103-112, As-O-S = 135°.

ANTIMONY OXIDE SULPHATE

$Sb_2O_3.2SO_3$

R. MERCIER, J. DOUGLADE and F. THEOBALD, 1975. Acta Cryst., B31, 2081-2085.

Tetragonal, $P4_12_12$, a = 6.59, c = 17.04 Å, D_m = 4.0, Z = 4. Mo radiation, R = 0.075 for 281 reflexions (film data).

The structure (Fig. 1) contains $Sb_2O_3.2SO_3$ molecular units, which consist of two SO_4 tetrahedra and two SbO_3 trigonal pyramids sharing corners; Sb-O = 1.92, 2.11, 2.17, S-O = 1.44-1.51 Å, O-Sb-O = 80, 90, 92, O-S-O = 108, 104, angles at bridging O = 116-141°. Intermolecular contacts correspond to van der Waals interactions.

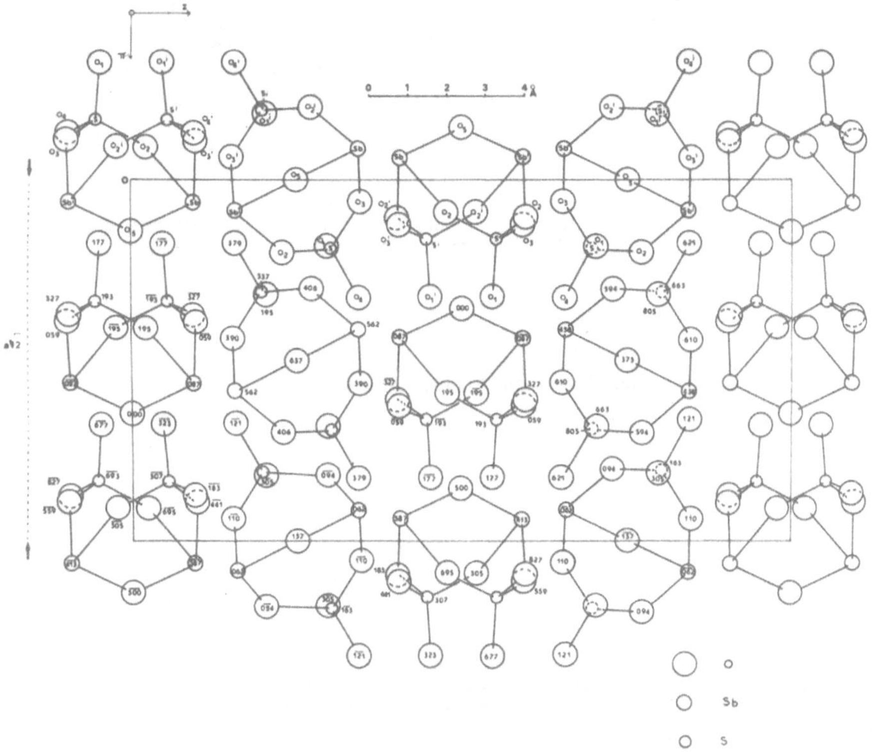

Fig. 1. Structure of $Sb_2O_3.2SO_3$.

HAFNIUM SULPHATE (BASIC)

$Hf_{18}O_{10}(OH)_{26}(SO_4)_{13}(H_2O)_{33}$

W. MARK and M. HANSSON, 1975. Acta Cryst., B31, 1101-1108.

Hexagonal, $P6_3/m$, a = 34.09, c = 17.664 Å, D_m = 3.19, Z = 6. Cu radiation, R = 0.075 for 2240 reflexions (film data).

The structure contains $Hf_{18}O_{10}(OH)_{26}(SO_4)_{13}(H_2O)_{20}$ building units (Fig. 1), which are joined along c by sulphate groups and hydrogen bonded water molecules. Hf atoms have 8-coordination (dodecahedra and square-antiprisms) and 7-coordination (capped trigonal prisms).

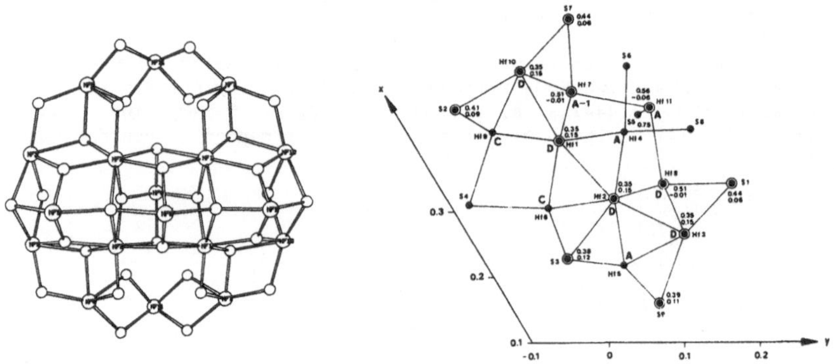

Fig. 1. The Hf-O framework (left side) and schematic outline of the building unit (right side) in $Hf_{18}O_{10}(OH)_{26}(SO_4)_{13}(H_2O)_{33}$; D = dodecahedron, A = antiprism, C = capped trigonal prism.

RUBIDIUM OXODISULPHATONIOBATE

$RbNbO(SO_4)_2$

V.Ja. KUZNECOV, D.L. ROGAČEV, M.A. PORAJ-KOŠIC and L.M. DIKAREVA, 1974. Izv. Akad. Nauk SSSR, Ser. Khim., 2167-2170.

Orthorhombic, Pnma, a = 17.10, b = 5.205, c = 8.76 Å, Z = 4. Cu radiation, R = 0.12 for 320 reflexions.

Isostructural with Cs compound (1). The structure contains Rb^+ ions and layers of polymeric $NbO(SO_4)_2^-$ anions. Nb has distorted octahedral coordination, Nb-OS = 1.96-2.26, Nb-O = 1.73 Å, the octahedra being linked by the bridging sulphate groups.

1. Structure Reports, 39A, 312.

POTASSIUM TETRASULPHATODIMOLYBDATE HYDRATES

$K_4Mo_2(SO_4)_4.2H_2O$ (I)
$K_3Mo_2(SO_4)_4.3{\cdot}5H_2O$ (II)

F.A. COTTON, B.A. FRENZ and T.R. WEBB, 1975. Inorg. Chem., 14, 391-398.

(I)
Monoclinic, C2/c, a = 17.206, b = 10.193, c = 10.061 Å, β = 94.92°, D_m not measured, Z = 4. Mo radiation, R = 0.028 for 1217 reflexions.

(II)
Monoclinic, C2/c, a = 30.654, b = 9.528, c = 12.727 Å, β = 97.43°, D_m = 2.72, Z = 8. Mo radiation, R = 0.026 for 2341 reflexions.

Both structures contain $Mo_2(SO_4)_4$ anions, in which four bidentate sulphate groups bridge across a strongly-bonded pair of Mo atoms; Mo-Mo = 2.111(1) Å in (I), and 2.164(2) Å in (II), Mo-O = 2.13-2.15 and 2.05-2.08 Å. Octahedral coordination at Mo is completed in (I) by sulphate oxygens of neighbouring anions at 2.59 Å, forming a polymeric structure, and in (II) by water molecules at 2.55 Å; one of these water molecules bridges between anions.

SODIUM TETRA-μ-SULPHATO-DIRHENATE(III) OCTAHYDRATE

$Na_2Re_2(SO_4)_4.8H_2O$

F.A. COTTON, B.A. FRENZ and L.W. SHIVE, 1975. Inorg. Chem., 14, 649-652.

Monoclinic, $P2_1/n$, a = 7.739, b = 13.212, c = 11.077 Å, β = 100.58°, D_m not given, Z = 2. Mo radiation, R = 0.042 for 1331 reflexions.

The structure contains $Re_2(SO_4)_4^{2-}$ ions, similar to those in related Mo compounds (1), with water molecules completing octahedral coordination. Re-Re = 2.214(1), Re-O = 2.01, Re-OH_2 = 2.28(1) Å.

1. Preceding report.

IRON(III) SULPHATE

$Fe_2(SO_4)_3$

P.C. CHRISTIDIS and P.J. RENTZEPERIS, 1975. Z. Kristallogr., 141, 233-245.

Monoclinic, $P2_1/n$, a = 8.296, b = 8.533, c = 11.630 Å, β = 90.25°, D_m = 3.20, Z = 4. Mo radiation, R = 0.038 for 4457 reflexions.

[The rhombohedral form is to described later (1), but both forms have been described previously and independently (2, 3). The present results for the monoclinic form are in agreement with those in 3.]

1. P.C. CHRISTIDIS and P.J. RENTZEPERIS, 1975. To be published.
2. Structure Reports, 39A, 313.
3. Ibid., 40A, 264.

RHOMBOCLASE

$(H_5O_2)[Fe(SO_4)_2(H_2O)_2]$

K. MEREITER, 1974. Miner. (Tschermaks) Petrogr. Mitt., 21, 216-232.

Orthorhombic, Pnma, a = 9.724, b = 18.333, c = 5.421 Å, D_m = 2.23, Z = 4. Mo radiation, R = 0.030 for 705 reflexions (film data for synthetic crystal).

The structure (Fig. 1) contains $[Fe(SO_4)_2(H_2O)_2^-]_n$ sheets parallel to (010), linked by $H_5O_2^+$ ions via hydrogen bonds. The $H_5O_2^+$ ions show disorder, 80% having cis- and 20% trans-conformation; O...O = 2.443(6) Å for the cis, and 2.42(2) Å for the trans. Fe-O = 1.964-2.063(4), S-O = 1.447-1.489(3), other O-H...O = 2.64-2.85 Å.

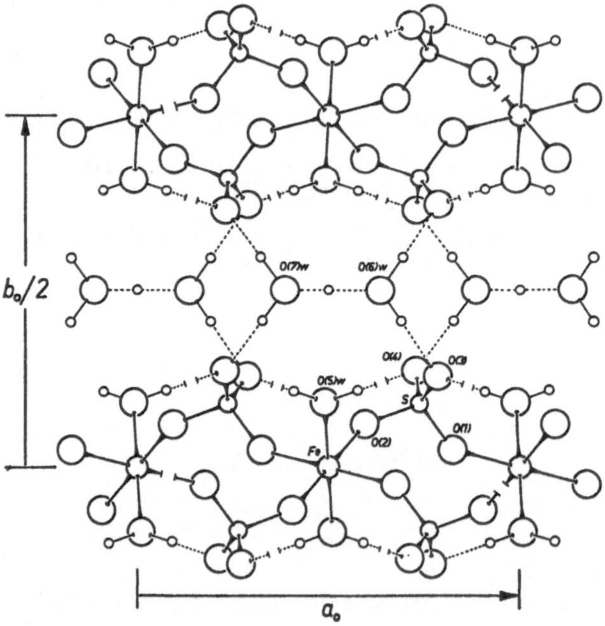

Fig. 1. Structure of rhomboclase.

SLAVIKITE

$NaMg_2Fe_5(SO_4)_7(OH)_6.33H_2O$

P. SÜSSE, 1975. Neues Jb. Miner., Mh., 27-40.

Rhombohedral, $R\bar{3}$, a = 13.67 Å, α = 53.03°, (hexagonal cell has a = 12.20, c = 35.13 Å), D_m = 1.90, Z = 1. Mo radiation, R = 0.12 for 1012 reflexions.

The structure contains infinite two-dimensional sheets of corner-sharing FeO_6 octahedra and sulphate tetrahedra; $Mg(H_2O)_6$ ions and Na ions (three oxygen neighbours) are located between the sheets. There is a system of hydrogen bonds, and one sulphate group is located at the unit-cell origin and is disordered. Fe-O = 1.98-2.05, Mg-O = 2.07-2.09, Na-O = 2.35, S-O = 1.47-1.52 Å.

POTASSIUM IRON(III) SULPHATE HYDROXIDE HYDRATE

(MAUS'S SALT)

$K_5Fe_3(SO_4)_6(OH)_2.nH_2O$

C. GIACOVAZZO, F. SCORDARI and S. MENCHETTI, 1975. Acta Cryst., B31, 2171-2173.

Hexagonal, $P6_3/m$, a = 9.71, c = 18.96 Å, D_m = 2.412, Z = 2. Mo radiation, R = 0.086 for 887 reflexions.

The structure consists of $Fe_3O(H_2O)_3(SO_4)_6$ groups (Fig. 1a), built up from three $FeO_5(H_2O)$ octahedra and six SO_4 tetrahedra. One type of K ion links these groups into sheets (Fig. 1b), with inter-sheet connections provided by other K ions. Three uncertain peaks are probably (K,H_2O), H_3O^+, and OH^-.

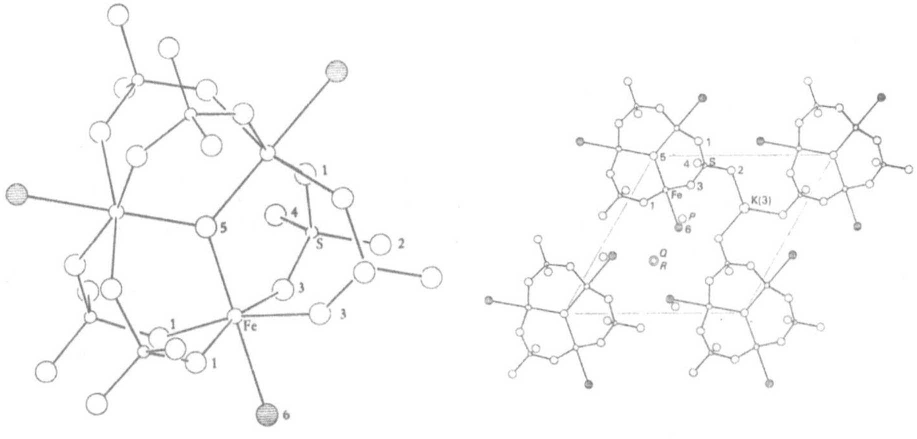

Fig. 1. Maus's salt, (a) the $Fe_3O(H_2O)_3(SO_4)_6$ group, and (b) a sheet parallel to (00.1).

trans-DICHLOROTETRAAMMINECOBALT(III) BISULPHATE

$[Co(NH_3)_4Cl_2]HSO_4$

J.A.P. BONAPACE and N.S. MANDEL, 1975. Acta Cryst., B31, 2540-2542.

Orthorhombic, $P2_12_12_1$, a = 8.815, b = 18.639, c = 5.934 Å, D_m = 2.01, Z = 4. Mo radiation, R = 0.039 for 779 reflexions.

Co has octahedral coordination (Fig. 1), Co-N = 1.946-1.973(6), Co-Cl = 2.251, 2.258(3) Å. In the HSO_4 tetrahedron, S-O(4)H = 1.569, S-O = 1.438-1.443(6) Å. The ions are joined by O-H...O, N-H...O, and N-H...Cl hydrogen bonds. The hydrogen atoms have been located, and those bonded to N(4) are disordered.

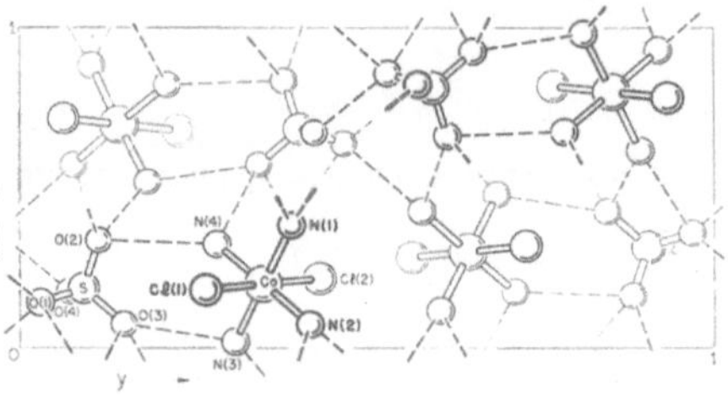

Fig. 1. Structure of trans-dichlorotetraamminecobalt(III) bisulphate, showing anion-anion and anion-cation hydrogen bonds.

COBALT BLÖDITE

(COBALT ASTRAKHANITE)

$Na_2Co(SO_4)_2.4H_2O$

V.I. BUKIN and Ju.Z. NOZIK, 1975. Kristallografija, 20, 293-296 [Soviet Physics - Crystallography, 20, 180-182].

Monoclinic, $P2_1/a$, a = 11.083, b = 8.426, c = 5.534 Å, β = 100.35°, Z = 2. Neutron diffraction data, R = 0.061 for 152 reflexions (hk0 and 0kℓ).

Isostructural with blödite (1) and zinc blödite (2). The structure contains SO_4 tetrahedra, $CoO_2(H_2O)_4$ octahedra, and $NaO_4(H_2O)_2$ distorted octahedra, linked by corner-sharing and hydrogen bonding. Mean S-O = 1.48, Co-O = 2.11, Na-O = 2.43 Å.

1. Structure Reports, 22, 461.
2. Ibid., 22, 464; 40A, 266.

COPPER SULPHATE PENTAHYDRATE

$CuSO_4.5H_2O$
$CuSO_4.5D_2O$

G.E. BACON and D.H. TITTERTON, 1975. Z. Kristallogr., 141, 330-341.

Triclinic, P$\bar{1}$, a = 6.141, b = 10.736, c = 5.986 Å, α = 82°16', β = 107°26', γ = 102°40', Z = 2. Neutron diffraction data, R = 0.037 and 0.059 for 725 and 1103 reflexions, respectively.

The structure is as previously described (1). The isotope effect is negligible; mean O-H = 0.967(5), O-D = 0.969(4) Å; H-O-H = 106.7-112.9(5), D-O-D = 104.7-112.6(3)°.

1. Strukturbericht, 3, 102, 449; Structure Reports, 27, 614.

KRÖHNKITE

$Na_2Cu(SO_4)_2.2H_2O$

F.C. HAWTHORNE and R.B. FERGUSON, 1975. Acta Cryst., B31, 1753-1755.

Monoclinic, $P2_1/c$, a = 5.807, b = 12.656, c = 5.517 Å, β = 108.32°, D_m = 2.913, Z = 2. Mo radiation, R = 0.045 for 1029 reflexions.

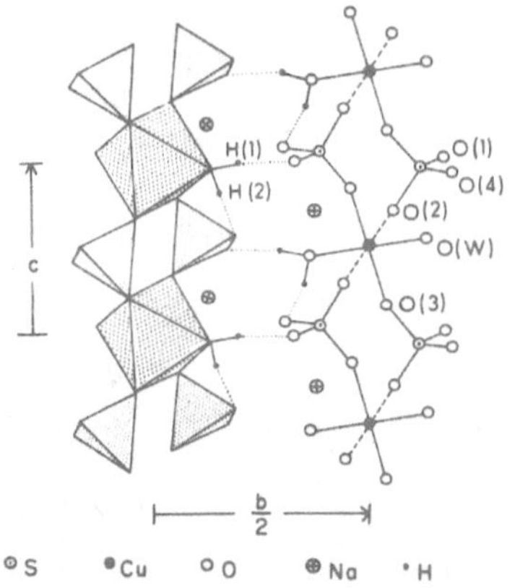

Fig. 1. Structure of kröhnkite.

The structure is as previously described (1). It consists of chains along c of alternating corner-sharing tetrahedra and octahedra, linked by Na ions and hydrogen bonds (Fig. 1). S-O = 1.465-1.504(3), Cu-O = 1.937, 1.975, 2.464 (each x 2), Na-O (7-coordination) = 2.386-2.566 Å.

1. Structure Reports, 16, 289; 17, 471; 18, 483; 26, 449.

RUBIDIUM COPPER(II) SULPHATE HEXAHYDRATE

POTASSIUM COPPER(II) SELENATE HEXAHYDRATE

POTASSIUM NICKEL SULPHATE HEXAHYDRATE

POTASSIUM ZINC SULPHATE HEXAHYDRATE

$Rb_2Cu(SO_4)_2.6H_2O$ (I)
$K_2Cu(SeO_4)_2.6H_2O$ (II)
$K_2Ni(SO_4)_2.6H_2O$ (III)
$K_2Zn(SO_4)_2.6H_2O$ (IV)

I. G. SMITH, F.H. MOORE and C.H.L. KENNARD, 1975. Cryst. Struct. Comm., 4, 407-412.

II. J. WHITNALL, C.H.L. KENNARD, J.K. NIMMO and F.H. MOORE, 1975. Ibid., 4, 709-712.

III. P.G. HODGESON, J. WHITNALL, C.H.L. KENNARD and F.H. MOORE, 1975. Ibid., 4, 713-716.

IV. J. WHITNALL, C.H.L. KENNARD, J. NIMMO and F.H. MOORE, 1975. Ibid., 4, 717-720.

Monoclinic, $P2_1/a$, Z = 2. Neutron diffraction data.

	I(77°K)	II	III	IV
a(Å)	9.28	9.165	8.985	9.034
b	12.15	12.224	12.167	12.184
c	6.23	6.336	6.128	6.148
β(deg.)	105.4	103.36	105.08	104.79
Reflexions	595	845	1477	797
R	0.075	0.028	0.048	0.058

The compounds have typical Tutton's salt structures (1, 2). The dimensions of the RbCu salt (Cu-O = 1.98, 2.00, 2.32 (1) Å) differ only slightly from those at room temperature (2). In the other three compounds, Cu-O = 1.94, 2.04, 2.30, Ni-O = 2.02, 2.08, 2.09, Zn-O = 2.03, 2.13, 2.13 Å.

1. Strukturbericht, 2, 93; Structure Reports, 27, 619; 29, 354.
2. Structure Reports, 38A, 338.

CADMIUM HYDROXIDE SULPHATE

β-$Cd_2(OH)_2SO_4$

J. LABARRE, D. LOUËR, M. LOUËR and D. GRANDJEAN, 1975. Cryst. Struct. Comm., 4, 657-659.

Monoclinic, C2/c, a = 11.546, b = 7.705, c = 6.486 Å, β = 116.49°, D_m = 4.56, Z = 4. Mo radiation, R = 0.065 for 762 reflexions.

Atomic positions

	x	y	z
Cd	0.2497	0.4021	0.3122
S	0	0.6875	1/4
O(1)	-0.1092	0.5755	0.0918
O(2)	-0.0432	0.7964	0.3897
O(3)	0.3186	0.6327	0.1983

The structure (Fig. 1) contains layers parallel to (100) of edge- and corner-sharing distorted CdO_6 octahedra, Cd-O = 2.20-2.40(1) Å, O-Cd-O = 74-118°. The layers are joined along a by corner sharing with SO_4 tetrahedra, S-O = 1.48-1.50(1) Å, O-S-O = 108.7-110.7°.

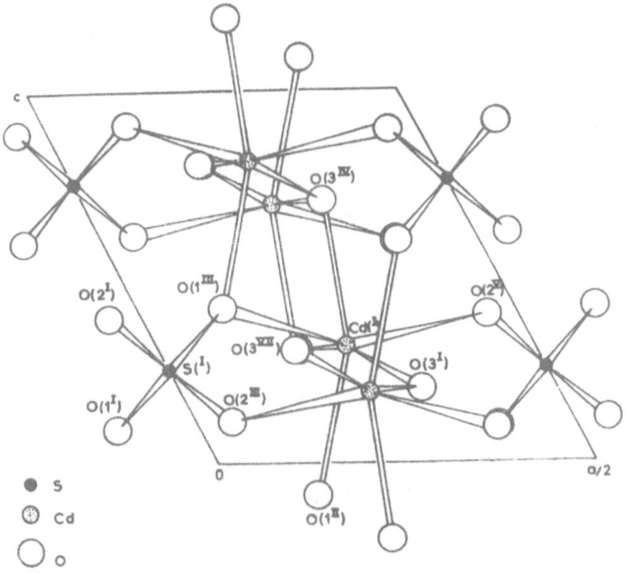

Fig. 1. Structure of β-$Cd_2(OH)_2SO_4$.

AMMONIUM CADMIUM SULPHATE

AMMONIUM CADMIUM MANGANESE SULPHATE

$(NH_4)_2Cd_2(SO_4)_3$
$(NH_4)_2CdMn(SO_4)_3$

H.N. NG and C. CALVO, 1975. Canad. J. Chem., 53, 1449-1455.

Cubic, $P2_13$, a = 10.362, 10.262 Å, D_m not given, Z = 4. Mo radiation, R = 0.016 and 0.051 for 257 reflexions.

Isostructural with langbeinite, $K_2Mg_2(SO_4)_3$ (1). In the manganese compound, Mn is substituted in both Cd sites in the ratio 3:1.

1. Structure Reports, 21, 362.

SODIUM DITHIONATE DIHYDRATE

$Na_2S_2O_6.2H_2O$

K. ZWOLL, 1974. Ber. Kernforsch.-Anlage Jülich, no. 1057, 1-141.

Orthorhombic, Pnma, a = 6.403, b = 10.750, c = 10.686 Å, D_m = 2.189, Z = 4 (all as in 1). Neutron diffraction data, R = 0.084 for 387 reflexions and 0.153 for 323 reflexions for hydrogen and deuterated samples, respectively.

The structure is as previously described (1).

1. Structure Reports, 20, 337; 28, 220; 32A, 323.

MAGNESIUM DITHIONATE HEXAHYDRATE

NICKEL DITHIONATE HEXAHYDRATE

ZINC DITHIONATE HEXAHYDRATE

$MgS_2O_6.6H_2O$
$NiS_2O_6.6H_2O$
$ZnS_2O_6.6H_2O$

W. BLACK, E.A.H. GRIFFITH and B.E. ROBERTSON, 1975. Acta Cryst., B31, 615-617.

Mg compound
Triclinic, $P\bar{1}$, a = 6.819, b = 6.747, c = 6.506 Å, α = 94.23, β = 96.76, γ = 101.72°, D_m = 1.8, Z = 1. Mo radiation, R = 0.036 for 1393 reflexions.

Ni compound
Triclinic, $P\bar{1}$, a = 6.751, b = 6.699, c = 6.451 Å, α = 94.53, β = 96.30, γ = 101.28°, D_m = 2.0, Z = 1. Mo radiation, R = 0.077 for 1478 reflexions.

Zn compound
Triclinic, $P\bar{1}$, a = 6.799, b = 6.752, c = 6.491 Å, α = 94.58, β = 96.20, γ = 101.47°, D_m = 2.0, Z = 1. Mo radiation, R = 0.031 for 1573 reflexions.

The structures are roughly as previously described for the Ni compound (1), but the previous study has errors in the reduced cell and in the atomic coordinates. The dithionate ions have S-S = 2.123-2.127(1), S-O = 1.446-1.454(3) Å, S-S-O = 104.2-104.7, O-S-O = 113.2-114.8°. The metal ions have nearly regular octahedral coordination, M-OH_2 = 2.042-2.055, 2.039-2.042, and 2.069-2.092 Å for M = Mg, Ni, and Zn, respectively. Each dithionate oxygen atom accepts hydrogen bonds from two water molecules, O-H...O = 2.79-2.85 Å.

1. Structure Reports, 35A, 380.

BARIUM THIOSULPHATE MONOHYDRATE

$BaS_2O_3.H_2O$

L. MANOJLOVIĆ-MUIR, 1975. Acta Cryst., B31, 135-139.

Orthorhombic, Pbcn, a = 20.07, b = 7.19, c = 7.37 Å, Z = 8 (1). Neutron diffraction data, R = 0.034 for 615 reflexions.

The structure is essentially as previously described (1), but with shifts up to 0.13 Å. Only one of the water hydrogen atoms is disordered (rather than both, as suggested in 1). O-H = 0.96-1.00(1) Å, H-O-H = 102°.

1. Structure Reports, 27, 630.

BIS[TRI-μ-HYDROXO-BIS(TRIAMMINECOBALT(III))] TRIS(DITHIONATE)

$[(NH_3)_3Co(OH)_3Co(NH_3)_3]_2(S_2O_6)_3$

U. THEWALT, 1975. Z. anorg. Chem., 412, 29-36.

Monoclinic, $P2_1/c$, a = 10.970, b = 9.412, c = 16.069 Å, β = 117.10°, D_m = 2.29, Z = 2. Mo radiation, R = 0.052 for 1334 reflexions.

The cation consists of two face-sharing octahedra, Co-O = 1.92-1.94, Co-N = 1.94-1.97(1) Å, O-Co-O = 81, N-Co-N = 91°. The two independent anions have staggered conformations (one lying on a centre of symmetry), S-S = 2.13, S-O = 1.45 Å.

SODIUM TRIHYDROGEN SELENITE

$NaH_2D(SeO_3)_2$

Ju.Z. NOZIK, 1975. Kristallografija, 20, 169-171 [Soviet Physics - Crystallography, 20, 98-99].

Unit-cell data and structure as in 1. Neutron diffraction data (R = 0.059 for 150 reflexions) indicate that D shows a preference for the H(2) position (as in the HD_2 compound, 2).

1. Structure Reports, 22, 473; 33A, 389; 37A, 315; 38A, 345.
2. Ju.Z. NOZIK, 1974. Geokhimija, 2, 228.

CAESIUM TRIHYDROGEN SELENITE

$CsH_3(SeO_3)_2$

R. TELLGREN and R. LIMINGA, 1974. Ferroelectrics, 8, 629-636.

Triclinic, $P\bar{1}$, a = 9.347, b = 6.540, c = 5.850 Å, α = 91.44, β = 105.34, γ = 91.63°, D_m not given, Z = 2. Mo radiation, R = 0.017 for 1829 reflexions.

The structure contains two types of O-Se-O-H...O-Se-O chains with disordered hydrogen positions. SeO_3 groups are pyramidal, Se-O = 1.683-1.735 Å (indicating the disordered hydrogens, since normal values are about 1.65 and 1.75 Å for Se-O and Se-OH, respectively). Cs-O bonds stabilize the three-dimensional network. An independent study has been reported (1).

1. S. SATO, 1972. J. Phys. Soc. Japan, 32, 1670.

SCANDIUM SELENATE PENTAHYDRATE

$Sc_2(SeO_4)_3.5H_2O$

J. VALKONEN, L. NIINISTÖ, B. ERIKSSON, L.O. LARSSON and U. SKOGLUND, 1975. Acta Chem. Scand., A29, 866-872.

Triclinic, P1, a = 11.225, b = 11.804, c = 5.766 Å, α = 91.35, β = 100.10, γ = 89.03°, D_m = 2.67, Z = 2. Cu radiation, R = 0.084 for 1730 reflexions.

In two of the four discrete ScO_6 octahedra (Fig. 1), two of the Sc-O contacts are to water molecules, while there are three water contacts in the other two. The Sc-O distances average 2.09 Å, but the Sc-water contacts are slightly longer than those to selenate oxygen. In the six SeO_4 tetrahedra, the average Se-O is 1.63 Å, though there are some small, but significant, differences.

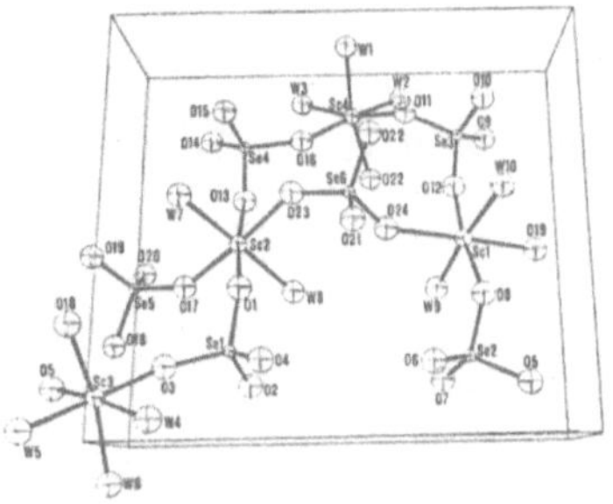

Fig. 1. A perspective view, in the c-direction, of the structure of $Sc_2(SeO_4)_3.5H_2O$.

LITHIUM TELLURATE

Li_2TeO_3

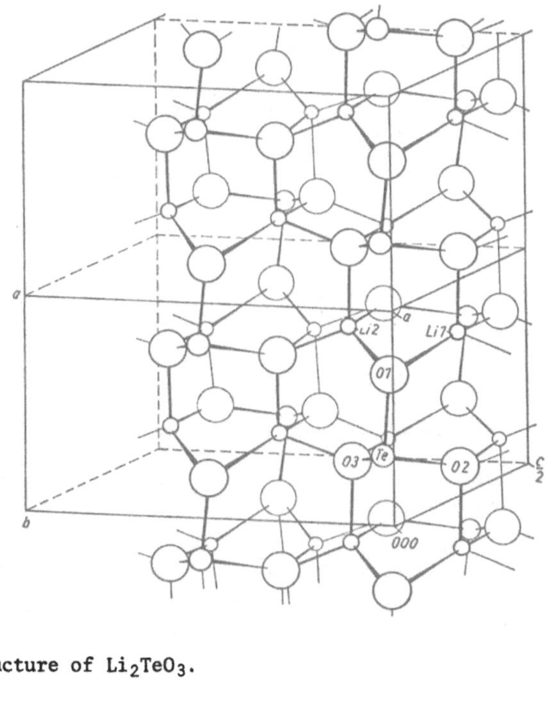

Fig. 1. Structure of Li_2TeO_3.

F. FOLGER, 1975. Z. anorg. Chem., 411, 103-110.

Monoclinic, C2/c, a = 5.069, b = 9.566, c = 13.727 Å, β = 95.4°, D_m = 3.77, Z = 8. Mo radiation, R = 0.10 for 772 reflexions (film data).

The structure (Fig. 1) contains double layers of TeO_3 trigonal pyramids and LiO_4 distorted tetrahedra; Te-O = 1.85-1.93(2) Å (next-nearest neighbour at 3.19 Å), Li-O = 1.89-2.02(8) Å. TeO_3 groups share corners with LiO_4 groups, which are linked to one another by common corners and edges.

MAGNESIUM DITELLURATE

$MgTe_2O_5$

M. TRÖMEL, 1975. Z. anorg. Chem., 418, 141-144.

Orthorhombic, Pbcn, a = 7.225, b = 10.64, c = 5.940 Å, Z = 4. Mo radiation, R = 0.123 for 402 reflexions.

Atomic positions

Atoms in 4(c) and 8(d)

	x	y	z
Te	0.1334	0.34534	0.4897
Mg	0	0.0570	1/4
O(1)	0.149	0.069	0.947
O(2)	0.190	0.204	0.320
O(3)	0	0.436	1/4

The structure contains $Te_2O_5^{2-}$ ions which consist of two TeO_3 pyramids sharing an oxygen atom, Te-O = 1.84 (terminal), 1.97(1) Å (bridging). The Te_2O_5 groups are linked into chains by longer Te-O contacts, 2.40 and 2.96 Å. Mg has distorted octahedral coordination, Mg-O = 2.08-2.12 Å. $MnTe_2O_5$ is isostructural.

MROSEITE

$CaTeO_2(CO_3)$

R. FISCHER, F. PERTLIK and J. ZEMANN, 1975. Canad. Miner., 13, 383-387.

Orthorhombic, Pbca, a = 6.988, b = 11.201, c = 10.566 Å, D_m = 4.35, Z = 8. Mo radiation, R = 0.066 for 807 reflexions.

The structure contains centrosymmetric Te_2O_4 groups which consist of two TeO_3 trigonal pyramids sharing an edge, Te-O = 1.85-2.05, Te...Te = 3.11 Å; two longer contacts to carbonate oxygens (Te-O = 2.31 and 2.55 Å) complete five-coordination for Te. Ca has eight-coordination, Ca-O = 2.38-2.73 Å, and the carbonate group has normal dimensions, C-O = 1.25-1.34(3) Å.

BARIUM TELLURATE

$BaTeO_3$

F. FOLGER, 1975. Z. anorg. Chem., 411, 111-117.

Monoclinic, $P2_1/m$, a = 4.633, b = 5.952, c = 7.308 Å, β = 111.2°, D_m = 5.51, Z = 2. Mo radiation, R = 0.070 for 1072 reflexions.

$KClO_3$-type structure (1). Te-O = 1.86-1.88, Ba-O = 2.83-3.30 Å.

1. Strukturbericht, 2, 66, 408; Structure Reports, 22, 484.

THALLIUM(III) TELLURATE(VI)

Tl_2TeO_6

B. FRIT, R. PRESSIGOUT and D. MERCURIO, 1975. Mater. Res. Bull., 10, 1305-1312.

Trigonal, P321, a = 9.070, c = 4.984 Å, D_m = 8.90, Z = 3. Mo radiation, R = 0.057 for 1505 reflexions.

Isostructural with Na_2SiF_6 (1). Each metal ion has octahedral coordination, Te-O = 1.95-1.99, Tl-O = 2.07-2.35(4) Å.

1. Structure Reports, 29, 264.

SODIUM CHLORITE TRIHYDRATE

$NaClO_2.3H_2O$

I. V. TAZZOLI, V. RIGANTI, G. GIUSEPPETTI and A. CODA, 1975. Acta Cryst., B31, 1032-1037.

II. C. TARIMCI, E. SCHEMPP and S.C. CHANG, 1975. Ibid., B31, 2146-2149.

III. V. TAZZOLI, V. RIGANTI, G. GIUSEPPETTI and A. CODA, 1975. Ibid., B31, 2750-2751.

Triclinic, $P\bar{1}$, a = 5.492, b = 6.412, c = 8.832 Å, α = 72.06, β = 87.73, γ = 70.88°, D_m = 1.72, Z = 2. Mo radiation, R = 0.040 for 2227 reflexions (I) and 0.037 for 1850 reflexions (II).

The structure (Fig. 1) contains chains along c of edge-sharing distorted Na octahedra, the chains being joined by hydrogen bonds. Only one of the chlorate oxygen atoms is bonded to Na. Cl-O = 1.557 and 1.564(1) Å, O-Cl-O = 108.2°, Na-O = 2.36-2.49, O-H...O = 2.73-3.01 Å.

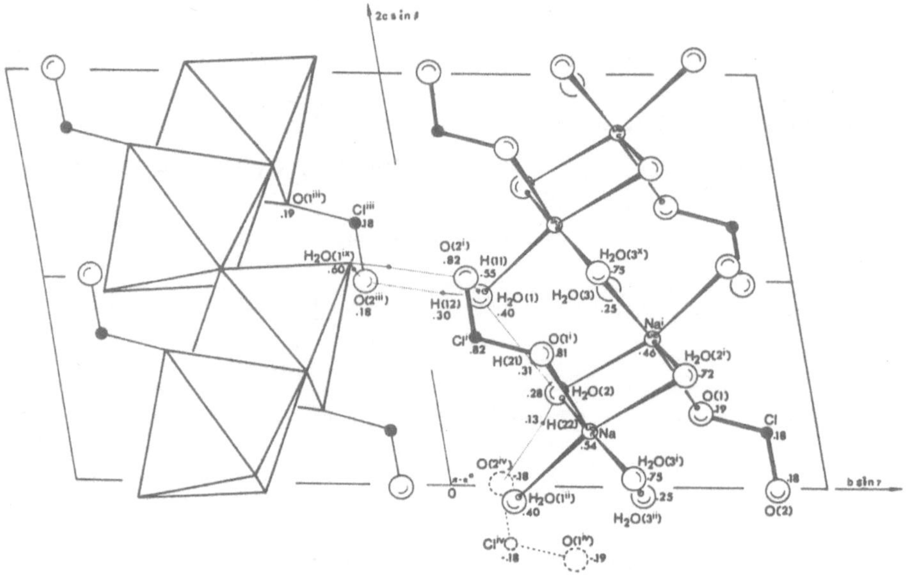

Fig. 1. Structure of sodium chlorite trihydrate (viewed down the a axis of a non-reduced cell used in I).

AMMONIUM PERCHLORATE

NH_4ClO_4

G. PEYRONEL and A. PIGNEDOLI, 1975. Acta Cryst., B31, 2052-2056.

Orthorhombic, $Pna2_1$, a = 9.227, b = 7.454, c = 5.819 Å, D_m not given, Z = 4. Mo radiation, R = 0.050 for 800 reflexions.

The structure is essentially as previously described in Pnam (1), the present parameters deviating only very slightly from the higher symmetry. The hydrogen atoms seem to be at least partially localized, and are all involved in bifurcated N-H...O bonds.

1. Strukturbericht, 1, 344, 372; 2, 413, 414; Structure Reports, 21, 357; 27, 639.

LITHIUM PERCHLORATE TRIHYDRATE

$LiClO_4.3H_2O$

A. SEQUEIRA, I. BERNAL, I.D. BROWN and R. FAGGIANI, 1975. Acta Cryst., B31, 1735-1739.

Hexagonal, $P6_3mc$, a = 7.719, c = 5.455 Å, D_m = 1.89, Z = 2. Mo radiation, R = 0.034 for 1075 reflexions; and neutron data, R = 0.087 for 247 reflexions.

The structure is as previously described (1). Cl-O = 1.458(2), Li-O = 2.133(4) Å.

1. Strukturbericht, 3, 117, 468; Structure Reports, 33A, 450.

SODIUM PERCHLORATE MONOHYDRATE

$NaClO_4.H_2O$

B. BERGLUND, J.O. THOMAS and R. TELLGREN, 1975. Acta Cryst., B31, 1842-1846.

Monoclinic, C2/c, a = 15.542, b = 5.540, c = 11.046 Å, β = 110.67°, D_m not given, Z = 8. Mo radiation, R = 0.025 for 1175 reflexions.

The structure contains a three-dimensional arrangement of ClO_4^- tetrahedra, linked by weak bifurcated O-H...O hydrogen bonds from the water molecules (Fig. 1) and by sodium ions, which are in special positions and octahedrally coordinated, Na-O = 2.36-2.42 Å.

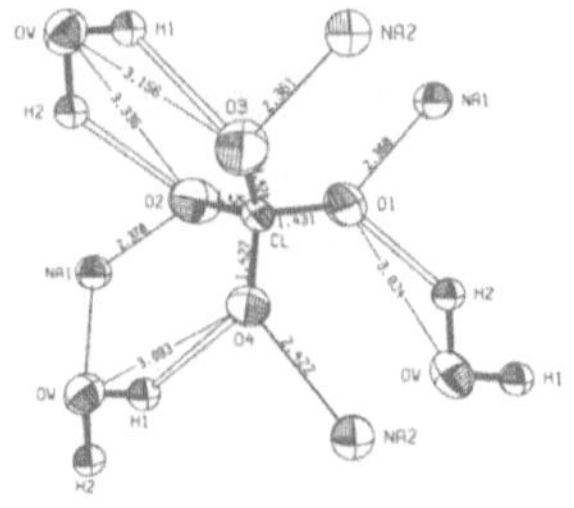

Fig. 1. The ClO_4^- environment in sodium perchlorate monohydrate.

BIS(IODOMERCURIO)-HYDROXOMERCURIO-OXONIUM PERCHLORATE

$[O(HgI)_2(HgOH)]ClO_4$

K. KÖHLER, G. THIELE and D. BREITINGER, 1975. Z. anorg. Chem., 418, 79-87.

Monoclinic, $P2_1/n$, a = 10.479, b = 13.022, c = 7.545 Å, β = 90.31°, D_m = 6.25, Z = 4. Mo radiation, R = 0.062 for 1135 reflexions.

The structure contains perchlorate anions (Cl-O = 1.40-1.57(5) Å) and planar oxonium cations with a central three-coordinate oxygen atom, O-HgI = 2.05(3), O-HgOH = 1.99(3), Hg-I = 2.57, Hg-OH = 1.98 Å, angles at O = 115-128, angles at Hg = 175-177°.

LITHIUM BROMITE MONOHYDRATE

$LiBrO_2.H_2O$

M.-T. Le BIHAN, B. GURTNER and A. KALT, 1975. Bull. Soc. Fr. Minér. Crist., 98, 223-226.

Orthorhombic, Pcab, a = 7.713, b = 13.336, c = 6.628 Å, D_m = 2.65, Z = 8. Cu radiation, R = 0.108 (films, densitometer intensities).

Atomic positions

	x	y	z
Li	0.2600	0.4416	0.3612
Br	0.0181	0.2891	0.1163
O(1)	0.0261	0.3737	-0.1000
O(2)	0.0601	0.3623	0.3438
O(W)	0.2717	0.5221	0.1112

The structure contains BrO_2^- ions (Br-O = 1.825 Å, O-Br-O = 108°) and Li^+ ions tetrahedrally coordinated to two oxygen atoms from two anions (Li-O = 1.87, 1.90) and two water oxygens (Li-O = 1.98, 1.99 Å); these form chains along c, which are linked by O-H...O hydrogen bonds (2.69 and 2.72 Å) and by a strong Br...Br interaction (3.49 Å).

ALUMINUM IODATE NITRATE HEXAHYDRATE

$Al(IO_3)_2NO_3.6H_2O$

P.D. CRADWICK and A.S. de ENDREDY, 1975. J. Chem. Soc., Dalton, 1926-1929.

Trigonal, P321, a = 6.764, c = 8.088 Å, D_m = 2.78, Z = 1. Mo radiation, R = 0.031 for 333 reflexions.

Atomic positions

	x	y	z
Al	0	0	1/2
I	2/3	1/3	0.11011
N	0	0	0
O(I)	0.5398	0.0611	0.2148
O(N)	0.1854	0	0
O(W)	0.1434	0.2694	0.3724

The structure contains discrete trigonal-pyramidal IO_3^- ions, planar NO_3^- ions, and distorted octahedral $Al(H_2O)_6^{3+}$ ions, linked by ionic forces and hydrogen bonds. Al-O = 1.887(4), I-O = 1.806(5), N-O = 1.25(1) Å.

POTASSIUM MESOPERIODATE

K_3IO_5

I. M. TRÖMEL and H. DÖLLING, 1975. Z. anorg. Chem., 411, 41-48.

II. Idem, 1975. Ibid., 411, 49-53.

Tetragonal, P4/ncc, a = 5.924, c = 18.13 Å, D_m not given, Z = 4. X-ray and neutron powder data.

The structure is closely related to that of $(NH_4)_3FeF_6$ (1), but with a doubled c-axis and IO_5 tetragonal prisms in place of the FeF_6 groups. I-O = 1.77 (x 4), 1.78 Å. The rubidium and caesium compounds are orthorhombic variants.

1. Strukturbericht, 1, 437, 449.

COPPER(II) IODATE

α-$Cu(IO_3)_2$

R. LIMINGA, S.C. ABRAHAMS and J.L. BERNSTEIN, 1975. J. Chem. Phys., 62, 4388-4399.

Monoclinic, $P2_1$, a = 5.5690, b = 5.1110, c = 9.2698 Å, β = 95.82°, at 298°K, Z = 2. Mo radiation, R = 0.063 and 0.035 for 1814 and 4272 reflexions (two data sets).

The two independent IO_3^- ions are trigonal pyramidal, I-O = 1.804-1.835(6) Å, O-I-O = 96.0-101.2°, with oxygen atoms from neighbouring anions completing 6- and 7-coordination, I...O = 2.61-3.21 Å; these polyhedra thus share corners to form a three-dimensional framework. Cu has tetragonally-elongated octahedral coordination, Cu-O = 1.958-1.988, 2.379, 2.405(6) Å.

NEODYMIUM IODATE MONOHYDRATE

$Nd(IO_3)_3.H_2O$

R. LIMINGA, S.C. ABRAHAMS and J.L. BERNSTEIN, 1975. J. Chem. Phys., 62, 755-763.

Monoclinic, $P2_1$, a = 10.201, b = 6.705, c = 7.354 Å, β = 113.11°, D_m = 4.80, Z = 2. Mo radiation, R = 0.058 for 2603 reflexions.

The three independent IO_3^- ions are trigonal pyramidal, I-O = 1.78-1.84(2) Å, O-I-O = 99°, with oxygen atoms from neighbouring anions and from water molecules completing 6-, 7-, and 7-coordination, I...O = 2.64-3.29 Å. The water molecule plays an important role in the structure by weak bonds to two I atoms and hydrogen bonds to O atoms. Nd has 8-coordination (trigonal prism with two rectangular faces capped), Nd-O = 2.38-2.52 Å.

HOLMIUM PERIODATE TETRAHYDRATE

$HoIO_5.4H_2O$

N.B. SHAMRAJ, M.B. VARFOLOMEEV, Ju.N. SAF'JANOV, E.A. KUZ'MIN and V.V. PLJUKHIN, 1975. Ž. Neorg. Khim., 20, 57-59 [Russ. J. Inorg. Chem., 20, 31-33].

Monoclinic, $P2_1/b$, a = 7.516, b = 10.40, c = 10.40 Å, γ = 118°41', D_m = 3.97, Z = 4. Mo radiation, R = 0.19 for 612 reflexions (film data).

The structure contains isolated $IO_4(OH)_2^{3-}$ octahedra (probably with trans OH groups), linked by HoO_8 polyhedra (trigonal prisms with two rectangular faces capped).

SODIUM MAGNESIUM SILICATE

$NaMg_4(Si_6O_{15}OH)(OH)_2$

V.A. DRITS, Ju.I. GONČAROV, V.A. ALEKSANDROVA, V.E. KHADŽI and A.L. DMITRIK, 1974. Kristallografija, 19, 1186-1193 [Soviet Physics - Crystallography, 19, 737-741].

Monoclinic, C2/c, a = 10.132, b = 27.12, c = 5.26 Å, β = 106.9°, Z = 4. Electron-diffraction patterns.

A structure for this fibrous material is proposed, based on strips of three pyroxene chains.

TRICALCIUM SILICATE

Ca_3SiO_5 $3CaO.SiO_2$

N.I. GOLOVASTIKOV, R.G. MATVEEVA and N.V. BELOV, 1975. Kristallografija, 20, 721-729 [Soviet Physics - Crystallography, 20, 441-445].

Triclinic, $P\bar{1}$, a = 11.67, b = 14.24, c = 13.72 Å, α = 105°30', β = 94°20', γ = 90°, Z = 18. Mo radiation, R = 0.114 for 7560 reflexions (films, visual intensities).

There is a pseudo-cell with a' = a/3, b' = b/4, c' = c/2, and the structure is quite similar to that previously proposed for the average structure in $R\bar{3}m$ (1). It consists of an array of Ca polyhedra with 6 and 7 vertices, linked mainly by common faces, and isolated SiO_4 tetrahedra which share corners with the Ca coordination polyhedra.

1. Structure Reports, 16, 336; 17, 548.

CALCIUM SULPHOSILICATE

$Ca_5(SiO_4)_2SO_4$

P.D. BROTHERTON, J.M. EPSTEIN, M.W. PRYCE and A.H. WHITE, 1974. Austral. J. Chem., 27, 657-660.

Orthorhombic, Pcmn, a = 10.182, b = 15.398, c = 6.850 Å, D_m = 2.95, Z = 4. Cu radiation, R = 0.042 for 570 reflexions.

Isostructural with silicocarnotite (1), except that there appears to be no disordering of the anions, the SO_4 ion lying on the mirror plane (S-O = 1.46-1.49(1) Å) and the SiO_4 ions in the general position (Si-O = 1.63-1.66(1) Å).

1. Structure Reports, 37A, 294.

SODIUM CALCIUM HYDROGEN SILICATE (Monoclinic form)

$NaCaHSiO_4$

V.F. KAZAK, V.I. LJUTIN, V.V. ILJUKHIN and N.V. BELOV, 1974. Kristallografija, 19, 954-957 [Soviet Physics - Crystallography, 19, 592-594].

Monoclinic, $P2_1/m$, a = 5.71, b = 5.45, c = 7.03 Å, γ = 122°, D_m not given, Z = 2. Mo radiation, R = 0.064 for 211 reflexions (films, visual intensities).

Atomic positions

	x	y	z
Ca	0	0	0
Na	0.641	0.375	3/4
Si	0.660	0.287	1/4
O(1)	0.765	0.066	1/4
O(2)	0.675	0.874	3/4
O(3)	0.792	0.500	0.059

The structure contains chains of CaO_6 octahedra, linked by SiO_4 tetrahedra (Fig. 1); Na has fivefold coordination. Si-O = 1.61-1.67, Ca-O = 2.34-2.37, Na-O = 2.29-2.84 Å. The orthorhombic form (1) has a very similar structure, except that Na has tetrahedral coordination.

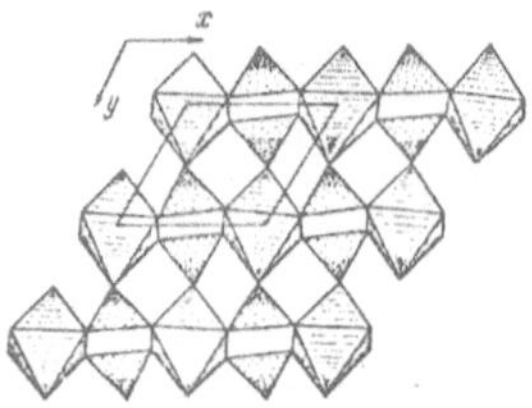

Fig. 1. CaO_6 octahedra and SiO_4 tetrahedra in $NaCaHSiO_4$.

1. Structure Reports, 38A, 385.

SODIUM ALUMINOBERYLLOSILICATE

$Na_3AlBeSi_2O_8$

Ju.A. KHARITONOV, V.M. GOLYŠEV, R.K. RASTSVETAEVA and N.V. BELOV, 1974. Kristallografija, 19, 1078-1080 [Soviet Physics - Crystallography, 19, 667-668].

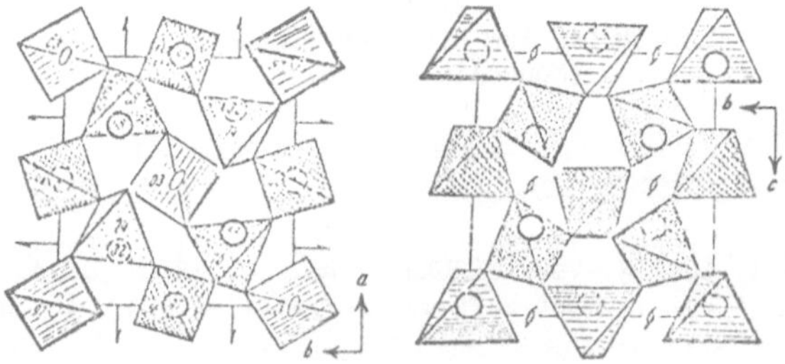

Fig. 1. Structure of $Na_3AlBeSi_2O_8$; Al tetrahedra are hatched, Si tetrahedra are dotted.

Orthorhombic, $P2_12_12$, a = 7.18, b = 6.82, c = 7.48 Å, D_m = 2.4, Z = 2. Mo radiation, R = 0.11 for 924 reflexions (films, visual intensities).

The structure (Fig. 1) contains SiO_4, AlO_4, and BeO_4 tetrahedra, joined into a framework of aluminosilicate, beryllosilicate, and aluminoberyllosilicate chains; the chains are linked by corner sharing. Na ions have 5- and 6-coordination. Si-O = 1.66, Al-O = 1.77, Be-O = 1.66, Na-O = 2.28-2.66 Å.

RUBIDIUM ALUMINUM SILICATE

$RbAlSiO_4$

R. KLASKA and O. JARCHOW, 1975. Z. Kristallogr., 142, 225-238.

Orthorhombic, $Pc2_1n$, a = 9.23, b = 5.34, c = 8.74 Å, D_m = 3.14, Z = 4. Cu radiation, R = 0.026 for 370 reflexions.

Atomic positions

	x	y	z
Rb	0.2037	0.5003	0.5009
Al	0.4166	-0.0170	0.3128
Si	0.0840	-0.0184	0.1939
O(1)	0.0848	-0.0176	0.0100
O(2)	-0.0311	-0.2247	0.2569
O(3)	0.0381	0.2566	0.2568
O(4)	0.2436	-0.0957	0.2535

The structure contains an ordered Si,Al framework with channels along c of six-membered rings of corner-sharing tetrahedra; these cavities contain the Rb ions. Mean Si-O = 1.62, Al-O = 1.74 Å; Rb-O = 2.91-3.54 Å.

STRONTIUM GALLIUM SILICATE

BARIUM GALLIUM SILICATE

STRONTIUM GALLIUM GERMANATE

$SrGa_2Si_2O_8$
$BaGa_2Si_2O_8$

I. M. CALLERI and G. GAZZONI, 1975. Acta Cryst., B31, 560-568.

Sr compound
Monoclinic, I2/c, a = 8.481, b = 13.142, c = 14.444 Å, β = 115.48°, D_m not given, Z = 8. Cu radiation, R = 0.078 for 1466 reflexions.

Ba compound
Monoclinic, I2/c, a = 8.727, b = 13.240, c = 14.608 Å, β = 115.00°, D_m not given, Z = 8. Mo radiation, R = 0.047 for 1463 reflexions.

The compounds are synthetic feldspars with a structure similar to that of celsian (1). There is complete Ga/Si order in the strontium compound (mean Si-O = 1.614, mean Ga-O = 1.821 Å) and probably also in the barium compound (although the mean bond lengths suggest some disorder, Si-O = 1.634, Ga-O = 1.805 Å). The divalent cations are seven-coordinate.

$SrGa_2Si_2O_8$
$SrGa_2Ge_2O_8$

II. M.W. PHILLIPS, H. KROLL, H. PENTINGHAUS and P.H. RIBBE, 1975. Amer. Min., 60, 659-666.

Monoclinic (pseudo-orthorhombic), $P2_1/a$, a = 9.001, 9.206, b = 9.484, 9.660, c = 8.399, 8.583 Å, β = 90.68, 90.43°, D_m not given, [Z = 4]. Mo radiation, R = 0.036 and 0.050 for 1706 and 1723 reflexions.

The materials are synthetic paracelsian analogues, and are isostructural with paracelsian (2) and hurlbutite (3). The T^{3+} and T^{4+} distributions are ordered (differently from danburite (4)).

1. Structure Reports, 24, 491.
2. Ibid., 17, 556; 24, 492.
3. Ibid., 24, 406; 40A, 238.
4. Ibid., 40A, 283.

NICKEL SILICATE

Ni_2SiO_4

C.-B. MA, 1975. Z. Kristallogr., 141, 126-137.

Cubic, Fd3m, a = 8.0424 Å, D_m not given, Z = 8. Cu radiation, R = 0.069 for 21 reflexions (powder data). (0.96 Ni + 0.04 Si) in 16(d); (0.92 Si + 0.08 Ni) in 8(a); O in 32(e): x = 0.367 (origin at 8(a)).

Spinel structure (1). The results agree with a previous study (2), except for slight differences in cation occupancies.

1. Strukturbericht, 1, 350.
2. Structure Reports, 40A, 280.

NICKEL ALUMINOSILICATE

$Ni_{10.3}Al_{11.4}Si_{2.3}O_{32}$ (Phase I)
$Ni_{17.0}Al_{13.9}Si_{5.1}O_{48}$ (Phase II)
$Ni_{12.5}Al_{7.0}Si_{4.5}O_{32}$ (Phase III)

I. C.-B. MA, K. SAHL and E. TILLMANNS, 1975. Acta Cryst., B31, 2137-2139.

II. C.-B. MA and E. TILLMANNS, 1975. Ibid., B31, 2139-2141.

III. C.-B. MA and K. SAHL, 1975. Ibid., B31, 2142-2143.

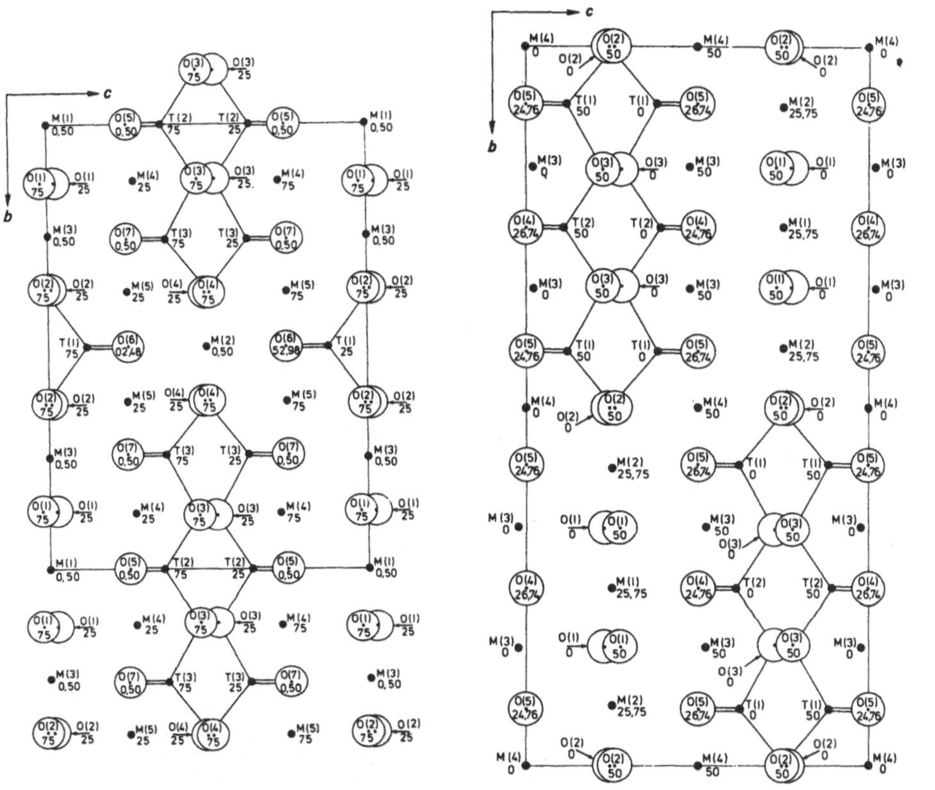

(a) (b)

Fig. 1. Structures of (a) phase I, and (b) phase II.

Phase I
Orthorhombic, Pmma, a = 5.666, b = 11.496, c = 8.098 Å, D_m not given, Z = 1. Mo radiation, R = 0.040 for 434 reflexions.

Phase II
Orthorhombic, Imma, a = 5.660, b = 17.298, c = 8.110 Å, D_m not given, Z = 1. Mo radiation, R = 0.036 for 348 reflexions.

Phase III
Orthorhombic, Imma, a = 5.665, b = 11.455, c = 8.101 Å, D_m not given, Z = 1. Mo radiation, R = 0.039 for 207 reflexions.

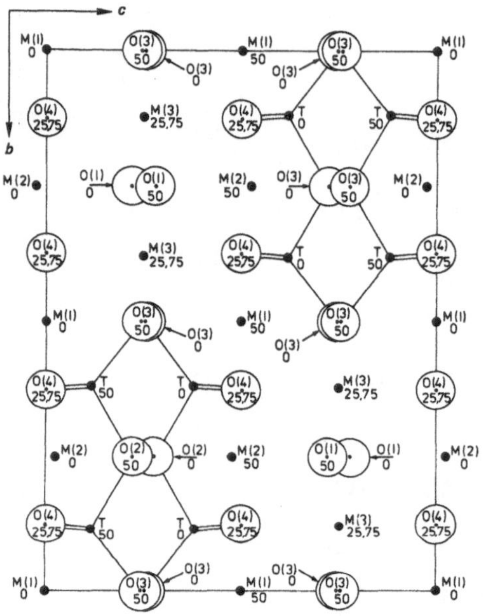

Fig. 2. Structure of phase III.

The materials were obtained at various pressures and temperatures in the SiO_2-NiO-$NiAl_2O_4$ system. All three structures are based on cubic close-packing of O atoms. In phase I (Fig. 1a) single and triple octahedral columns run parallel to a, cross-linked by single octahedral columns parallel to b. The tetrahedra are either isolated or linked in T_3O_{10} groups. Partial ordering of cations occurs in octahedral sites (Ni and Al) and in tetrahedral sites (Si, Al, and possibly some Ni).

Phase II (Fig. 1b) is isostructural with manganostibite (1), and contains triple octahedral columns running parallel to a, cross-linked by single octahedral columns parallel to b. The tetrahedra are linked in T_3O_{10} groups, and partial ordering of cations occurs in octahedral (Ni and Al) and tetrahedral (Si and Al) sites.

Phase III (Fig. 2) is isostructural with β-Co_2SiO_4 (2), and contains double octahedral columns running parallel to a, cross-linked by single octahedral columns parallel to b. The tetrahedra are linked in T_2O_7 groups, and slight ordering of Ni and Al may occur in octahedral sites.

1. Structure Reports, 35A, 443.
2. Ibid., 40A, 281, 312.

NICKEL YTTERBIUM BERYLLIUM SILICATE

$NiYb_2Be_2Si_2O_{10}$

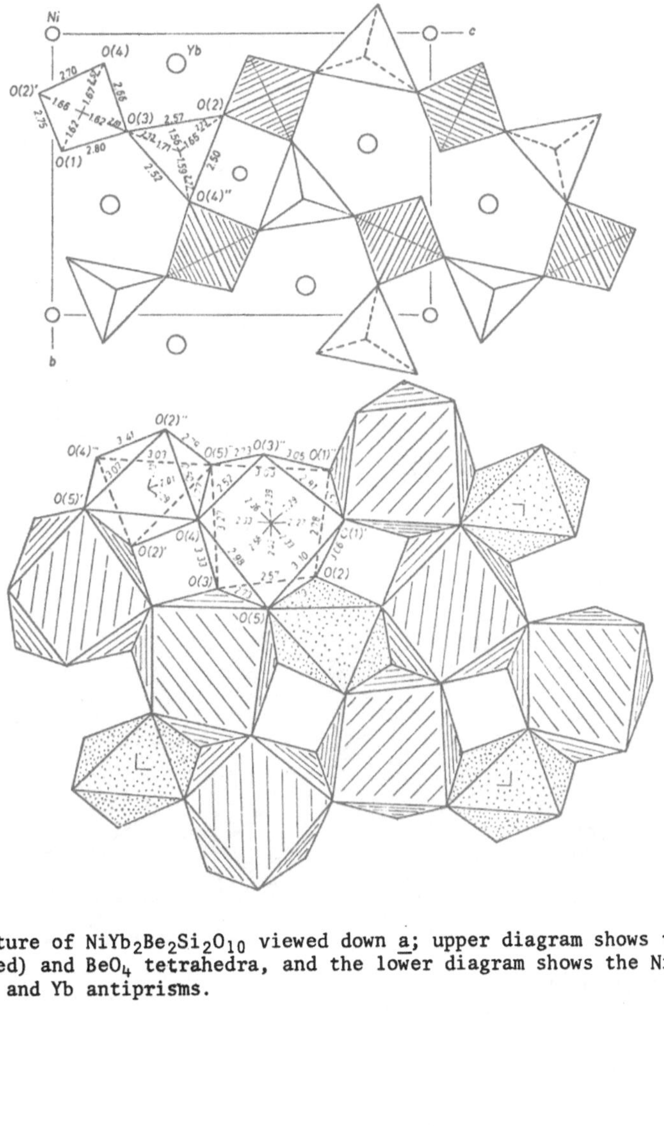

Fig. 1. Structure of $NiYb_2Be_2Si_2O_{10}$ viewed down a; upper diagram shows the SiO_4 (shaded) and BeO_4 tetrahedra, and the lower diagram shows the Ni octahedra and Yb antiprisms.

F.F. FOIT and G.V. GIBBS, 1975. Z. Kristallogr., 141, 375-386.

Monoclinic, $P2_1/c$, a = 4.664, b = 7.385, c = 9.866 Å, β = 90.02°, Z = 2. Mo radiation, R = 0.073 for 595 reflexions.

The material is synthetic and is isostructural with gadolinite (1), the structure being similar to that of datolite (1, 2), with an additional (Ni) atom at the unit cell origin. The structure (Fig. 1) is based on sheets of four- and eight-membered rings of alternating SiO_4 and BeO_4 tetrahedra; these sheets are held together by Ni and Yb ions which form a sheet of edge-sharing NiO_6 octahedra and YbO_8 antiprisms.

1. Structure Reports, 17, 554; 21, 383; 23, 470.
2. Ibid., 32A, 425; 38A, 365; 39A, 339.

ZINC METASILICATE

MAGNESIUM ZINC METASILICATE

$ZnSiO_3$
$ZnMgSi_2O_6$

N. MORIMOTO, Y. NAKAJIMA, Y. SYONO, S. AKIMOTO and Y. MATSUI, 1975. Acta Cryst., B31, 1041-1049.

	$ZnSiO_3$		$ZnMgSi_2O_6$
Crystal system	Monoclinic	Orthorhombic	Orthorhombic
Space group	C2/c	Pbca	Pbca
a (Å)	9.787	18.204	18.201
b	9.161	9.087	8.916
c	5.296	5.278	5.209
β	111.42°		
Z	8	16	8
Radiation	Mo	Mo	Mo
R	0.057	0.073	0.101
No. of reflexions	1101	1263	935

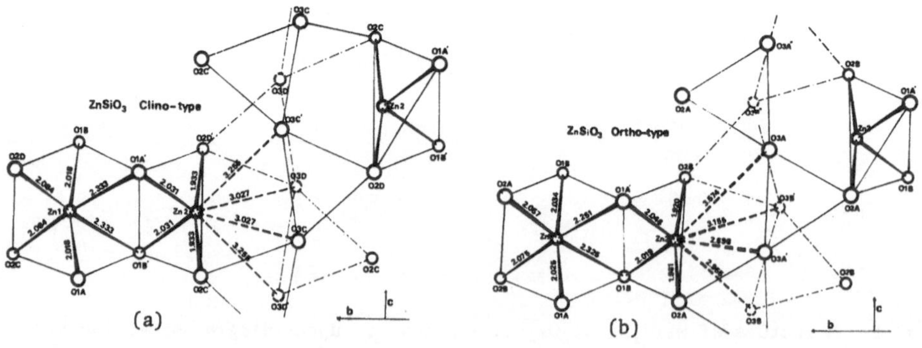

Fig. 1. Part of the structures of (a) monoclinic and (b) orthorhombic $ZnSiO_3$.

The materials have the monoclinic pyroxene (diopside, 1) and orthorhombic pyroxene (enstatite, 2) structures, with partial ordering of cations in $ZnMgSi_2O_6$ (36 and 64% Zn in M(1) and M(2) sites, respectively). The refined structure of monoclinic $ZnSiO_3$ differs in detail from that of diopside. Zn atoms have octahedral (M(1)) and tetrahedral (M(2)) coordination, with the M(2) site not coordinated to the bridging oxygen, O(3), of the SiO_3 chains (Fig. 1a). In the orthorhombic pyroxenes, M(2) sites have irregular octahedral coordination, including O(3) atoms (Fig. 1b), which results in a different shape for the SiO_3 chains.

1. Strukturbericht, 2, 130, 528.
2. Ibid., 2, 134, 529; Structure Reports, 34A, 365.

SODIUM ZINC SILICATE

$Na_2Zn_3(SiO_4)_2$

G.F. PLAKHOV, M.A. SIMONOV and N.V. BELOV, 1975. Kristallografija, 20, 46-51 [Soviet Physics - Crystallography, 20, 24-27].

Triclinic, P1, a = 5.124, b = 8.830, c = 13.504 Å, α = 72.59, β = 101.76, γ = 89.11°, D_m = 3.64, Z = 3. Mo radiation, R = 0.038 for 3370 reflexions.

The structure (Fig. 1) is characterized by alternate layers of Na-O polyhedra (octahedra and tetrahedra) and Zn-O tetrahedra which form $[Zn_9O_{24}]_n$ bands and are linked by isolated SiO_4 tetrahedra. Si-O = 1.58-1.66, Zn-O = 1.89-2.05, Na-O = 2.21-2.90 Å.

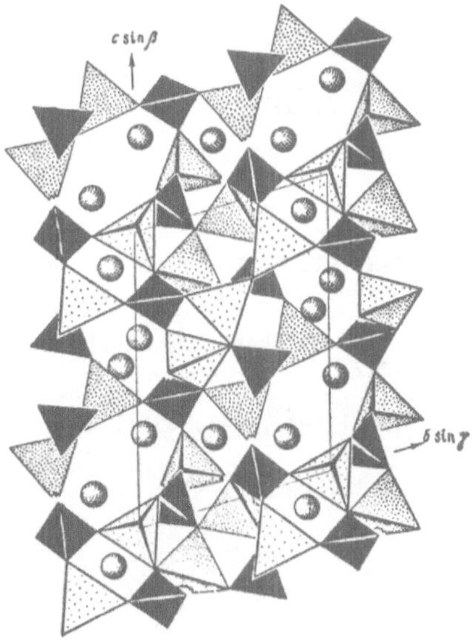

Fig. 1. Structure of $Na_2Zn_3(SiO_4)_2$.

SAMARIUM SILICON OXYNITRIDE

$Sm_5Si_3O_{12}N$

J. GAUDÉ, P. L'HARIDON, C. HAMON, R. MARCHAND and Y. LAURENT, 1975. Bull. Soc. Fr. Minér. Crist., 98, 214-217.

Hexagonal, $P6_3$, a = 9.517, c = 6.981 Å, D_m = 6.13, Z = 2. Mo radiation, R = 0.051 for 844 reflexions.

Apatite-type structure, but with no mirror plane, and possibly with an oxygen rather than a nitrogen atom unique.

SODIUM URANYL SILICATE

$Na_2[(UO_2)SiO_4]$

D.P. ŠAŠKIN, E.A. LUR'E and N.V. BELOV, 1974. Kristallografija, 19, 958-963 [Soviet Physics - Crystallography, 19, 595-597].

Tetragonal, $I4_1/acd$, a = 12.718, c = 13.376 Å, D_m = 4.85, Z = 16. No details of data, R = 0.105, but the oxygen atoms refine to unreasonable positions.

A structure is derived, mainly from packing considerations, in which U has octahedral, Si tetrahedral, and Na ions distorted cubic coordinations.

AXINITE

$(Fe,Mn)Ca_2Al_2BSi_4O_{15}(OH)$

I. T. ITO, Y. TAKÉUCHI, T. OZAWA, T. ARAKI, T. ZOLTAI and J.J. FINNEY, 1969. Proc. Japan Acad., 45, 490-494.

II. Y. TAKÉUCHI, T. OZAWA, T. ITO, T. ARAKI, T. ZOLTAI and J.J. FINNEY, 1974. Z. Kristallogr., 140, 289-312.

III. Y. TAKÉUCHI, 1975. Z. Kristallogr., 141, 471-472.

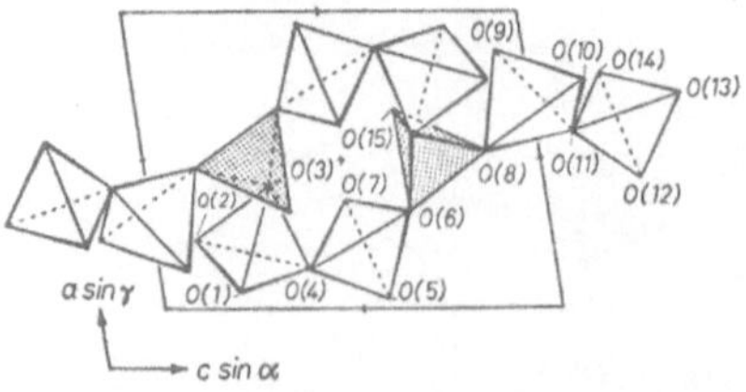

Fig. 1. The $B_2Si_8O_{30}$ group in axinite (BO_4 tetrahedra shaded).

Triclinic, $P\bar{1}$, a = 7.157, b = 9.200, c = 8.959 Å, α = 91.8, β = 98.1, γ = 77.3°, Z = 2. Mo radiation, R = 0.042 for 3315 reflexions.

The previous structure (1) is incorrect. The revised structure contains a $B_2Si_8O_{30}$ group of four pairs of SiO_4 tetrahedra connected by two BO_4 tetrahedra (Fig. 1), and finite octahedral chains of four Al and two Fe(Mn) connected into sheets by elongated Ca antiprisms (Fig. 2); the sheets are linked by the $B_2Si_8O_{30}$ groups. The OH group is bonded to 2 Al, 1 Ca, and hydrogen bonded to a silicate oxygen atom.

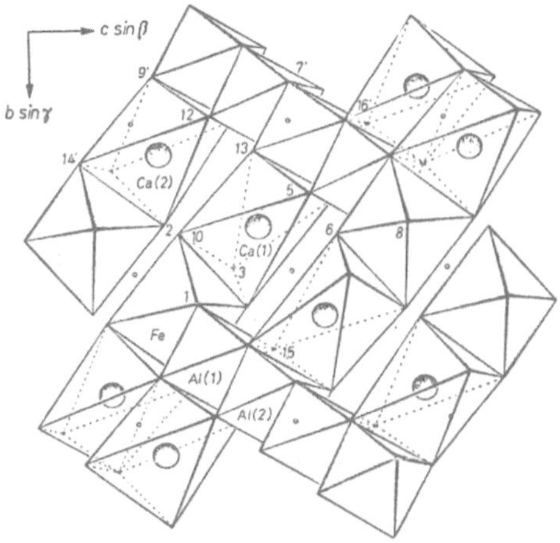

Fig. 2. The polyhedral sheet in axinite.

1. Structure Reports, 16, 349.

BABINGTONITE

$Ca_2Fe_2Si_5O_{14}(OH)$

A.L. KOSOI, 1975. Kristallografija, 20, 730-739 [Soviet Physics - Crystallography, 20, 446-451].

Triclinic, P1 [compare 1], a = 12.18, b = 6.68, c = 7.50 Å, α = 93°59', β = 112°19', γ = 86°3', Z = 2. Mo radiation, R = 0.059 for 3000 reflexions. [Independent study in 1.]

The unit cell parameters are similar to those of rhodonite (2), but there are quite large differences in atomic parameters. The structure is nearly centrosymmetrical, with maximum displacement of about 0.1 Å from $P\bar{1}$. It contains chains of corner-sharing SiO_4 tetrahedra, with a repeat pattern of five tetrahedra; Fe^{2+}, Fe^{3+}, and one Ca ion have octahedral coordination, and the other Ca ion has eight-coordination.

1. Structure Reports, 38A, 362.
2. Ibid., 22, 506; 23, 476; 28, 258.

1M-BIOTITE

$2M_1$-BIOTITE

$K(Mg,Fe)_3AlSi_3O_{10}(OH)_2$

H. TAKEDA and M. ROSS, 1975. Amer. Min., 60, 1030-1040.

1M
Monoclinic, C2/m, a = 5.317, b = 9.22, c = 10.09 Å, β = 100.2°, Z = 2. Mo radiation, R = 0.044 for 649 reflexions.

$2M_1$
Monoclinic, C2/c, a = 5.315, b = 9.22, c = 19.95 Å, β = 95.1°, Z = 4. Mo radiation, R = 0.056 for 875 reflexions.

The 1M structure is the same as that of 1M-phlogopite (1). The $2M_1$ structure is quite similar, but with differences in the position of two oxygen atoms as a result of relative shifting of the upper and lower triads of octahedral oxygens. The deformations of the octahedral and tetrahedral sheets in $2M_1$-biotite are different from those in $2M_1$-muscovite (2).

1. Structure Reports, 38A, 372; 39, 337.
2. Ibid., 37A, 339.

BRAUNITE

Mn_7SiO_{12}

J.P.R. de VILLIERS, 1975. Amer. Min., 60, 1098-1104.

Tetragonal, $I4_1/acd$, a = 9.432, c = 18.703 Å, D_m = 4.77, Z = 8. Mo radiation, R = 0.025 for 486 reflexions.

The structure is very similar to that described previously (1), but in a different space group; $CaMn_6SiO_{12}$ (2) is isostructural. Mn^{2+} has distorted cubic coordination (Mn-O = 2.17, 2.51 Å), Si^{4+} has tetrahedral coordination (Si-O = 1.62 Å), and the non-equivalent Mn^{3+} ions have distorted octahedral coordinations (Mn-O = 1.86-2.28 Å).

1. Structure Reports, 9, 251.
2. Ibid., 31A, 217.

CARPHOLITE

$MnAl_2(Si_2O_6)(OH)_4$

I.S. NAUMOVA, E.A. POBEDIMSKAJA and N.V. BELOV, 1974. Kristallografija, 19, 1155-1160 [Soviet Physics - Crystallography, 19, 718-721].

Orthorhombic, Ccca, a = 13.831, b = 20.296, c = 5.121 Å, D_m not given, Z = 4. R = 0.06.

Isostructural with ferrocarpholite (1).

1. Structure Reports, 20, 406.

CAVANSITE (DEHYDRATED)

$Ca(VO)(Si_4O_{10}).H_2O$

R. RINALDI, J.J. PLUTH and J.V. SMITH, 1975. Acta Cryst., B31, 1598-1602.

Orthorhombic, Pcmn, a = 9.368, b = 12.808, c = 9.550 Å, D_m not given, Z = 4. Cu radiation, R = 0.036 for 571 reflexions.

The structure (Fig. 1) is similar to that of the hydrated mineral (1), with removal of water from between the silicate layers. The V position is disordered, with a small (10%) occupancy of an alternate position; both positions have square-pyramidal geometry. Ca has 5-fold coordination in the dehydrated sample.

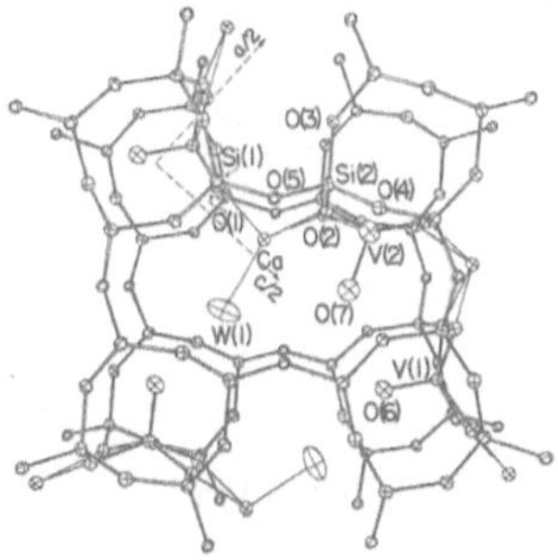

Fig. 1. Structure of dehydrated cavansite; V(2) is one of the 0.1 occupancy V positions.

1. Structure Reports, 39A, 338.

CHLORITOID

$(Fe,Mg)_2(Al,Fe)_4Si_2O_{10}(OH)_4$

R.H. HANSCOM, 1975. Acta Cryst., B31, 780-784.

Monoclinic, C2/c, a = 9.482, b = 5.484, c = 18.182 Å, β = 101.74°, D_m = 3.56, Z = 4 [not 8]. Mo radiation, R = 0.041 for 3047 reflexions.

The structure and cation occupancies are as previously suggested (1). The structure contains two octahedral layers connected by isolated silicon tetrahedra. Substitution of Mg for Fe^{2+} and Fe^{3+} for Al occurs in the trioctahedral layer, with mean M-O = 2.163 and 1.935 Å for these sites, respectively. Both sites in the Al octahedral layer have mean M-O = 1.896 Å, and no substitution occurs in these sites. The mean tetrahedral bond distance is 1.643 Å. The hydroxyl groups in one octahedral layer form hydrogen bonds with oxygen atoms in the adjacent octahedral layer, O-H...O = 2.666 and 2.814 Å.

1. Structure Reports, 16, 364; 21, 463.

CLINOPTILOLITE

$(Ca,Na,K,Mg)Al_2Si_7O_{18}.6H_2O$

A. ALBERTI, 1975. Miner. (Tschermaks) Petrogr. Mitt., 22, 25-37.

Monoclinic, C2/m, a = 17.646, 17.641, b = 17.898, 18.031, c = 7.397, 7.402 Å, β = 116°22', 116°26', for specimens from Agoura, California and Siusi, Italy, respectively, Z = 4. Mo radiation, R = 0.051 and 0.057 for 1515 and 1141 reflexions.

Isostructural with heulandite (1), with partial ordering of Al in the Si(2) site, but with a new partially-occupied cation site at (0,0,0).

1. Structure Reports, 33A, 484; 38A, 368.

CUPROSKLODOWSKITE

$Cu[(UO_2)_2(SiO_3OH)_2].6H_2O$

A. ROSENZWEIG and R.R. RYAN, 1975. Amer. Min., 60, 448-453.

Triclinic, P$\bar{1}$, a = 7.052, b = 9.267, c = 6.655 Å, α = 109.23, β = 89.84, γ = 110.01°, D_m = 3.83, Z = 2. Mo radiation, R = 0.030 for 1143 reflexions.

The structure is as previously described (jachimovite, 1) except that an additional oxygen atom is found, an oxygen being moved from the previous special to a general position. The fundamental structural unit is a sheet of composition $[(UO_2)_2(SiO_3OH)_2]^{2-}$ (Fig. 1). Cu lies on a centre of symmetry and has square-planar coordination to four H_2O molecules at 1.96 Å, with two uranyl oxygen atoms at 2.48 Å completing a distorted octahedron. U-O = 1.77 (uranyl), 2.29-2.43, Si-O = 1.59, 1.60, 1.61, 1.67 Å.

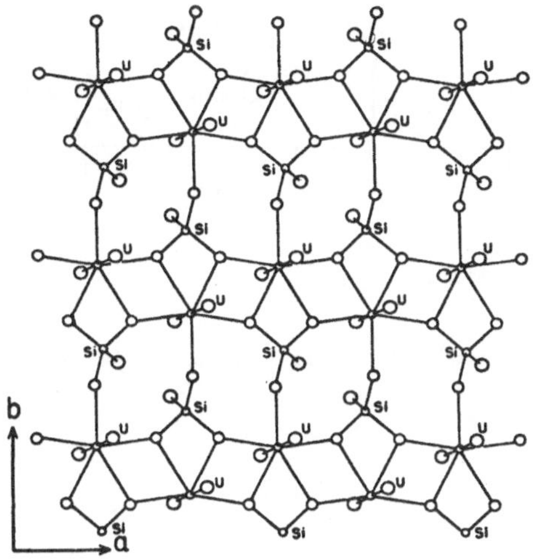

Fig. 1. The uranyl silicate layer in cuprosklodowskite.

1. Structure Reports, 28, 277.

CYMRITE

$BaSi_2Al_2O_8.H_2O$

V.A. DRITS, A.A. KAŠAEV and G.V. SOKOLOVA, 1975. Kristallografija, 20, 280-286 [Soviet Physics - Crystallography, 20, 171-175].

Monoclinic, $P2_1$, a = 5.33, b = 36.6, c = 7.67 Å, β = 90°, D_m not given, Z = 8. Mo radiation, R = 0.14 (film data).

The mineral is from California, a specimen from Alaska having been studied previously (1). The structure (Fig. 1) contains double layers of tetrahedra in which Si and Al are disordered (mean T-O = 1.68 Å). Each Ba ion has 9 oxygen near-neighbours, and the four water molecules in the asymmetric unit are distributed over eight positions.

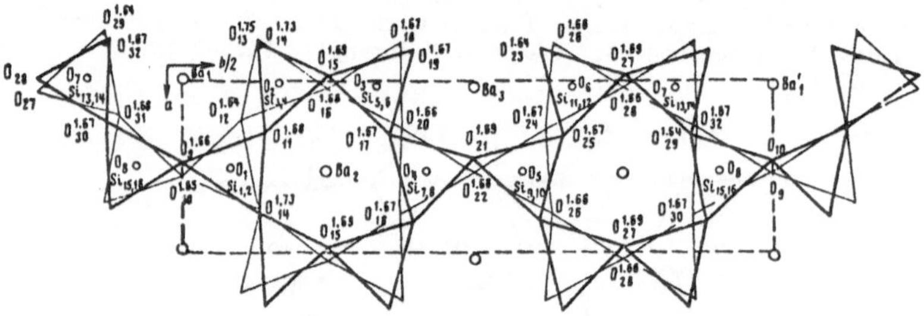

Fig. 1. Structure of cymrite.

1. V.A. DRITS and A.A. KAŠAEV, 1968. Kristallografija, 13, 809.

DIOPSIDE

(Mg,Fe,Al)(Mg,Fe,Ca,Na)Si_2O_6

R.H. McCALLISTER, L.W. FINGER and Y. OHASHI, 1974. Carnegie Inst. Washington, Year Book, 73, 518-522.

Monoclinic, [C2/c], a = 9.699, b = 8.871, c = 5.251 Å, β = 107.03°, D_m not given, [Z = 4]. Mo radiation, R = 0.063 for 631 reflexions.

Diopside structure (1). Site occupancies are: M(1) = 0.85 Mg + 0.03 Fe^{2+} + 0.03 Fe^{3+} + 0.09 Al, M(2) = 0.23 Mg + 0.10 Fe^{2+} + 0.55 Ca + 0.12 Na, T = 0.99 Si + 0.01 Al.

1. Strukturbericht, 2, 130, 528.

FAYALITE

Fe_2SiO_4

J.R. SMYTH, 1975. Amer. Min., 60, 1092-1097.

Orthorhombic, Pbnm, a = 4.818-4.860, b = 10.471-10.559, c = 6.086-6.150 Å, at 20-900°C, Z = 4. Mo radiation, R = 0.041-0.055 for 625-306 reflexions at 20, 300, 600, 900°C.

Olivine structure as previously described (1). With increasing temperature the Fe-O distances become more regular and the central bond angles of the octahedra show increasing distortion.

1. Structure Reports, 28, 265; 33A, 468.

STRONTIUM FELDSPAR

$SrAl_2Si_2O_8$

G. CHIARI, M. CALLERI, E. BRUNO and P.H. RIBBE, 1975. Amer. Min., 60, 111-119.

Monoclinic, I2/c, a = 8.388, b = 12.974, c = 14.263 Å, β = 115.2°, D_m not given, Z = 8. Mo radiation, R = 0.047 for 1316 reflexions.

The material is synthetic. The structure is partially disordered and similar to that of I2/c celsian (1). Mean (Al,Si)-O = 1.626, 1.630, 1.732, 1.735 Å; Sr can be considered to be 7- or 9-coordinate.

1. Structure Reports, 24, 491.

GILLESPITE II

$BaFeSi_4O_{10}$

R.M. HAZEN and C.W. BURNHAM, 1975. Amer. Min., 60, 937-938.

Correction for absorption effects in the high-pressure cell, and further refinement (R now 0.063) results in minor changes in the atomic positions (1).

1. Structure Reports, 40A, 284.

GROSSULARITE

PYROPE

$Ca_3Al_2Si_3O_{12}$
$Mg_3Al_2Si_3O_{12}$

E.P. MEAGHER, 1975. Amer. Min., 60, 218-228.

Cubic, Ia3d, a = 11.846, 11.880, 11.917 Å for grossularite at 25, 368, 675°C; 11.459, 11.490, 11.507, 11.530 Å for pyrope at 25, 350, 550, 750°C; Z = 8. Mo radiation, R = 0.18-0.29 for 7 data sets. Garnet structure (1).

The Si-O bond distances do not change significantly with temperature (1.647-1.654(6) Å for grossularite, 1.635-1.636(6) Å for pyrope). The other distances increase slightly with temperature: Al-O = 1.921-1.937 and 1.887-1.897, Ca-O = 2.319-2.333, 2.491-2.505, Mg-O = 2.197-2.210, 2.342-2.373(6) Å. In pyrope the SiO_4 tetrahedra rotate with increasing temperature to allow the shared octahedral edge to lengthen at a greater rate than the unshared edge; in grossularite the tetrahedra do not rotate.

1. Strukturbericht, 1, 363, 411; 2, 520, 521; Structure Reports, 22, 500; 26, 416; 28, 246; 30A, 426; 31A, 228; 37A, 334.

HOLMQUISTITE

$Li_2(Mg,Fe)_3Al_2Si_8O_{22}(OH)_2$

M.C. IRUSTETA and E.J.W. WHITTAKER, 1975. Acta Cryst., B31, 145-150.

Crystal data as in 1. Mo radiation, R = 0.056 for 1160 reflexions.

The results are essentially as previously given (1), with Li preferentially occupying site M(4) and Fe^{2+} in M(3). Some details differ from those in 2.

1. Structure Reports, 34A, 390.
2. A.L. LITVIN, I.V. GINZBURG, L.N. EGOROVA and S.S. OSTAPENKO, 1973. Konst. Sviosta Miner., 7, 18.

IFTISITE

$(Y,Ln)_4(F,OH)_6TiO(SiO_4)_2$

V.P. BALKO and V.V. BAKAKIN, 1975. Ž. Strukt. Khim., 16, 837-842.

Orthorhombic, Cmcm, a = 14.949, b = 10.626, c = 7.043 Å, Z = 4. Mo radiation, R = 0.067 for 385 reflexions.

Atomic positions

	x	y	z
Ti	0	0	1/2
Y(1)	0.1464	0.4368	1/4
Y(2)	0.1313	0.7933	1/4
Si	0.1559	0.1200	1/4
O(1)	0.0954	0.1262	0.0553
O(2)	0.2952	0.4854	1/4
O(3)	0.2201	0.2487	1/4
O(4)	1/2	0.4337	1/4
F(1)	0.3582	0.1181	0.0610
F(2)	0	0.4204	1/4
F(3)	0	0.6987	1/4

The structure is built up from SiO_4 tetrahedra, TiO_6 octahedra, and Y ions with 7- and 8-coordination. Si-O = 1.60-1.67, Ti-O = 1.90, 2.00, Y-O = 2.27-2.47, Y-F = 2.20-2.34 Å.

ILVAITE

$CaFe_3(OH)O(Si_2O_7)$

A. BERAN and H. BITTNER, 1974. Miner. (Tschermaks) Petrogr. Mitt., 21, 11-29.

The results are essentially as previously described by the same authors (1). Fe^{2+} is in the (B) site, and (Fe^{2+} + Fe^{3+}) in the (A) site.

1. Structure Reports, 38A, 369.

JOAQUINITE

$NaFeBa_2Ln_2Ti_2Si_8O_{26}(OH).H_2O$

E. DOWTY, 1975. Amer. Min., 60, 872-878.

Monoclinic, C2, a = 10.516, b = 9.686, c = 11.833 Å, β = 109.67°, Z = 2. R = 0.086 for 2630 reflexions.

The main features of the structure are essentially as described in an independent study (1), except that the space group is C2 (rather than C2/m) and three additional atoms (Na, OH, and H_2O) are found [however there are some anomalies in the final difference synthesis of the present study].

1. Structure Reports, 38A, 369.

OXY-KAERSUTITE

$Ca_2(Na,K)(Mg,Fe)_4Ti(Si_6Al_2)O_{22}(OH,F)_2$

M. KITAMURA, M. TOKONAMI and N. MORIMOTO, 1975. Contrib. Mineral. Petrol., 51, 167-172.

The cation distributions are determined from neutron diffraction data. The results agree with X-ray studies (1, 2), with Ti preferentially in site M1.

1. M. KITAMURA and M. TOKONAMI, 1971. Sci. Rep. Tohoku Univ., Ser 3, 11, 125.
2. Structure Reports, 39A, 342.

LEVYNITE

$(Ca,Na,K)Al_2Si_4O_{12}.6H_2O$ (idealized)

S. MERLINO, E. GALLI and A. ALBERTI, 1975. Miner. (Tschermaks) Petrogr. Mitt., 22, 117-129.

Rhombohedral, $R\bar{3}m$, a = 10.87 Å, α = 75°42' (hexagonal cell has a = 13.338, c = 23.014 Å), Z = 3. Cu radiation, R = 0.092 for 613 reflexions.

The framework of layers of single and double six-membered rings of $(Si,Al)O_4$ tetrahedra is as previously described (1). The cations are distributed over five sites on the threefold axis; one site is fully occupied by Ca, with the other four sites partially occupied. Three of the four water sites are also only partially occupied.

1. Structure Reports, 23, 491.

LITIDIONITE

$CuNaKSi_4O_{10}$

J.M. MARTIN POZAS, G. ROSSI and V. TAZZOLI, 1975. Amer. Min., 60, 471-474.

Triclinic, $P\bar{1}$, a = 9.80, b = 8.01, c = 6.97 Å, α = 114.1, β = 99.5, γ = 105.6°, D_m = 2.75, Z = 2. Mo radiation, R = 0.032 for 1450 reflexions.

Isostructural with fenaksite, $FeNaKSi_4O_{10}$ (1). The structure (Fig. 1) contains tubular chains of silicate tetrahedra, formed by condensation of vlasovite-type chains, and interconnected by fivefold coordinated (square pyramid) Cu and Na ions. K ions are in large cavites within the chains. Cu-O = 1.96-1.99 (4 distances), 2.55 (next closest at 3.24), Na-O = 2.39-2.58 (5 distances, next closest at 2.85, 2.93), Si-O = 1.57-1.64, K-O = 2.67-3.17 Å (8 distances).

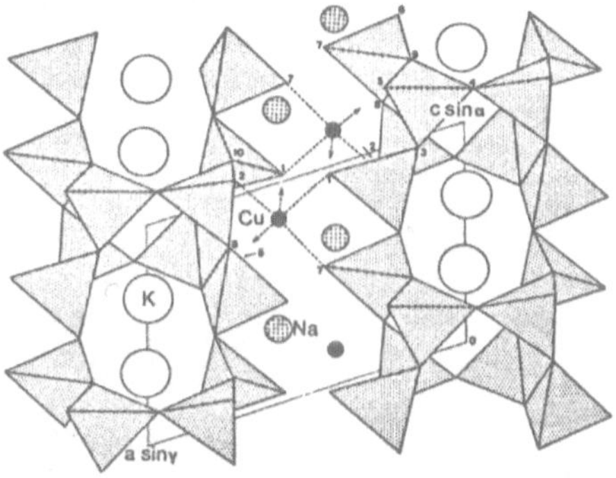

Fig. 1. Structure of litidionite.

1. Structure Reports, 35A, 480.

BETALOMONOSOVITE

$Na_7Ti_4Si_4P_2O_{23}(OH)_3$

R.K. RASTSVETAEVA, M.I. SIROTA and N.V. BELOV, 1975. Kristallografija, 20, 259-264 [Soviet Physics - Crystallography, 20, 158-160].

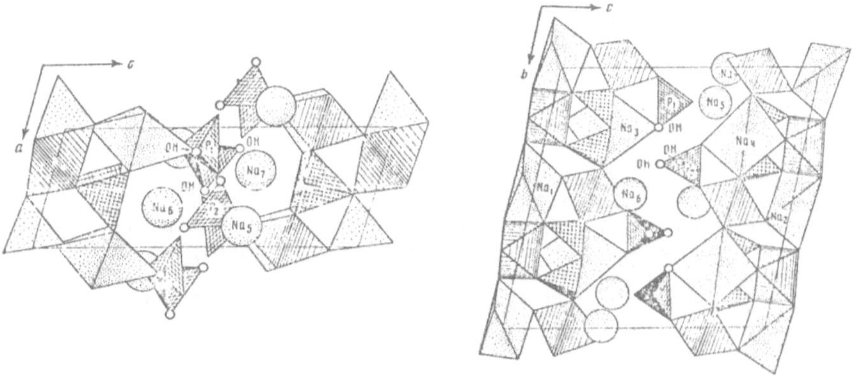

Fig. 1. Two views of the structure of betalomonosovite.

Triclinic, $P\bar{1}$, a = 5.34, b = 14.26, c = 14.23 Å, α = 102°33', β = 105°52', γ = 89°06', Z = 2. Mo radiation, R = 0.154 for 1091 reflexions (film data).

The structure (Fig. 1) is related to that described briefly for lomonosovite (1). It contains phosphate and silicate tetrahedra, TiO_6 octahedra, and Na ions with 5-, 6-, and 8-coordination.

1. Structure Reports, 37A, 353.

MARGARITE

$CaAl_2(Si,Al)_4O_{10}(OH)_2$

S. GUGGENHEIM and S.W. BAILEY, 1975. Amer. Min., 60, 1023-1029.

Monoclinic, Cc, a = 5.104, b = 8.829, c = 19.148 Å, β = 95.46°, Z = 4. Mo radiation, R = 0.075 for 1071 reflexions.

The mineral is a $2M_1$ dioctahedral mica. The structure is essentially as previously described in C2/c (1), but with nearly complete ordering of tetrahedral Si and Al resulting in the lower symmetry. Refinement of the muscovite-$2M_1$ structure with the neutron diffraction data of 2 indicated a disordered C2/c structure.

1. Y. TAKÉUCHI, 1965. Clays Clay Miner., 13, 1.
2. R. ROTHBAUER, 1971. Neues Jb. Miner., Mh., 143; Structure Reports, 37A, 339.

MAZZITE (DEHYDRATED)

$Na_{0.3}K_{2.5}Ca_{1.4}Mg_{2.1}(Al_{9.9}Si_{26.5}O_{72}).7H_2O$

R. RINALDI, J.J. PLUTH and J.V. SMITH, 1975. Acta Cryst., B31, 1603-1608.

Hexagonal, $P6_3/mmc$, a = 18.007, c = 7.608 Å, D_m not given, Z = 1. Cu radiation, R = 0.06 for 353 reflexions.

The aluminosilicate framework of mazzite (1) consists of gmelinite-type cages, cross-linked to form 12-rings perpendicular to c, which are connected to form channels whose walls consist of alternating ladders of 4- and 5-rings. This framework undergoes only minor distortion on dehydration, but the cations migrate to maintain a stable configuration (Fig. 1), and all the water in the 12-ring channels is removed. No evidence of Si,Al ordering is found.

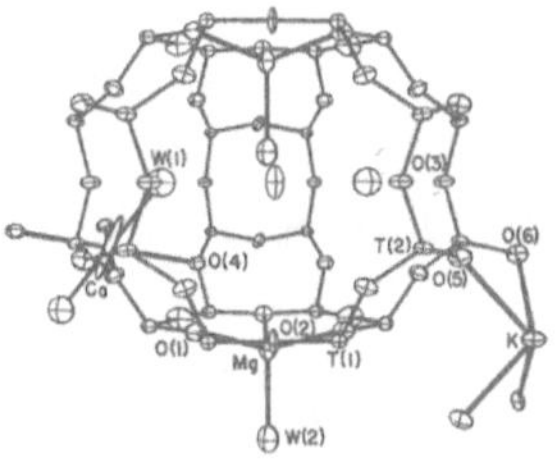

Fig. 1. Structure of dehydrated mazzite.

1. Structure Reports, 40A, 286.

DIOCTAHEDRAL 1M MICA

$KAl_2Si_4O_{10}(OH)_2$

O.V. SIDORENKO, B.B. ZVJAGIN and S.V. SOBOLEVA, 1975. Kristallografija, 20, 543-549 [Soviet Physics - Crystallography, 20, 332-335].

Monoclinic, C2, a = 5.186, b = 8.952, c = 10.12 Å, β = 101°50', Z = 2. Electron diffraction, R = 0.109 for 588 reflexions. Previous study in 1.

Mica structure, with an ordered Si/Al distribution in the tetrahedra and a reduction in symmetry from C2/m [in e.g. 2], average rotation 9.3° for the bases of the tetrahedra and 6.9° for those of the octahedra.

1. S.V. SOBOLEVA and B.B. ZVJAGIN, 1968. Kristallografija, 13, 605.
2. Structure Reports, 39A, 343.

DIOCTAHEDRAL $2M_2$ MICA

$KAl_2(Si_3Al)O_{10}(OH)_2$

A.P. ZHOUKHLISTOV, B.B. ZVYAGIN, S.V. SOBOLEVA and A.F. FEDOTOV, 1973. Clays Clay Miner., 21, 465-470.

Monoclinic, C2/c, a = 8.965, b = 5.175, c = 20.31 Å, β = 100°40', D_m not given, Z = 4. Electron diffraction patterns.

Muscovite-type structure with cubic packing of oxygens in the layers, which are stacked $\sigma_4\sigma_5\sigma_4\sigma_5$. Interlayer K ions have trigonal prismatic coordination, and there is partial ordering of Si and Al.

NAMBULITE

$(Li,Na)Mn_4Si_5O_{14}(OH)$

H. NARITA, K. KOTO, N. MORIMOTO and M. YOSHII, 1975. Acta Cryst., B31, 2422-2426.

Triclinic, $P\bar{1}$, a = 7.621, b = 11.761, c = 6.731 Å, α = 92°46', β = 95°5', γ = 106°52', D_m not given, Z = 2. Mo radiation, R = 0.086 for 2233 reflexions.

The structure (Fig. 1) contains infinite silicate chains with a repeat of five tetrahedra, and Mn polyhedral bands, both parallel to [110]. Mn ions have octahedral and seven coordination; (Li,Na) has distorted square-antiprismatic coordination. Si-O = 1.58-1.68, Mn-O = 2.09-2.85, (Li,Na)-O = 2.16-2.87 Å. The structure is more closely related to that of babingtonite, $Ca_2Fe_2Si_5O_{14}(OH)$ (1), than to that of rhodonite, $CaMn_4Si_5O_{15}$ (2).

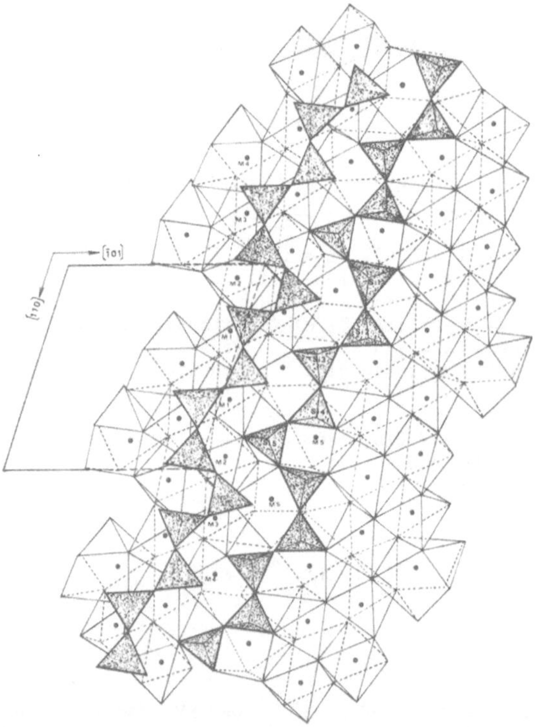

Fig. 1. Structure of nambulite.

1. Structure Reports, 38A, 362.
2. Ibid., 22, 506; 23, 476; 28, 258.

NAUJAKASITE

$Na_6FeAl_4Si_8O_{26}$

R. BASSO, A. DAL NEGRO, A. DELLA GIUSTA and L. UNGARETTI, 1975. Bull. Grønlands Geol. Unders., 116, 11-24.

Monoclinic, C2/m, a = 15.025, b = 7.991, c = 10.486 Å, β = 113°40', D_m = 2.622, Z = 2. Mo radiation, R = 0.041 for 1496 reflexions.

The structure (Fig. 1) contains double layers of corner-sharing tetrahedra; each single layer contains rings of six tetrahedra, and the linkage between two single layers gives rise to rings of four and six tetrahedra. One tetrahedral site is occupied by Si (mean Si-O = 1.62 Å) and the other two by (0.5Si + 0.5Al) (mean T-O = 1.68 Å). The double layers are linked by FeO_6 tetragonal bipyramids (Fe-O = 2.06, 2.72 Å) and irregular Na coordination polyhedra (Na-O = 2.26-3.20 Å).

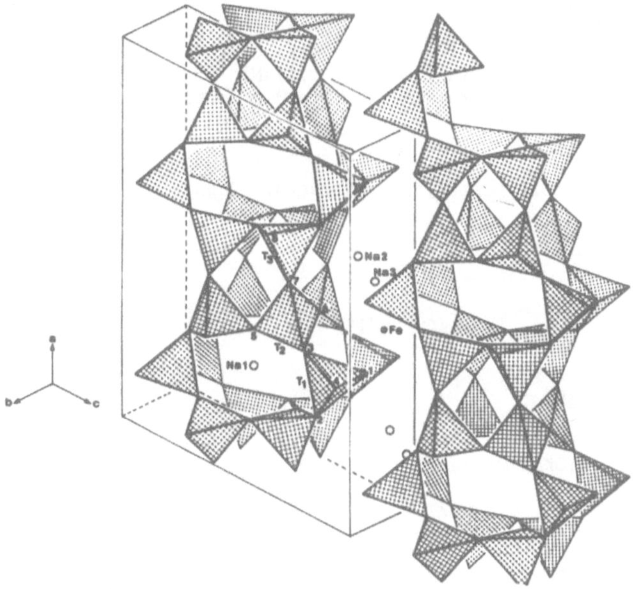

Fig. 1. Structure of naujakasite.

COBALT OLIVINE

$Co_{1.1}Mg_{0.9}SiO_4$

S. GHOSE and C. WAN, 1974. Contrib. Miner. Petr., 47, 131-150.

Olivine structure (1) with (0.73 Co + 0.27 Mg) in M(1) and (0.37 Co + 0.63 Mg) in M(2). R = 0.044 for 612 reflexions.

1. Strukturbericht, 1, 352.

NICKEL-MAGNESIUM OLIVINE

$NiMgSiO_4$

V. RAJAMANI, G.E. BROWN and C.T. PREWITT, 1975. Amer. Min., 60, 292-299.

Orthorhombic, [Pbnm], a = 4.7366, b = 10.1716, c = 5.9374 Å, Z = 4. Mo radiation, R = 0.029 for 650 reflexions.

Olivine structure (1). In spite of the small difference in size (0.03 Å) there is significant cation ordering: M(1) = 0.77 Ni + 0.23 Mg, M(2) = 0.26 Ni + 0.74 Mg. This probably results from crystal field effects, and is in accord with previous predictions.

1. Strukturbericht, 1, 352, 401; 2, 121; Structure Reports, 15, 306; 28, 245; 39A, 348.

OMPHACITE

$(Ca,Na)(Mg,Fe,Al)(Si,Al)_2O_6$

T. MATSUMOTO, M. TOKONAMI and N. MORIMOTO, 1975. Amer. Min., 60, 634-641.

Monoclinic, P2/n, a = 9.585, b = 8.776, c = 5.260 Å, β = 106.85°, Z = 4. Mo radiation, R = 0.058 for 955 reflexions.

The mineral is from Bessi, Japan, and the structure is the same as that of a sample from California (1), which is now shown to have space group P2/n, rather than P2 as previously reported (1). The structure contains one kind of SiO_3 chain in which two crystallographically different Si atoms alternate, the chain being different from those of C2 spodumene, $P2_1/c$ enstatite, and C2/c diopside (2). Mg and Al atoms are ordered in the M1 and M1(2) sites, respectively; Na and Ca atoms are partially ordered in the M2 and M2(1) sites.

1. Structure Reports, 33A, 459.
2. J.R. CLARK, D.E. APPLEMAN and J.J. PAPIKE, 1969. Min. Soc. Amer. Spec. Pap., 2, 31; this volume, p. 394.

TITANIAN FERRO-OMPHACITE

$(Na,Ca)(Al,Fe,Ti)Si_2O_6$

L. CURTIS, J. GITTINS, V. KOCMAN, J.C. RUCKLIDGE, F.C. HAWTHORNE and R.B. FERGUSON, 1975. Canad. Miner., 13, 62-67.

Monoclinic, P2/n, a = 9.622, b = 8.826, c = 5.279 Å, β = 106.92°, D_m = 3.42, Z = 4. Mo radiation, R = 0.047 and 0.041 for 843 and 922 reflexions (two independent data sets).

The structure can be derived by modifying the C2/c jadeite structure (1) and it bears a close resemblance to P2 omphacite (2). Ordering of the cations is similar to that reported for the P2 omphacite (2).

1. Structure Reports, 31A, 228.
2. Ibid., 33A, 459.

PHLOGOPITE

$KMg_3AlSi_3O_{10}(OH)_2$

J.H. RAYNER, 1974. Miner. Mag., 39, 850-856.

Monoclinic, C2/m, a = 5.322, b = 9.206, c = 10.24 Å, β = 100.03°, D_m not given, Z = 2. Neutron diffraction data, R = 0.066 for 293 reflexions.

Structure as in 1. The OH group is perpendicular to the plane of the silicate sheets.

1. Structure Reports, 27, 696; 39A, 337.

FLUOROPHLOGOPITE

$KMg_3AlSi_3O_{10}F_2$

H. TAKEDA and B. MOROSIN, 1975. Acta Cryst., B31, 2444-2452.

Monoclinic, [C2/m], a = 5.307, 5.337, b = 9.195, 9.240, c = 10.134, 10.253 Å, β = 100.08, 100.00°, at 25 and 700°C, respectively, [Z = 2]. Mo radiation, R = 0.043 and 0.095 for 451 and 410 reflexions.

A geometrical model structure for trioctahedral micas is applied to predict structural changes as a function of temperature; these changes are in agreement with those observed in the refinement of the synthetic fluorophlogopite at 700°C. The structure is similar to that of a synthetic lithium fluorophlogopite (1), and the room-temperature results agree with those of an independent study (2).

1. Structure Reports, 31A, 231.
2. Ibid., 39A, 343.

PIGEONITE

(Ca,Mg,Fe)SiO_3

Y. OHASHI and L.W. FINGER, 1974. Carnegie Inst. Washington, Year Book, 73, 525-531.

Monoclinic, $P2_1/c$, a = 9.683, b = 8.900, c = 5.228 Å, β = 108.50°, D_m not given, Z = 8. Mo radiation, R = 0.047 for 1532 reflexions (0.042 for 711 (h + k) odd reflexions), lunar material.

Structure as previously described (1). The role of Ca is still not fully established.

1. Structure Reports, 24, 465; 35A, 475; 37A, 342; 39A, 349.

CLINOPYROXENES

I. J.R. CLARK, D.E. APPLEMAN and J.J. PAPIKE, 1969. Min. Soc. Amer. Spec. Pap., 2, 31-50.

Refinements of the structures of spodumene, $LiFeSi_2O_6$, synthetic ureyite, acmite, diopside, augite, omphacite (C2/c), and omphacite (P2). The structures are very similar throughout the range of chemical compositions, with ordered cation distributions in the end-members.

(Ca,Fe)SiO_3

II. Y. OHASHI, C.W. BURNHAM and L.W. FINGER, 1975. Amer. Min., 60, 423-434.

Monoclinic, for Fe:Ca = 65:35, 75:25, 80:20, C2/c, a = 9.812, 9.781, 9.760, b = 9.049, 9.072, 9.057, c = 5.233, 5.246, 5.234 Å, β = 105.34, 106.55, 106.28°, Z = 8; for Fe:Ca = 85:15, $P2_1/c$, a = 9.779, b = 9.088, c = 5.258 Å, β = 107.39°, Z = 8. Mo radiation, R = 0.043-0.062 for 967-1144 reflexions.

The major structural change between the $P2_1/c$ phases (e.g. 1) and the C2/c form (2), is in the size and shape of the M(2) polyhedron, the M(1) polyhedron remaining essentially unchanged. The M(2) polyhedron decreases in size and causes a kinking of the tetrahedral chain, with eventual change of space group from $P2_1/c$ to C2/c. Large anisotropic thermal factors for intermediate compositions indicate positional disorder.

1. Structure Reports, 24, 465.
2. Strukturbericht, 2, 130, 528.

SANIDINE

$KAlSi_3O_8$

Y. OHASHI and L.W. FINGER, 1974. Carnegie Inst. Washington, Year Book, 73, 539-544.

Monoclinic, C2/m, a = 8.543, 8.603, b = 13.021, 13.011, c = 7.183, 7.175 Å, β = 115.98, 115.90°, at 25 and 400°C, respectively, Z = 4. Mo radiation, R = 0.034 and 0.068 for 1471 and 621 reflexions.

Structure as previously determined (1).

1. Strukturbericht, 3, 161, 547; Structure Reports, 38A, 375.

SCHORL TOURMALINE

$(Na,Ca)(Fe,Al)_3(Al,Fe)_6B_3Si_6O_{27}(F,OH)_4$

S. FORTIER and G. DONNAY, 1975. Canad. Miner., 13, 173-177.

Rhombohedral, R3m, a = 15.992, c = 7.190 Å, D_m = 3.20, Z = 3. Mo radiation, R = 0.034 for 3961 reflexions.

The material is an Andreasberg schorl tourmaline, and has a typical tourmaline structure (1), but with small but significant parameter differences from other tourmalines, e.g. dravite (1), buergerite (2), and elbaite (3).

1. Structure Reports, 15, 310; 27, 712.
2. Ibid., 34A, 391; 37A, 330.
3. Ibid., 39A, 340.

VERMICULITE-PIPERIDINE

$Mg_3Si_4O_{10}(OH)_2.xH_2O.yC_5H_{11}N$

J.E. IGLESIAS and H. STEINFINK, 1974. Clays Clay Miner., 22, 91-95.

Monoclinic, C2/m, a = 5.346, b = 9.256, c = 17.57 Å, β = 96.29°. Cu radiation, R = 0.17 for 453 reflexions.

The silicate structure was taken from that of 1, modified for C2/m and c = 17.5 Å. The organic molecules are statistically distributed over a large number of positions between the silicate layers.

1. Structure Reports, 31A, 237.

VESUVIANITE

$Ca_{19}FeAl_4(Al,Mg,Fe)_8Si_{18}O_{70}(OH)_8$

I. A. CODA, A. DELLA GIUSTA, G. ISETTI and F. MAZZI, 1970. Atti Accad. Sci. Torino, Cl. Sci. Fis. Mat. Nat., 105, 63-84.

II. J.C. RUCKLIDGE, V. KOCMAN, S.H. WHITLOW and E.J. GABE, 1975. Canad. Miner., 13, 15-21.

I. Tetragonal, P4/nnc, a = 15.565, c = 11.816 Å, D_m = 3.38, for a sample from Italy, Z = 2. Mo radiation, R = 0.079 for 1004 reflexions.

II. Tetragonal, P4/nnc, a = 15.516, 15.558, 15.543, c = 11.769, 11.810, 11.791 Å, for three samples from localities in Canada, Z = 2. Mo radiation, R = 0.039, 0.047, 0.050 for 1727, 1701, 3390 reflexions.

The structure is similar to that previously described (1), except that the Ca ion position on the fourfold axis is only half-occupied and there is an additional (Fe) atom position on this axis, also half-occupied, with an additional oxygen atom which completes five-coordination around the new Fe site.

1. Strukturbericht, 2, 126, 525.

WERNERITE

$(Ca,Na,K)_4(Si,Al)_{12}O_{24}(Cl,SO_4,CO_3)$

I. S.B. LIN and B.J. BURLEY, 1974. Miner. (Tschermaks) Petrogr. Mitt., 21, 196-215.

II. Idem, 1975. Acta Cryst., B31, 1806-1814.

Tetragonal, $P4_2/n$, a = 12.116, c = 7.581 Å, D_m = 2.69, Z = 2. Mo radiation, R = 0.083 for 1730 reflexions.

The material is a scapolite, with a structure (Fig. 1) similar to that of marialitic scapolite (1). Al is concentrated in the T(2) sites.

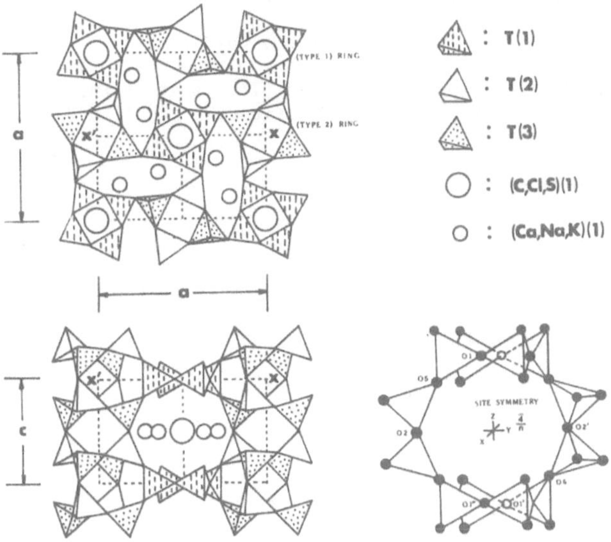

Fig. 1. Structure of wernerite.

1. Structure Reports, 39A, 350.

HYDRATED ZEOLITES

$M_{12}Al_{12}Si_{12}O_{48}.xH_2O$

W. THÖNI, 1975. Z. Kristallogr., 142, 142-160.

Cubic, Pm3m, a = 12.35, 12.24, 12.30 Å, x = 18, 28, 23, D_m = 3.68, 2.00, not given, for M = Tl, Ca, Ag. Single-crystal (Mo) data for Tl, Ca, powder data (Cu) for Ag.

The framework structure for zeolites Tl-A, Ca-A, and Ag-A is similar to that of Na-A (1). The cations are mainly distributed among two positions on the three-fold axis to either side of the six-membered rings; the two positions are too close together to be occupied simultaneously. Tl-A has been studied previously (2).

1. Structure Reports, 37A, 346.
2. Ibid., 38A, 376.

ZEOLITE A SODIUM NITRATE

$Na_{12}Al_{12}Si_{12}O_{48}.9NaNO_3.7H_2O$

R.M. BARRER and H. VILLIGER, 1975. Z. Kristallogr., 142, 82-98.

Cubic, Pm3m, a = 12.39 Å, Z = 1. Cu radiation, powder data.

The framework is as previously described (1). Most of the nitrates occupy positions in the large cavity, and two are inside the sodalite cage.

1. Structure Reports, 24, 480; 37A, 346.

POTASSIUM-EXCHANGED ZEOLITE A

CAESIUM-EXCHANGED ZEOLITE A

$K_{12}Al_{12}Si_{12}O_{48}.xH_2O$
$Cs_7Na_5Al_{12}Si_{12}O_{48}.xH_2O$

I. P.C.W. LEUNG, K.B. KUNZ, K. SEFF and I.E. MAXWELL, 1975. J. Phys. Chem., 79, 2157-2162.

II. T.B. VANCE and K. SEFF, 1975. Ibid., 79, 2163-2167.

Cubic, Pm3m, a = 12.301, 12.309, 12.320, 12.158 Å, for K-exchanged hydrated and dehydrated and seven-twelfths Cs-exchanged hydrated and dehydrated, respectively, Z = 1. Mo radiation, R = 0.114, 0.057, 0.098, 0.090 for 374, 214, 323, 123 reflexions, respectively.

The frameworks are similar to those in related zeolites (1). The cations occupy several positions (see I and II for details); in the dehydrated potassium-exchanged sample one K^+ ion is not coordinated to any atom or ion.

1. Structure Reports, 38A, 376; 40A, 291.

ZEOLITE 4A (MANGANESE(II)-EXCHANGED ACETYLENE SORPTION COMPLEX)

ZEOLITE 4A (COBALT(II)-EXCHANGED ACETYLENE SORPTION COMPLEX)

$Mn_{4.5}Na_3Al_{12}Si_{12}O_{48}.4{\cdot}5C_2H_2$
$Co_4Na_4Al_{12}Si_{12}O_{48}.4C_2H_2$

P.E. RILEY and K. SEFF, 1975. Inorg. Chem., 14, 714-721.

Cubic, Pm3m, a = 12.205, 12.171 Å, Z = 1. Mo radiation, R = 0.065 and 0.062 for 205 and 297 reflexions, respectively.

Structure as in 1. Mn and Co are on a threefold axis in the large central cavity, close to three equivalent trigonally arranged zeolite framework oxygen atoms (Mn-O = 2.18, Co-O = 2.19 Å), and symmetrically to both carbons of a C_2H_2 molecule; M...C interactions are weak, 2.63 and 2.54 Å. Na ions occupy similar threefold axial positions, somewhat recessed into the small sodalite cages, but with no interactions with the C_2H_2 molecules.

1. Structure Reports, 40A, 291.

COBALT(II)-EXCHANGED ZEOLITE A ETHYLENE SORPTION COMPLEX

$Co_4Na_4Al_{12}Si_{12}O_{48}.4C_2H_4$

P.E. RILEY, K.B. KUNZ and K. SEFF, 1975. J. Amer. Chem. Soc., 97, 537-542.

Cubic, Pm3m, a = 12.135 Å, Z = 1. Mo radiation, R = 0.062 for 312 reflexions.

The structure is similar to that of related zeolites (1). Co lies in the large zeolite cage on a threefold axis, bonded to three framework oxygen atoms at 2.15(1) Å. The C_2H_4 molecule is weakly π-bonded to Co, Co...C = 2.51(6) Å.

1. Preceding report.

COBALT(II)-EXCHANGED ZEOLITE A

$Co_4Na_4Al_{12}Si_{12}O_{48}.35H_2O$

P.E. RILEY and K. SEFF, 1975. J. Phys. Chem., 79, 1594-1601.

Cubic, Pm3m, a = 12.267 Å, Z = 1. Mo radiation, R = 0.072 for 291 reflexions.

The framework is as in related materials (e.g. 1). One Co ion is in a sodalite unit, surrounded octahedrally by six water molecules, Co-O = 2.11(3) Å. The other three Co ions are distributed about equivalent sides on threefold axes, have tetrahedral coordination, and appear to be responsible for the addition of three water molecules to the zeolite framework, which gives fivefold coordination at one Si,Al (probably Al) site.

1. Structure Reports, 38A, 376.

COPPER(II)-EXCHANGED FAUJASITE

$Cu_7Al_{14}Si_{34}O_{96}.xH_2O$

I.E. MAXWELL and J.J. de BOER, 1975. J. Phys. Chem., 79, 1874-1879.

Cubic, Fd3m, a = 24.713, 24.643 Å, for hydrated and dehydrated samples, respectively, Z = 4. Cu radiation, R = 0.046 and 0.050 for 218 and 212 reflexions, respectively.

The overall structure is similar to that of natural faujasite (1). In the dehydrated form Cu ions are distributed over several sites and are strongly bound to the zeolite framework; in the hydrated form the Cu ions are rather mobile and bound mainly to water molecules.

1. Structure Reports, 33A, 498.

PTILOLITE (MORDENITE)

(DEHYDRATED CALCIUM-EXCHANGED)

$Ca_{3.33}Al_{7.82}Si_{40.2}O_{96}$

I. W.J. MORTIER, J.J. PLUTH and J.V. SMITH, 1975. Mater. Res. Bull., 10, 1037-1046.

Orthorhombic, Cmcm (the true symmetry may be monoclinic, C2/m), a = 18.01, b = 20.27, c = 7.465 Å, Z = 1. Mo radiation, R = 0.075 for 1425 reflexions.

The alumino-silicate framework contains columns of 5-rings of tetrahedra, as previously described (1). Ca partially occupies four sites: one each in twisted and near-circular 8-rings, one on the wall of the main channel, and one in a boat-shaped packet in a zigzag channel.

DEHYDRATED HYDROGEN-PTILOLITE

$(K,Na,NH_4)AlSi_5O_{12}.xH_2O$

II. W.J. MORTIER, J.J. PLUTH and J.V. SMITH, 1975. Mater. Res. Bull., 10, 1319-1326.

Orthorhombic, Cmcm, a = 18.223, b = 20.465, c = 7.531 Å, Z = 8. Mo radiation, R = 0.048 for 1427 reflexions.

The material was obtained from natural ptilolite (mordenite) by ion-exchange with NH_4^+ and heating to 320°C. The general structure is similar to that of mordenite (1), with some distortions of the framework. Two per cent of the silicate rings are not correctly stacked in Cmcm.

1. Structure Reports, 26, 515.

STELLERITE

$CaAl_2Si_7O_{18}.7H_2O$

E. GALLI and A. ALBERTI, 1975. Bull. Soc. Fr. Minér. Crist., 98, 11-18.

Orthorhombic, Fmmm, a = 13.599, b = 18.222, c = 17.863 Å, Z = 8. Cu radiation, R = 0.078 for 843 reflexions.

The framework is similar to that in monoclinic stilbite (1), with disorder of Si and Al in the tetrahedra, (Si,Al)-O = 1.64 Å. Ca is coordinated only to water molecules, the sites for some of which have low occupancies.

1. Structure Reports, 37A, 343.

BARRERITE

$(Ca,Mg,Na,K)_{15}(Al,Si)_{72}O_{144}.52H_2O$

E. GALLI and A. ALBERTI, 1975. Bull. Soc. Fr. Minér. Crist., 98, 331-340.

Orthorhombic, Amma, a = 13.643, b = 18.200, c = 17.842 Å, Z = 1. Mo radiation, R = 0.068 for 2483 reflexions.

The framework is topologically identical to that of stilbite (1) and stellerite (2), with a lowering of symmetry from the Fmmm of stellerite to Ammm resulting from the presence of cations in sites that are vacant in stilbite and stellerite, and which impose rotational displacements within the framework.

1. Structure Reports, 37A, 343.
2. Preceding report.

LANTHANUM Y ZEOLITES

J. SCHERZER, J.L. BASS and F.D. HUNTER, 1975. J. Phys. Chem., 79, 1194-1199.

Cubic, a ∿ 25 Å, for various specimens. Powder data suggest a non-framework Al atom.

ZIRCON

$ZrSiO_4$

L.W. FINGER, 1974. Carnegie Inst. Washington, Year Book, 73, 544-547.

Tetragonal, $I4_1/amd$, a = 6.612, c = 5.994 Å, D_m not given, Z = 4. Mo radiation, R = 0.028 for 220 reflexions. x(O) = 0.0660, z(O) = 0.1941 (origin at centre).

Structure as previously determined (1, 2). z(O) differs from that in 2 by 2.5σ, Si-O being 1.630(3) Å in the present analysis.

1. Strukturbericht, 1, 345, 408; Structure Reports, 22, 314; 29, 411.
2. Structure Reports, 37A, 349.

TABLE I

Some structural information has also been given for the following materials (listed with abbreviated 1975 reference).

Compound	Structure	Reference
AgI, 4H 8H 12H 16H	Ždanov symbol = 22 $(211)_{\bar{2}}$ $(2111\bar{1})_2$ $(112111\bar{1})_{\bar{2}}$	Acta Cryst., A30, 369 (1974)
γ-Vanadium oxide, $VO_{0.53}$	NaCl, with oxygen vacancies	Ibid., A31, 63
Rutile, TiO_2	Rutile, z = 0.30493	Ibid., B31, 1981
γ-$BiTaO_4$	$LaTaO_4$ (Dokl. Akad. Nauk SSSR, 201, 1095(1971))	Ibid., B31, 2748
Frankdicksonite, BaF_2	Fluorite	Amer. Min., 59, 885 (1974)
Cookeite, $(Li,Al)_3Al_2Si_4O_{10}(OH)_8$	Di,trioctahedral chlorite	Ibid., 60, 1041
$CaPbO_{3-x}$	CaF_2	Ann. Chim., 10, 63
K_3TaOF_6	K_3NbOF_6	Ibid., 10, 75
$CaPb(OH)_6$ $CdPb(OH)_6$	Cubic, Pn3	Ibid., 10, 105
$CdPbO_3$	C-Ln_2O_3	
$(Fe,Cr)_3O_4$ $(Fe,Sn)_3O_4$	Spinel Spinel	Ibid., 10, 159

TABLE I

Compound	Structure	Reference
$(Fe,Co)_2SnO_4$	Spinel	Ann. Chim., 10, 301
$xMg_2TiO_4.(1-x)MgFe_2O_4$	Spinel	Appl. Phys., 7, 77
$(Pb,Zn)_2(Nb,Zn)_2O_{6.5}$	Pyrochlore, x(O) = 0.44	Bull. Soc. Chim. Fr., 1989
Samarium bimolybdate, $Sm_2Mo_2O_7$	Pyrochlore, x(O) = 0.41, possibly with O(2) disordered in 32(e)	Ibid., 2463
$Na_3Cd_2(SO_4)_3Cl$ $Na_3Pb_2(SO_4)_3Cl$	Apatites	Bull. Soc. Fr. Minér. Crist., 98, 254
$BaCrO_{2.90}$	$Ba_5Ta_4O_{15}$	Chem. Letters, 557
Silver iodide, AgI	7-layer polytype	Cryst. Lattice Defects, 5, 235
$DyAsO_4$ (low-temp.) $DyVO_4$ (low-temp.) $TbVO_4$ (low-temp.)	Imma Imma C2/c	Internat. J. Magnetism, 3, 123 (1972)
$(Fe,Li)_3(O,F)_4$	Spinel	Izv. Akad. Nauk SSSR, Neorg. Mater., 10, 1743
$FeCo_2O_4$	Spinel	Jap. J. Appl. Phys., 13, 1891 (1974)
$CoAl_6O_{10}$	Spinel	J. Ceram. Soc. Japan, 83, 87
Copper(I) hexafluoroarsenate(V), $CuAsF_6$	$LiSbF_6$	J. Fluor. Chem., 6, 379
FeV_2O_4	Spinel (partially inverted, u = 0.385)	J. Phys., C, 8, 370
Copper(II) fluoride, CuF_2	Structure as previously described (Structure Rep-Reports, 21, 205)	J. Phys. Chem. Solids, 35, 1683 (1974)
Silver fluoride, AgF	Three phases, B1, B2, and probably inverse-NiAs [$B8_1$]	Ibid., 36, 939
Caesium tribromoferrate-(II), $CsFeBr_3$	$BaNiO_3$	J. Phys. Soc. Japan, 37, 276 (1974)
$MnGeO_3$	Ilmenite	Ibid., 37, 1242 (1974)
Cobalt molybdate, $CoMoO_4$	Disordered modifications	J. Solid State Chem., 14, 117

TABLE I

Compound	Structure	Reference
Vanadium suboxide, V_7O_3	C2/m superstructure of the metal sublattice	J. Solid State Chem., 14, 219
$Sr_3Cr_2WO_9$	Perovskite, x(O) = 0.242	Ibid., 14, 354
$Ca_3Cr_2WO_9$	Tetragonally-distorted perovskite	
$Ba_3Cr_2WO_9$	Hexagonal perovskite	
Calcium lead hydroxy-apatite, $Ca_2Pb_3(PO_4)_3(OH)$	Apatite; Ca shows a preference for position 4(f), and Pb for 6(h)	Ibid., 15, 117
Gd_2O_3	CaF_2	Kristallografija, 20, 192 [Soviet Physics - Crystallography, 20, 114]
Tb_2O_3	CaF_2	
Potassium titanium tungstate hydrate, $K_2Ti(WO_4)_3 \cdot 2H_2O$	Pyrochlore, x = 0.30	Ibid., 20, 314 [Ibid., 20, 194]
HfO_2	ZrO_2 (see Structure Reports, 40A, 300)	Ibid., 20, 392 [Ibid., 20, 239]
Hf_2O	Cuprite	
$Li_{0.5}(Mg,Fe)_{2.5}(O,F)_4$	Spinel	Ibid., 20, 1188 [Ibid., 20, 720]
$Tl(Nb,W)O_6$	Pyrochlore	Mater. Res. Bull., 9, 1053 (1974)
$Nd(Al,Ga)O_3$	Distorted perovskite	Ibid., 10, 481
$(Tl,Ag)NbO_3$	Pyrochlore	Ibid., 10, 933
$(Tl,Pb)NbO_3$	Pyrochlore	
$K(TaW)O_6$	Pyrochlore	Ibid., 10, 949
$K(TaW)O_6.H_2O$	Pyrochlore	
K_2CuBr_3	K_2CuCl_3	Ibid., 10, 1151
$Bi_2Mo_2O_9$	Bi, Mo, and approximate O positions are given	Ibid., 10, 1163
$MnFe_2O_4$	Spinel	Mineral J., Japan, 7, 202 (1973)
Fe_3O_4-Fe_2TiO_4	Spinel	Ibid., 7, 472 (1974)
β-Cristobalite, SiO_2	Disordered structures fit the data better than do previous models	Philos. Mag., 31, 1391
Aluminum phosphate, $AlPO_4$		

TABLE I

Compound	Structure	Reference
$Cd_2(Sb,Nb)_2O_7$ $Cd_2(Sb,Ta)_2O_7$ $(Cd,Bi)_2(Sb,Sn)_2O_7$	Pyrochlores	Rev. Chim. Minér., 12, 247
$Na_{0.5}La_{0.5}UO_4$	$CaUO_4$	Ibid., 12, 382
γ-$Na_3Ln(VO_4)_2$	Na_2CrO_4	Ibid., 12, 448
K_3NbO_4	Elpasolite	Ibid., 12, 454
FeI_2	FeI_2-type (Strukturbericht, 2, 247), $z \sim 1/4$	Solid State Comm., 14, 187 (1974)
Cs_2CrCl_4 (low-temp.)	K_2NiF_4-type (Structure Reports, 17, 332; 19, 323). z(Cs) = 0.352, z(Cl) = 0.145	Ibid., 15, 313 (1974)
Ba_2MnUO_6 Ba_2CoUO_6 Ba_2NiUO_6	Perovskite, x(O) = 0.2537, 0.2520, 0.2500	Ibid., 15, 1831 (1974)
Yttrium iron garnet, $Y_3(Al,Fe)_5O_{12}$	Garnet	Ibid., 16, 987
Talc, $Mg_3Si_4O_{10}(OH)_2$	Space group P1; structure proposed	Sprechsaal Keram. Glas Baustoffe, 108, 234
$(Mn,Al)_3O_4$	Spinel	Z. anorg. Chem., 415, 69
$Eu_3Ln(PO_4)_3$	Eulytite	Ibid., 417, 81
Lead(II) iodide, PbI_2	Twelve-layer polytype, $(11)_3 2112$	Z. Kristallogr., 141, 59
Cadmium iodide, CdI_2	$20H_4$ $2112(11)_7$ $20H_5$ 222221122211 $20H_6$ $(22)_3 211211$ $30H_2$ 222221121122222211 $30H_3$ $(22)_7 11$ $30H_4$ $(22)_4 2112221111$	Ibid., 141, 67
	$16H_6$ $(22)_2 112211$ $26H_3$ $[(22)_2 11]_2 2112$ $34H_1$ $[(22)_2 11]_3 22$	Ibid., 141, 451
	$10H_2$ $(221)_2$ $14H_4$ $(22)_2(11)_3$ $18H_5$ $(22)_3 1221$	Ibid., 141, 458
	$10H_3$ $22(11)_3$ $10H_4$ 21111211 $12H_7$ $22(11)_4$ $12H_8$ $2112(11)_3$ $16H_7$ 12322222	Ibid., 142, 121

TABLE I

Compound	Structure	Reference
Cadmium iodide, CdI_2 (continued)	$24H_2$ $221(22)_4 3$ $28H_4$ $(22)_2(11)_4 22(11)_4$	Z. Kristallogr., 142, 121
Ca_2PCl Sr_2PCl	NaCl NaCl	Z. Naturforsch., 30B, 165
Ca_2PBr Sr_2PBr	NaCl NaCl	Ibid., 30B, 378
Ca_2PBr Sr_2PBr Ba_2PBr	Rhombohedral, anti-$NaFeO_2$ type (Strukturbericht, 3, 75, 392), z ∿ 0.23	

ELECTRON DIFFRACTION

The following compounds have been studied by electron diffraction of the vapours (listed with abbreviated 1975 reference). Bond lengths are in Å, angles in degrees.

Tungsten oxide tetrachloride, $WOCl_4$	Square-pyramid		Bull. Chem. Soc. Japan, 47, 1393 (1974)
	W-Cl	2.28	
	W-O	1.69	
	Cl-W-O	102	
Boron triiodide, BI_3	B-I	2.12	Ibid., 47, 2337 (1974)
Molybdenum oxide tetrachloride, $MoOCl_4$	Square pyramid		Ibid., 48, 666
	Mo-Cl	2.279	
	Mo-O	1.658	
Tetrafluorodiphosphine, P_2F_4	Trans conformer		Inorg. Chem., 14, 599
	P-P	2.281	
	P-F	1.587	
	P-P-F	95.4	
	F-P-F	99.1	
Vanadyl chloride, $VOCl_3$	C_{3v} symmetry		Inorg. Chim. Acta, 13, 113
	V-Cl	2.142	
	V-O	1.570	
	Cl-V-Cl	111.3	
1-Silylpentaborane, $H_3SiB_5H_8$	Si-B	1.981	J. Amer. Chem. Soc., 97, 1074
2-Silylpentaborane, $H_3SiB_5H_8$	Si-B	2.006	

Silylsulphinylamine, H_3SiNSO	Si-N	1.76	J. Chem. Soc., Dalton, 805
	N-S	1.52	
	S-O	1.44	
	Si-N-S	130	
	N-S-O	119	
Vanadium hexacarbonyl, $V(CO)_6$	O_h symmetry		J. Molec. Struct., 24, 1
	V-C	2.015	
	C-O	1.138	
Aluminum chloride-ammonia, $Cl_3Al.NH_3$	Al-Cl	2.100	Ibid., 24, 27
	Al-N	2.00	
	Cl-Al-Cl	116.9	
Cs_2SO_4	Cs-O	2.60	Ibid., 25, 357
Tl_2SO_4	Tl-O	2.41	
Cs_2MoO_4	Cs-O	2.80	
Cs_2WO_4	Cs-O	2.78	

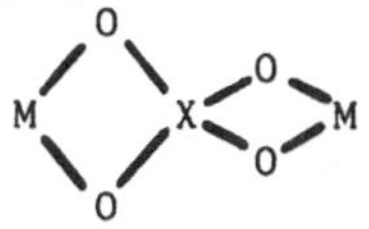

Germanium tetrabromide, $GeBr_4$	T_d symmetry, Ge-Br = 2.272 Å. Previous studies in Strukturbericht, 6, 70; Structure Reports, 8, 245, 247		Ibid., 25, 442
Manganese(II) chloride, $MnCl_2$	Linear molecule assumed		Ibid., 26, 116
	Mn-Cl	2.205	
Hexafluorodisilane, F_3SiSiF_3	Si-Si	2.324	Ibid., 27, 438
	Si-F	1.569	
	Si-Si-F	110.6	
	Conformation is 35° from the eclipsed position		
Bis(difluorophosphine) oxide, F_2POPF_2	P-F	1.568	Ibid., 28, 205
	P-O	1.631	
	P-O-P	135	
	O-P-F	98	
	F-P-F	99	
	Independent study in Structure Reports, 39A, 360		
Decaborane, $B_{10}H_{14}$	Structure and dimensions close to those previously given [Structure Reports, 28, 360]		Ž. Strukt. Khim., 16, 128 [J. Struct. Chem., 16, 110]
Rubidium metaborate, $RbBO_2$	B-O	1.26	Ibid., 16, 662 [Ibid., 16, 611]
	Rb-O	2.57	
	[see Structure Reports, 38A, 382]		
Caesium metaborate, $CsBO_2$	B-O	1.25	Ibid., 16, 899
	Cs-O	2.63	

MICROWAVE SPECTRA

Phosphine-trifluoroborane, H_3PBF_3	Staggered conformation assumed		Inorg. Chem., 14, 2837
	P-B	1.92	
	B-F	1.37	
Diphosphine, P_2H_4	P-P	2.219	J. Amer. Chem. Soc., 96, 2688 (1974)
	P-H	1.416	
	H-P-H	92.0	
	H-P-P	94.3, 99.1	
	Dihedral angle 74° (for electron diffraction study, see Structure Reports, 38A, 381)		
Copper(I) chloride, CuCl	r_e = 2.051177		J. Chem. Phys., 62, 1040
Copper(I) iodide, CuI	r_e = 2.338316		Ibid., 62, 4796
Silylphosphine, $H_3Si\text{-}PH_2$	Si-P	2.250	Ibid., 63, 915
Copper(I) bromide, CuBr	r_e = 2.173435		Ibid., 63, 2724
Hydrogen sulphide, H_2S and D_2S	S-H	1.3518	J. Molec. Struct., 28, 237
	S-D	1.3474	
	H-S-H	92.13	
	D-S-D	92.11	
	r_e	1.3362	
	θ_e	92.06	

PAPERS REFERRED TO LATER YEARS

Many preliminary notes have not been reported, since fuller accounts will appear at a later date. The compounds studied, and abbreviated 1975 references, are listed below.

Tin(II) bromide, $SnBr_2$	Acta Chem. Scand., A29, 956
Margarite (see this volume, p. 388) Amesite Cronstedtite Cookeite (see this volume, p. 402)	Amer. Min., 60, 175
Tellurium iodides, α-TeI and β-TeI	Angew. Chem., 86, 411 (1974)
Hexamanganato(VII)-manganic(IV) acid, $(H_3O)_2[Mn(MnO_4)_6].11H_2O$	Ibid., 86, 647 (1974)
Bis(iodomercury) hexafluorotitanate, $(HgI)_2TiF_6$	Ibid., 86, 863 (1974)
Ammonium dimolybdate, $(NH_4)_2Mo_2O_7$	Ibid., 86, 894 (1974) [this volume, p. 246]
$Cs_8Sn_{10}O_4S_{20}.13H_2O$	Ibid., 87, 451

Ammonium decamolybdate, $(NH_4)_8Mo_{10}O_{34}$	Angew. Chem., 87, 634
Holmquistite, $Li_2(Mg,Fe)_3(Al,Fe)_2Si_8O_{22}$-$(OH,F)_2$	Carnegie Inst. Washington, Year Book, 73, 535 (1974) [see also this volume, p. 384]
Di-μ-iodo-bis(tetracarbonylmolybdenum)-(Mo-Mo), $[(CO)_4MoI]_2$	Chem. Ber., 108, 260
2,2'-Thiaborane, $(B_9H_8S)_2$	Chem. Comm., 629 (1974) [this volume, p. 137]
α-$Ba_3P_2W_{18}O_{62}.29H_2O$	Ibid., 691
Li_2SO_4 Na_2SO_4 Rb_2SO_4	Chem. Communic., Sweden, No. 12, 1-44 (1974)
Potassium dihydrogen phosphate, KD_2PO_4	Chem. Phys. Letters, 34, 175
$Bi_2Ru_2O_7$	C.R. Acad. Sci., Paris, C, 280, 279
MoSBr	Ibid., C, 280, 949; 281, 23
Fe_2TeO_5	Ibid., C, 280, 1141
$FeAl_2O_4$	Ibid., C, 280, 1367
$CuTeO_3$	Ibid., C, 280, 1463
Ba_2CuUO_6	Ibid., C, 281, 19
Caesium uranyl fluoride hydrate, $Cs_2(UO_2)_2F_6.2H_2O$	Ibid., C, 281, 593
Indium tellurate, In_2TeO_6	Ibid., C, 281, 769
Sodium thorium phosphate, $Na_2Th(PO_4)_2$	Croat. Chem. Acta, 47, 41
Solongoite, $Ca_2[B_3O_4(OH)_4]Cl$	Dokl. Akad. Nauk SSSR, 216, 1281 (1974)
$Na_2InSiO_4(OH)$	Ibid., 217, 86 (1974)
Lemoynite	Ibid., 217, 326 (1974)
$Na_{11}Nb_2Ti(Si_2O_7)_2(PO_4)_2O_3F$	Ibid., 217, 569 (1974)
Samarium oxygermanate, $Sm(OH)_3$.-$6Sm_2[GeO_4][O,(OH,F)_2]$	Ibid., 217, 824 (1974)
$Na_3Sc(SO_4)_3.5H_2O$	Ibid., 217, 1073 (1974)
Na_2BeSiO_4	Ibid., 218, 335 (1974)
Sakhaite, $Ca_3Mg(BO_3)_2CO_3.xH_2O$	Ibid., 218, 576 (1974)
$Na_3HZr(GeO_4)_2$	Ibid., 218, 830 (1974)
Ytterbium perrhenate, $Yb(ReO_4)_3.4H_2O$	Ibid., 218, 1086 (1974)

$Na_2Pr_6(Ge_4O_{12})(Ge_2O_7)_2$	Dokl. Akad. Nauk SSSR, 219, 91 (1974)
Tricalcium silicate trihydrate	Ibid., 219, 340 (1974)
Dioctahedral mica $2M_2$	Ibid., 219, 704 (1974) [this volume, p. 389]
$Na(Fe,Zn)PO_4$	Ibid., 219, 860 (1974)
$KIO_3.HIO_3$	Ibid., 219, 1108 (1974)
Potassium bisulphate iodate, $K_2H(SO_4)(IO_3)$	Ibid., 219, 1352 (1974)
Zinc triploidite, $(Fe,Zn)_2PO_4OH$	Ibid., 220, 89
Caesium lanthanum sulphate hydrate, $Cs_2SO_4.La_2(SO_4)_3.8H_2O$	Ibid., 220, 346
Phosphuranylite, $Ca[(UO_2)_3(PO_4)_2(OH)_2]$.-$6H_2O$	Ibid., 220, 1161
Francevillite, $Ba[(UO_2)_2(VO_4)_2].5H_2O$	Ibid., 220, 1410
Calcium borate, $Ca_2[B_8O_{13}(OH)_2]$	Ibid., 221, 87
Muirite, $Ba_9(Ca,Ba)(Ca,Ti)_4(OH)_8[Si_8O_{24}]$-$(Cl,OH)_8$	Ibid., 221, 343
Sodium decatungstouranate hydrate, $Na_8UW_{10}O_{36}.30H_2O$	Ibid., 221, 351
$Na_2LiFe(Si_2O_5)_3$	Ibid., 221, 842
$NaFe(MoO_4)_2$	Ibid., 221, 1322
Roweite, $Ca_2(Mn,Mg)_2[B_4O_7(OH)_2](OH)_4$	Ibid., 221, 1326
Potassium ytterbium germanate, $K_4Yb_2(OH,F)_2[Ge_8O_{20}]$	Ibid., 222, 87
Bi_2WO_6	Ibid., 222, 94
Sodium samarium germanate, $NaSm_3[GeO_4]_2(OH)_2$	Ibid., 222, 343
Dysprosium perrhenate tetrahydrate, $Dy(ReO_4)_3.4H_2O$	Ibid., 222, 1097
Holmium periodate tetrahydrate, $HoIO_5.4H_2O$	Ibid., 222, 1335 [this volume, p. 366]
Hypersthene	Ibid., 223, 192
Djerfisherite, $K_6Mg(Fe,Cu)_{24}S_{26}Cl$	Ibid., 223, 343
Thoreaulite, $SnTa_2O_6$	Ibid., 223, 1115
Lanthanum perrhenate, $La(ReO_4)_3.4H_2O$	Ibid., 224, 94

Caesium indium selenate dihydrate, $CsIn(SeO_4)_2.2H_2O$	Dokl. Akad. Nauk SSSR, 224, 335
Sodium dichromate dihydrate, $Na_2Cr_2O_7.2H_2O$	Ibid., 224, 580
$K_2Nd_4(Ge_4O_{13})(OH,F)_4$	Ibid., 224, 817
Potassium hydrogen iodate, $KIO_3.HIO_3$	Ibid., 224, 1066
Sodium scandium germanate, $Na_2ScGeO_4(OH)$	Ibid., 224, 1069
$Cs_4\{Rh[SnF_2(H_2O)_2]_2[Sn_4F_{15}]\}.4H_2O$	Ibid., 224, 1323
Barcarite, $Ca_4Mg[B_4O_6(OH)_6](CO_3)_2$	Ibid., 225, 823
Ckalovite, $Na_2BeSi_2O_6$	Ibid., 225, 1319
$CaU_2O_6F_2$	Inorg. Nucl. Chem. Letters, 11, 207
$RbCoCl_3.2H_2O$	Ibid., 11, 813
$NaCrO_2$	Ibid., 11, 817
Rubidium lead(II) hexanitrocuprate(II), $Rb_2PbCu(NO_2)_6$	J. Amer. Chem. Soc., 97, 444
Tetradecaborane(20), $B_{14}H_{20}$	Ibid., 97, 1621
Osmium carbonyl hydride, $H_2Os_3(CO)_{11}$	Ibid., 97, 4145
Potassium lanthanum molybdate, $K_5La(MoO_4)_4$	Kristallografija, 19, 989 (1974) [Soviet Physics - Crystallography, 19, 613]
Hulsite, $(Fe,Mg,Sn)_3BO_3O_2$	Ibid., 20, 156 [Ibid., 20, 89]
$BaLu_2F_8$	Ibid., 20, 642 [Ibid., 20, 393]
V_3O_5	Mater. Res. Bull., 10, 861
Inesite, $Ca_2Mn_7Si_{10}O_{28}(OH)_2.5H_2O$	Naturwissenschaften, 62, 96
Sodium oxoferrate(III), Na_5FeO_4	Ibid., 62, 138
Sodium oxopyroberyllate, $Na_6Be_2O_5$	Ibid., 62, 236
Lanthanum niobate, $LaNb_3O_9$	Ibid., 62, 296
γ-Platinum(IV) iodide, PtI_4	Ibid., 62, 297
Strontium aluminogermanate, $SrAl_2Ge_2O_8$	Ibid., 62, 485
Eglestonite, $Hg_6Cl_3O_2$	Österr. Akad. Wiss., Kl. Mat.-Nat., No. 10, 1
Potassium dihydrogen pyrophosphate hemihydrate, $K_2H_2P_2O_7.0{\cdot}5H_2O$	Phosphorus, 2, 159 (1972) [compare Structure Reports, 39A, 290]
Trisodium hydrogen pyrophosphate nonahydrate, $Na_3HP_2O_7.9H_2O$	Ibid., 3, 75 (1973)

Tetrasodium hypophosphate decahydrate, $Na_4P_2O_6.10H_2O$	Phosphorus, 3, 131 (1973)
Trisodium hydrogen hypophosphate nonahydrate, $Na_3HP_2O_6.9H_2O$	Ibid., 4, 207 (1974)
Prussian Blue, $Fe_4[Fe(CN)_6]_3.14H_2O$	Z. Phys. Chem., 92, 354 (1974) [see also Structure Reports, 38A, 387]

ADDITIONAL PAPERS

The following papers were omitted from previous volumes, mainly Volume 34A (1969).

TETRAAMMINEZINC OCTAHYDROOCTABORATE(-2)

$Zn(NH_3)_4B_8H_8$

L.J. GUGGENBERGER, 1969. Inorg. Chem., 8, 2771-2774.

Tetragonal, $P4_2/nmc$, a = 7.503, c = 10.784 Å, D_m not measured, Z = 2. Mo radiation, R = 0.078 for 363 reflexions.

Atomic positions

Zn in 2(b); other atoms in 8(g)

	x	y	z
Zn	1/4	3/4	3/4
N	1/4	0.9747	0.8585
B(1)	1/4	0.5813	0.2170
B(2)	1/4	0.6461	0.3697
H(1)	1/4	0.415	0.207
H(2)	1/4	0.548	0.446

The structure contains tetrahedral $Zn(NH_3)_4^{2+}$ and distorted dodecahedral $B_8H_8^{2-}$ ions, the latter geometry being similar to that found in B_8Cl_8 (1) and $B_6H_6C_2(CH_3)_2$ (2). Zn-N = 2.05, B-B = 1.56-1.93, B-H = 1.11, 1.25 Å.

1. Structure Reports, 23, 303.
2. Ibid., 33B, 262.

BIS(PENTACHLOROANTIMONY(V))DISULPHUR DINITRIDE

$S_2N_2(SbCl_5)_2$

R.L. PATTON and K.N. RAYMOND, 1969. Inorg. Chem., 8, 2426-2431.

Tetragonal, I$\bar{4}$2d, a = 14.933, c = 15.547 Å, D_m = 2.64, Z = 8. Mo radiation, R = 0.038 for 768 reflexions.

The structure contains discrete molecules, which consist of a planar four-membered S_2N_2 ring of alternating S and N atoms, with Sb atoms bonded to each N; Cl-Sb-N...N-Sb-Cl lies along a crystallographic twofold axis. S-N = 1.62(1), Sb-N = 2.28(1), Sb-Cl = 2.30(1) Å, S-N-S = 95°.

PROSOPITE

$CaAl_2(F,OH)_8$

C. GIACOVAZZO and S. MENCHETTI, 1969. Atti Accad. Sci. Lincei, R.C., Cl. Sci. Fis. Mat. Nat., 47, 55-68.

Monoclinic, C2/c, a = 6.70, b = 11.13, c = 7.33 Å, β = 95.0°, D_m = 2.88, Z = 4. (all as in 1). Cu radiation, R = 0.067 for 326 reflexions (films, densitometer intensities).

Atomic positions

	x	y	z
Ca	1/2	0.5395	1/4
Al(1)	3/4	1/4	1/2
Al(2)	0	0.3588	1/4
F,O(1)	0.9755	0.3468	0.5047
F(2)	0.5769	0.3811	0.4614
F(3)	0.1830	0.4758	0.2865
O(4)	0.7990	0.2365	0.2506

Ca has distorted antiprismatic coordination, Ca-O,F = 2.28-2.57 Å, the antiprisms sharing edges to form zigzag chains along c. Both Al atoms have octahedral coordination, Al-O,F = 1.79-1.92 Å; the octahedra share edges to form chains along [10$\bar{1}$], and each antiprism shares corners and edges with seven octahedra.

1. Structure Reports, 34A, 215.

COPPER(II) HEXAFLUOROSILICATE·TETRAHYDRATE

$CuSiF_6.4H_2O$

M.J.R. CLARK, J.E. FLEMING and H. LYNTON, 1969. Canad. J. Chem., 47, 3859-3861.

Monoclinic, $P2_1/a$, a = 7.22, b = 9.64, c = 5.36 Å, β = 105.2°, D_m = 2.56, Z = 2. Cu radiation, R = 0.083 for 574 reflexions (films, visual intensities).

The structure contains chains of trans corner-sharing $Cu(H_2O)_4F_2$ and SiF_6 octahedra; Cu-O = 1.97, Cu-F = 2.34, Si-F = 1.69(1) Å, Cu-F-Si = 152°. The chains are linked by possible O-H...O and O-H...F hydrogen bonds.

POTASSIUM HEXAFLUOROMOLYBDATE(III)

K_3MoF_6

L.M. TOTH, G.D. BRUNTON and G.P. SMITH, 1969. Inorg. Chem., 8, 2694-2697.

Cubic, Fm3m, a = 8.784 Å, D_m = 3.23, Z = 4. Mo radiation, R = 0.10 for 142 reflexions. Mo in 4(a); K(1) in 4(b); K(2) in 8(c); F in 24(e): x = 0.226.

Isostructural with $(NH_4)_3FeF_6$ (1). Mo-F = 2.00(2), K(1)-F = 2.40 (x 6), K(2)-F = 3.11 Å (x 12).

1. Strukturbericht, 1, 437, 449.

SODIUM TRIFLUOROCOBALTATE(II)

SODIUM TRIFLUORONICKELATE

$NaCoF_3$
$NaNiF_3$

F. POMPA and F. SICILIANO, 1969. Ric. Sci., 39, 370-385.

Orthorhombic, Pnma, a = 5.607, 5.529, b = 7.790, 7.695, c = 5.428, 5.369 Å. Orthorhombic perovskites.

HEXAAQUOALUMINUM HEXACHLORORUTHENATE(III) TETRAHYDRATE

$Al(H_2O)_6RuCl_6.4H_2O$

T.E. HOPKINS, A. ZALKIN, D.H. TEMPLETON and M.G. ADAMSON, 1969. Inorg. Chem., 8, 2421-2425.

Monoclinic, $P2_1/n$, a = 10.492, b = 11.415, c = 7.069 Å, β = 92.69°, D_m not given, Z = 2. Mo radiation, R = 0.036 for 1238 reflexions.

The structure contains slightly-distorted octahedral $Al(H_2O)_6{}^{3+}$ and $RuCl_6{}^{3-}$ ions, which are joined by O-H...O and O-H...Cl hydrogen bonds, which also involve the free water molecules and some of which are bifurcated. Al-O = 1.873-1.883(4), Ru-Cl = 2.370-2.384(2) Å.

CAESIUM TRICHLOROCOBALTATE(II) DIHYDRATE

$CsCoCl_3.2H_2O$

N. THORUP and H. SOLING, 1969. Acta Chem. Scand., 23, 2933-2934.

Orthorhombic, Pcca, a = 8.914, b = 7.174, c = 11.360 Å, D_m ≐ 3.01, Z = 4. Mo radiation, R = 0.080 for 559 reflexions.

Isostructural with $CsMnCl_3.2H_2O$ (1) and α-$RbMnCl_3.2H_2O$ (2). Co-Cl = 2.444 and 2.494(5), Co-O = 2.07(1), O-H...Cl = 3.17 Å.

1. Structure Reports, 27, 456.
2. Ibid., 32A, 188.

MOLYBDENUM(II) BROMIDE HYDRATE

$Mo_6Br_{12}.2H_2O$

L.J. GUGGENBERGER and A.W. SLEIGHT, 1969. Inorg. Chem., 8, 2041-2049.

Tetragonal, I4/m, a = 9.437, c = 11.729 Å, D_m = 4.995, Z = 2. Mo radiation, R = 0.060 for 559 reflexions.

Atomic positions

		x	y	z
Mo(1)	in 8(h)	0.1752	0.0917	0
Mo(2)	4(e)	0	0	0.1580
Br(1)	8(h)	0.4205	0.2410	0
Br(2)	16(i)	0.0820	0.2621	0.1580
O	4(e)	0	0	0.3448

The structure contains discrete $Mo_6Br_{12}.2H_2O$ molecules, which consist of an octahedron of Mo atoms, with each face triply bridged by a Br atom, 4 terminal Br atoms, and 2 terminal trans H_2O molecules. Mo-Mo = 2.630 and 2.640(2), Mo-Br = 2.587 (terminal), 2.591-2.606 (bridging), Mo-O = 2.19 Å.

SILVER THALLIUM HALIDES

$AgTl_2X_3$ (X = Cl, Br, I)
$AgTlI_2$

P. MESSIEN, 1969. Bull. Soc. Roy. Sci., Liège, 38, 490-495.

$AgTl_2X_3$
Rhombohedral, space group not determined, a = 8.26, 8.56, 8.97 Å, for X = Cl, Br, I, α = 70°36' for all three, Z = 3.

$AgTlI_2$
Tetragonal, I4/mcm, a = 8.34, c = 7.66 Å, Z = 4. Powder data. Tl in 4(a); Ag in 4(b); I in 8(h): x = 0.182.

LITHIUM OXYFLUORONIOBATE(V)

Li_2NbOF_5

J. GALY, S. ANDERSSON and J. PORTIER, 1969. Acta Chem. Scand., 23, 2949-2954.

Hexagonal, P$\bar{3}$1m, a = 4.966, c = 4.572 Å, D_m = 3.8, Z = 1. Cu radiation, R = 0.047 for powder data. Nb in 1(a); Li in 2(d); O,F in 6(k): x = 0.325, z = 0.240.

Isostructural with Li_2ZrF_6 (1).

1. Structure Reports, 24, 277; 39A, 154.

SELENINYL DIFLUORIDE - NIOBIUM PENTAFLUORIDE

$SeOF_2.NbF_5$

A.J. EDWARDS and G.R. JONES, 1969. J. Chem. Soc. (A), 2858-2861.

Orthorhombic, Pmcn, a = 8.02, b = 7.13, c = 11.09 Å, D_m not measured, Z = 4. Cu radiation, R = 0.10 for 445 reflexions (films, photometer intensities).

The structure contains discrete F_2Se-O-NbF_5 units, in which selenium has trigonal pyramidal and niobium octahedral coordination; Se-F = 1.68, Se-O = 1.60, Nb-F = 1.85, Nb-O = 2.13(2) Å, Se-O-Nb = 144°. The units are linked by three long Se...F contacts (2.69, 2.69, and 2.88 Å) which complete very distorted octahedral geometry around Se.

MOLYBDENUM(V) OXIDE TRICHLORIDE

$MoOCl_3$

G. FERGUSON, M. MERCER and D.W.A. SHARP, 1969. J. Chem. Soc. (A), 2415-2418.

Monoclinic, $P2_1/c$, a = 5.74, b = 13.51, c = 6.03 Å, β = 92.9°, D_m not measured, Z = 4. Cu radiation, R = 0.13 for 516 reflexions (films, densitometer intensities).

Atomic positions

	x	y	z
Mo	-0.0251	0.1610	0.2007
Cl(1)	-0.2825	0.3152	0.3852
Cl(2)	0.1823	0.1923	0.5544
Cl(3)	-0.2958	0.0628	0.3550
O	0.1520	0.0792	0.1189

The structure contains chains of $MoCl_5O$ octahedra which share two Cl-Cl edges, leaving cis-terminal Cl and O atoms. Mo-Cl = 2.28 (terminal), 2.37-2.82(1) (shared), Mo-O = 1.60(2) Å. Independent study in 1.

1. Structure Reports, 35A, 184.

BARIUM TETRACYANOPLATINATE(II) TETRAHYDRATE

$BaPt(CN)_4.4D_2O$

L. DUPONT, O. DIDEBERG and E. LEGRAND, 1969. Bull. Soc. Roy. Sci., Liège, 38, 503-508.

Monoclinic, C2/c, a = 12.052, b = 13.851, c = 6.635 Å, β = 103°2'. Neutron diffraction study of D positions from hk0 data, to assist in interpretation of the infrared spectrum.

RUBIDIUM TETRACYANOPLATINATE(II) SESQUIHYDRATE

$Rb_2Pt(CN)_4.1{\cdot}5H_2O$

L. DUPONT, 1969. Bull. Soc. Roy. Sci., Liège, 38, 509-515.

Monoclinic, C2/c, a = 12.67, b = 12.78, c = 13.57 Å, β = 112°11'. Refinement of Rb and Pt parameters with two-dimensional data.

TOHDITE

$5Al_2O_3.H_2O$

G. YAMAGUCHI, M. OKUMIYA and S. ONO, 1969. Bull. Chem. Soc. Japan, 42, 2247-2249.

Hexagonal, $P6_3mc$, a = 5.575, c = 8.761 Å. Refinement with powder data, R = 0.08.

Structure as previously described (1), with P31c not definitely ruled out.

1. Structure Reports, 29, 299.

PLATINUM DIOXIDE

β-PtO_2

S. SIEGEL, H.R. HOEKSTRA and B.S. TANI, 1969. J. Inorg. Nucl. Chem., 31, 3803-3807.

Orthorhombic, Pnnm, a = 4.488, b = 4.533, c = 3.138 Å, D_m = 11.36, Z = 2. Cu radiation, powder data. Pt in 2(a); O in 4(g): x = 0.281, y = 0.348.

Isostructural with $CaCl_2$ (1), as previously proposed (2); Pt-O = 1.98 (x 4), 2.02 (x 2) Å (σ = 0.02 Å).

1. Strukturbericht, 3, 30, 278; Structure Reports, 9, 145.
2. Structure Reports, 33A, 270.

SODIUM GALLATE

POTASSIUM GALLATE

$NaGaO_2$
$KGaO_2$

E. VIELHABER and R. HOPPE , 1969. Z. anorg. Chem., 369, 14-32.

$NaGaO_2$
Orthorhombic, $P2_1nb$, a = 5.30, b = 5.52, c = 7.20 Å, D_m = 3.91, Z = 4. Cu, Mo radiations, two-dimensional film data.

$KGaO_2$
Orthorhombic, Pbca, a = 5.52, b = 11.08, c = 15.82 Å, D_m = 3.78, Z = 16. Mo radiation, R = 0.112 for 360 reflexions (film data).

$NaGaO_2$ has the β-$NaFeO_2$-type structure (1), but the value of the x-parameter of one of the oxygen atoms is uncertain. $KGaO_2$ is a stuffed derivative of a distorted β-cristobalite-type structure. In both structures Ga has tetrahedral coordination, Ga-O = 1.72-1.98 Å.

1. Structure Reports, 18, 422; 28, 133.

MANGANESE(II) GERMANATE

$MnGeO_3$

J.H. FANG, W.D. TOWNES and P.D. ROBINSON, 1969. Z. Kristallogr., 130, 139-147.

Orthorhombic, Pbca, a = 19.267, b = 9.248, c = 5.477 Å, D_m not given, Z = 16. Mo radiation, R = 0.08 for 694 reflexions.

Isostructural with enstatite (1); Ge-O = 1.69-1.80, Mn-O = 2.10-2.33(2) Å. Independent study in 2.

1. Strukturbericht, 2, 134.
2. Structure Reports, 37A, 231.

STRONTIUM PLUMBATE

CADMIUM STANNATE

CALCIUM STANNATE

Sr_2PbO_4
Cd_2SnO_4
Ca_2SnO_4

M. TRÖMEL, 1969. Z. anorg. Chem., 371, 237-247.

Orthorhombic, Pbam, a = 6.159, 5.546, 5.748, b = 10.078, 9.888, 9.694, c = 3.502, 3.193, 3.264 Å, Z = 2. Fe radiation, powder data.

Atomic positions (for Sr_2PbO_4, closely similar parameters for the other structures)

		x	y	z
Pb	in 2(a)	0	0	0
Sr	4(h)	0.077	0.319	1/2
O(1)	4(h)	0.22	0.05	1/2
O(2)	4(g)	0.36	0.31	0

The structures contain chains of edge-sharing PbO_6 or SnO_6 octahedra; the M^{2+} ions have 7-coordination.

POTASSIUM SCANDATE

RUBIDIUM SCANDATE

$KScO_2$
α-$RbScO_2$
β-$RbScO_2$ (high-temperature)

R. HOPPE and H. SABROWSKY, 1965. Z. anorg. Chem., 339, 144-154.

$KScO_2$, α-$RbScO_2$
Rhombohedral, R$\bar{3}$m, a = 3.22, 3.25, c = 18.3, 19.2 Å, D_m = 3.46, 4.26, Z = 3. Cu radiation, powder data, R(I) = 0.16 and 0.12. K(Rb) in 3(a); Sc in 3(b); O in 6(c): z = 0.22, 0.22.

β-$RbScO_2$
Hexagonal, P$\bar{6}$m2, a = 3.25, c = 12.8 Å, D_m = 4.23, Z = 2. Powder data. Rb in 1(a) and 1(f); Sc in 2(h): z = 0.25; O in 2(i): z = 0.17, and 2(g): z = 0.34.

$KScO_2$ and α-$RbScO_2$ have the α-$NaFeO_2$-type structure (1), and the β-$RbScO_2$ is a stacking variant of this structure.

1. Strukturbericht, 3, 75, 392.

LITHIUM ZIRCONATE

LITHIUM HAFNATE

Li_2ZrO_3
Li_2HfO_3

G. DITTRICH and R. HOPPE, 1969. Z. anorg. Chem., 371, 306-317.

Monoclinic, Cc, a = 5.43, 5.42, b = 9.03, 8.98, c = 5.43, 5.39 Å, β = 112.8, 112.9°, D_m = 4.09, 6.54, Z = 4. Mo radiation, R = 0.068 for 115 reflexions for Zr compound.

The structure is described [but the Li position is not well established].

TETRAAMMINEZINC TETRAPEROXOMOLYBDATE(VI)

$[Zn(NH_3)_4][Mo(O_2)_4]$

R. STOMBERG, 1969. Acta Chem. Scand., 23, 2755-2763.

Tetragonal, I$\bar{4}$, a = 8.523, c = 7.024 Å, D_m = 2.30, Z = 2. Cu radiation, R = 0.075 for 153 reflexions (film data, visual intensities for a crystal at -15°C).

Atomic positions

			x	y	z
Mo	in	2(a)	0	0	0
Zn		2(c)	0	1/2	1/4
O(1)		8(g)	0.089	0.217	-0.007
O(2)		8(g)	0.051	0.144	-0.203
N		8(g)	0.057	0.316	0.424

The structure contains tetrahedral $Zn(NH_3)_4^{2+}$ and distorted dodecahedral $Mo(O_2)_4^{2-}$ ions; Zn-N = 2.05(3), Mo-O = 1.97(3), O-O = 1.55(5) Å.

AMMONIUM TETRATHIOANTIMONATE(V)

$(NH_4)_3SbS_4$

H. GRAF, H. SCHÄFER and A. WEISS, 1969. Z. Naturforsch., B24, 1345-1346.

Cubic, I$\bar{4}$3m, a = 7.940 Å, D_m = 2.05, Z = 2. R = 0.11 [no details]. Sb in 2(a); NH_4 in 6(b); S in 8(c): x = 0.171.

Sb has tetrahedral coordination, Sb-S = 2.35 Å, and NH_4 has eight S neighbours, four at 3.24 and four at 3.94 Å.

MACALLISTERITE

$Mg_2[B_6O_7(OH)_6]_2.9H_2O$ $2MgO.6B_2O_3.15H_2O$

A. DAL NEGRO, C. SABELLI and L. UNGARETTI, 1969. Atti Accad. Naz. Lincei, R.C., Cl. Sci. Fis. Mat. Nat., 47, 353-364.

Rhombohedral, R$\bar{3}$c, a = 11.549, c = 35.567 Å, D_m = 1.868, Z = 6. Cu radiation, R = 0.048 for 461 reflexions (films, densitometer intensities).

Atomic positions

	x	y	z
Mg	0	0	0.1409
OH(1)	0.6957	0.4917	0.9400
$H_2O(2)$	0.5241	0.3378	0.0085
O(3)	0.1033	0.2367	0.0463
O(4)	0.2388	0.1345	0.0517
$H_2O(5)$	0	0.1950	1/4
OH(6)	0.3217	0.3530	0.9741
O(7)	0	0	0.0584
B(1)	0.1222	0.1372	0.0659
B(2)	0.2157	0.2381	0.9591

The borate anion contains three tetrahedra sharing a vertex, with three additional bridging BO_3 triangles; all the unshared corners are OH groups. The OH corners of the three tetrahedra are shared with the face of an $Mg(OH)_3(H_2O)_3$ octahedron. The octahedra are linked by hydrogen bonding via an uncoordinated water molecule.

GAYLUSSITE

PIRSSONITE

$CaNa_2(CO_3)_2.5H_2O$
$CaNa_2(CO_3)_2.2H_2O$

B. DICKENS and W.E. BROWN, 1969. Inorg. Chem., 8, 2093-2103.

Gaylussite
Monoclinic, C2/c, a = 14.361, b = 7.781, c = 11.209 Å, β = 127.84°, D_m = 1.99, Z = 4. Mo radiation, R = 0.054 for 2632 reflexions (synthetic sample).

Pirssonite
Orthorhombic, Fdd2, a = 11.340, b = 20.096, c = 6.034 Å, D_m = 2.35, Z = 8. Mo radiation, R = 0.044 for 1079 reflexions.

The structures are essentially as previously described (1, 2).

1. Structure Reports, 33A, 435.
2. Ibid., 32A, 416.

trans-DINITROTETRAAMMINECOBALT(III) NITRATE MONOHYDRATE

$[Co(NH_3)_4(NO_2)_2]NO_3.H_2O$

I. OONISHI, F. MUTO and Y. KOMIYAMA, 1969. Bull. Chem. Soc. Japan, 42, 2791-2796.

Orthorhombic, $P2_12_12_1$, a = 10.02, b = 6.02, c = 16.84 Å, D_m = 1.865, Z = 4. Ni radiation, R = 0.103 for 998 reflexions (film data, visual intensities).

Co has slightly distorted octahedral coordination, Co-NO_2 = 1.94, Co-NH_3 = 1.97(1) Å. The cations are joined into layers by N-H...O hydrogen bonds, the layers being linked by O-H...O bonds involving the water molecules.

CADMIUM COPPER HYDROXIDE NITRATE HYDRATE

$CdCu_3(OH)_6(NO_3)_2.H_2O$

H.R. OSWALD, 1969. Helv. Chim. Acta, 52, 2369-2380.

Trigonal, $P\bar{3}m1$, a = 6.522, c = 7.012 Å, D_m = 3.50, Z = 1. Cu radiation, R = 0.11 for 804 reflexions (films, densitometer intensities).

The true unit cell is triclinic, with the triclinic domains grown together at 120°, or with statistically disordered nitrate groups. Mixed layers normal to c of Cu and Cd atoms are imbedded between layers of OH and one nitrate O atom; the nitrate groups are nearly perpendicular to the layers and connect them by hydrogen bonding to OH groups. Cd has octahedral coordination to 6 OH at 2.27 Å, and Cu has distorted octahedral coordination to 4 OH at 2.03 Å and 2 nitrate O at 2.43 Å.

HYDROXYAPATITE

$Ca_5(PO_4)_3(OH)$

K. SUDARSANAN and R.A. YOUNG, 1969. Acta Cryst., B25, 1534-1543.

Hexagonal, $P6_3/m$, a = 9.424, c = 6.879 Å, Z = 2. Mo radiation, R = 0.022-0.028 for 500-726 reflexions (three data sets).

Structure as previously determined (1); z(OH) = 0.196.

1. Structure Reports, 22, 411; 29, 369.

POTASSIUM ZIRCONIUM PHOSPHATE

$KZr_2(PO_4)_3$

M. ŠLJUKIĆ, B. MATKOVIĆ, B. PRODIĆ and D. ANDERSON, 1969. Z. Kristallogr., 130, 148-161.

Rhombohedral, $R\bar{3}c$, a = 9.41 Å, α = 55°08' (hexagonal cell has a = 8.71, c = 23.89 Å), D_m = 3.23, Z = 2. Cu radiation, R = 0.085 for 397 reflexions (films, densitometer intensities).

Atomic positions (hexagonal axes)

		x	y	z
Zr	in 12(c)	0	0	0.1497
K	6(b)	0	0	0
P	18(e)	-0.2870	0	1/4
O(1)	36(f)	0.3067	0.4685	0.2628
O(2)	36(f)	0.1680	0.2112	0.1994

Zr has octahedral coordination, the octahedra being linked by corner-sharing with PO_4 tetrahedra. K ions are in holes with sixfold trigonal prismatic coordination. Zr-O = 2.06, P-O = 1.52, K-O = 2.81(1) Å. [The Na compound (1) is isostructural.]

1. Structure Reports, 33A, 397.

TRIPLITE

$(Mn,Fe)_2PO_4F$

L. WALDROP, 1969. Z. Kristallogr., 130, 1-14.

Monoclinic, I2/a, a = 12.065, b = 6.454, c = 9.937 Å, β = 107.09°, Z = 8. Fe radiation, R = 0.049 for 348 reflexions.

The structure contains PO_4 tetrahedra which share corners with distorted MO_4F_2 octahedra; the octahedra share edges to form chains. The F atom is statistically disordered over two sites separated by 0.62 Å, and the metal atoms (Mn, Fe, Mg, Ca) are statistically distributed in the sample examined (from Mica Lode, Colorado). P-O = 1.54, M-O = 2.08-2.16(1), M-½F = 1.99-2.81 Å.

HOPEITE

$Zn_3(PO_4)_2.4H_2O$

A. KAWAHARA, Y. TAKANO and M. TAKAHASHI, 1973. Mineral. J., Japan, 7, 289-297.

Orthorhombic, Pnma, a = 10.553, b = 18.199, c = 5.031 Å, Z = 4.

[The structure has been determined independently (1).]

1. Structure Reports, 26, 462; 27, 586; 28, 190; this volume, p. 323.

PARAHOPEITE

$Zn_3(PO_4)_2.4H_2O$

G.Y. CHAO, 1969. Z. Kristallogr., 130, 261-266.

Triclinic, $P\bar{1}$, a = 5.77, b = 7.55, c = 5.28 Å, α = 93.4, β = 91.1, γ = 91.4°, Z = 1. Cu radiation, R = 0.097 for 2097 reflexions (film data).

The structure is essentially as previously described (1). P-O = 1.53-1.57, Zn-O = 1.93-2.00 (tetrahedral), 2.10-2.12 Å (octahedral).

1. Structure Reports, 33A, 395.

STRANSKIITE

$CuZn_2(AsO_4)_2$

C. CALVO and K.Y. LEUNG, 1969. Z. Kristallogr., 130, 231-233.

Refinement of the previous data (1). R is now 0.063 for 146 reflexions. As-O = 1.63-1.72, Zn-O = 2.03-2.17, Cu-O = 1.89, 2.01(2) Å.

1. Structure Reports, 32A, 389.

SODIUM SULPHITE

Na_2SO_3

L.O. LARSSON and P. KIERKEGAARD, 1969. Acta Chem. Scand., 23, 2253-2260.

Trigonal, $P\bar{3}$, a = 5.459, c = 6.179 Å, Z = 2. Cu radiation, R = 0.064 for 135 reflexions (twinned crystal, film data). x(Na(3)) = 0.6667, x(S) = 0.1730, O at (0.1313, 0.3820, 0.2683).

The structure is essentially as previously described (1); S-O = 1.504(3) Å, O-S-O = 105.7°. See also 2.

1. Strukturbericht, 2, 74, 409.
2. Structure Reports, 20, 335.

SILVER SULPHITE

Ag_2SO_3

L.O. LARSSON, 1969. Acta Chem. Scand., 23, 2261-2269.

Monoclinic, $P2_1/c$, a = 4.651, b = 7.891, c = 11.173 Å, β = 120.7°, D_m = 5.4, Z = 4. Cu radiation, R = 0.095 for 448 reflexions.

Atomic positions

	x	y	z
Ag(1)	0.1524	0.4800	0.3038
Ag(2)	0.2733	0.6230	0.0599
S	0.4324	0.2100	0.1218
O(1)	0.2778	0.2266	0.2112
O(2)	0.1440	0.2920	0.4703
O(3)	0.5917	0.3800	0.1301

The structure contains pyramidal SO_3^{2-} ions (S-O = 1.52 Å, O-S-O = 105°), joined by tetrahedrally coordinated Ag ions (AgO_3S and AgO_4), Ag-S = 2.47, Ag-O = 2.23-2.50 Å (next shortest Ag-O = 2.73, 3.04 Å).

HAFNIUM(IV) HYDROXIDE SULPHATE MONOHYDRATE

$Hf(OH)_2SO_4.H_2O$

M. HANSSON, 1969. Acta Chem. Scand., 23, 3541-3554.

Monoclinic, C2/c, a = 6.465, b = 12.448, c = 6,797 Å, β = 96.20°, D_m = 3.94, Z = 4. Mo radiation, R = 0.066 for 1467 reflexions.

Atomic positions

Atoms in 4(e) and 8(f)

	x	y	z
Hf	0	0.9584	1/4
S	0	0.3809	1/4
OH(1)	0.0139	0.0887	0.4473
O(2)	0.0601	0.3156	0.4229
O(3)	0.1771	0.4540	0.2178
$H_2O(4)$	1/2	0.2831	1/4

The structure contains almost planar, infinite, hydroxy-bridged $[Hf(OH)_2{}^{2+}]_n$ chains along c. Hf has pentagonal bipyramidal coordination to 2 O, 4 OH, and 1 H_2O; Hf-O = 2.08-2.18(1) Å. The chains are joined into layers parallel to the ac plane by sulphate groups, and the layers are held together by hydrogen bonds.

AMMONIUM PERSULPHATE

$(NH_4)_2S_2O_8$

B.K. SIVERTSEN and H. SORUM, 1969. Z. Kristallogr., 130, 449-460.

Monoclinic, $P2_1/c$, a = 7.830, b = 8.008, c = 9.505 Å, β = 139.93°, D_m = 1.982, Z = 2. Cu radiation, R = 0.086 for 707 reflexions.

Atomic positions

	x	y	z
N	0.0899	0.8839	0.2331
S	0.4242	0.3572	0.2869
O(1)	0.4059	0.5186	0.3818
O(2)	0.3430	0.2170	0.3158
O(3)	0.2414	0.4109	0.0667
O(4)	0.6985	0.3510	0.4181

The structure is as previously described (1, where the setting $P2_1/n$ was used). In the $[O_3S\text{-}O\text{-}O\text{-}SO_3]^{2-}$ ion S-O = 1.42-1.44 (terminal), 1.64, O-O = 1.50(1) Å, S-O-O = 106°. NH_4 has 12 O neighbours at 2.90-3.62 Å.

1. Strukturbericht, 3, 136, 499.

POTASSIUM THIOSULPHATE HYDRATE

$K_2S_2O_3.1/3H_2O$

L. CSORDÁS, 1969. Acta Chim. Acad. Sci. Hungar., 62, 371-393.

Monoclinic, $P2_1/c$, a = 9.389, b = 6.006, c = 30.98 Å, β = 98°22', Z = 12. R = 0.13 for 2500 reflexions.

The dimensions of the three independent $S_2O_3^{2-}$ ions are normal. K ions have irregular coordination, and H_2O molecules form chains around the screw axes.

POTASSIUM TRIHYDROGEN SELENITE

$KH_3(SeO_3)_2$

F. HANSEN, R.G. HAZELL and S.E. RASMUSSEN, 1969. Acta Chem. Scand., 23, 2561-2566.

Orthorhombic, Pbcn, a = 16.152, b = 6.249, c = 6.307 Å, D_m not measured, Z = 4. Mo radiation, R = 0.039 for 653 reflexions.

Atomic positions

	x	y	z
Se	0.1515	0.1881	0.2133
K	1/2	0.1885	1/4
O(1)	0.1111	0.3880	0.0715
O(2)	0.0673	0.1119	0.3583
O(3)	0.2069	0.3228	0.4065

The structure contains pyramidal SeO_3 groups, probably $H_{1.5}SeO_3^{-1/2}$, since Se-O = 1.669, 1.707, 1.730(5) Å. These ions are joined into double sheets by two strong hydrogen bonds, O...O = 2.57 and 2.60 Å, and K lies in the middle of a double layer, with eight oxygen neighbours at 2.75-2.97 Å.

SODIUM POTASSIUM TELLURATE

$Na_2K_4[Te_2O_8(OH)_2].14H_2O$

O. LINDQVIST, 1969. Acta Chem. Scand., 23, 3062-3070.

Monoclinic, I2/m, a = 8.06, b = 7.05, c = 21.40 Å, β = 93.0°, D_m not given, Z = 2. Cu radiation, R = 0.078 for 805 reflexions (films, visual intensities).

The structure contains $Te_2O_8(OH)_2^{6-}$ ions, which consist of two octahedra sharing an O...O edge; Te-O = 2.04 (bridging), 1.91 (terminal), Te-OH = 2.03(2) Å. Na has 6, and K 8 and 9 oxygen neighbours, and O...O distances of 2.60-2.85 Å indicate hydrogen bonding.

MACKAYITE

$Fe(OH)(Te_2O_5)$

F. PERTLIK, 1969. Miner. (Tschermaks) Petrogr. Mitt., 13, 219-232.

Tetragonal, $I4_1/acd$, a = 11.704, c = 14.984 Å, Z = 16. R = 0.072 for 380 reflexions (film data).

The structure contains Te_2O_5 groupings; each Te is bonded strongly to three O at 1.90-1.95 Å, with a fourth O from a neighbouring group at 2.37 Å. Fe had distorted octahedral coordination, the $FeO_4(OH)_2$ octahedra sharing edges.

SILVER HYDROXYTELLURATE

$Ag_2TeO_2(OH)_4$

R. FISCHER, 1969. Mh. Chem., 100, 1809-1822.

Orthorhombic, Fdd2, a = 18.72, b = 6.48, c = 8.94 Å, D_m not given, Z = 8. Cu, Ag radiations, R = 0.085 for 177 reflexions (films, photometer intensities).

Atomic positions

Te in 8(a); other atoms in 16(b)

	x	y	z
Te	0	0	0
Ag	0.0798	0.5279	0.9822
O(1)	0.0401	0.8153	0.8489
O(2)	0.0458	0.8327	0.1376
O(3)	0.0851	0.1860	0.0037

Two distorted AgO_4 tetrahedra are linked to a TeO_6 octahedron to form units which are further joined into a three dimensional array by sharing oxygens and by hydrogen bonding. Te-O = 1.85-2.00(8), Ag-O = 2.23-2.50(8) Å.

SODIUM NICKEL(IV) HEXAOXIDOIODATE(VII)

$NaNiIO_6$

I.D. BROWN, 1969. Canad. J. Chem., 47, 3779-3782.

Trigonal, P312, a = 4.962, c = 5.148 Å, D_m not given, Z = 1. Powder data.

Isostructural with $KNiIO_6$ (1); O at (0.608, -0.013, 0.312).

1. Structure Reports, 30A, 447; 31A, 215.

CALCIUM SILICATE

(PSEUDOWOLLASTONITE)

$CaSiO_3$ (high-pressure)

F.J. TROJER, 1969. Z. Kristallogr., 130, 185-206.

Triclinic, $P\bar{1}$, a = 6.695, b = 9.257, c = 6.666 Å, α = 86°38', β = 76°08', γ = 70°23', D_m = 3.05, Z = 6. Cu radiation, R = 0.065 for 1051 reflexions.

The structure contains CaO layers and Si_3O_9 rings. The layers contain two types of Ca (6- and 8-coordinate), and are connected by a third type of Ca (6-coordinate) and by the Si_3O_9 rings.

STRONTIUM TITANIUM SILICATE

$Sr_3TiSi_4O_{12}(OH).2H_2O$

T. MIZATO, M. KOMATSU and K. CHIHARA, 1973. Mineral. J., Japan, 7, 302-305.

Monoclinic, $P2_1/m$, a = 10.958, b = 7.778, c = 7.799 Å, β = 100.9°.

Silicate chains similar to those in haradaite (1) run along b. Ti has octahedral coordination (4 O + 2 OH) and Sr ions have 7-9 oxygen neighbours.

1. Structure Reports, 32A, 441.

GADOLINIUM OXIDE SILICATE

$7Gd_2O_3.9SiO_2$

Ju.I. SMOLIN and Ju.F. ŠEPELEV, 1969. Izv. Akad. Nauk SSSR, Neorg. Mater., 5, 1823-1825 [Inorganic Materials, 5, 1551-1553].

Hexagonal, $P6_3/m$, a = 9.45, c = 6.87 Å, D_m not given, Z = 2/3. Mo radiation, R = 0.089 for 1100 reflexions.

Atomic positions

				x	y	z
6	Gd(1)	in	6(h)	0.2400	0.2332	3/4
$3\frac{1}{3}$	Gd(2)		4(f)	2/3	1/3	0
6	Si		6(h)	0.4001	0.3721	1/4
6	O(1)		6(h)	0.3178	0.4872	1/4
6	O(2)		6(h)	0.6002	0.4740	1/4
12	O(3)		12(i)	0.3418	0.2497	0.0575
2	O(4)		2(a)	0	0	1/4

Apatite structure, as previously found for related compounds (1).

1. E.A. KUZ'MIN and N.V. BELOV, 1965. Dokl. Akad. Nauk SSSR, 165, 88.

YTTERBIUM ORTHOSILICATE

Yb_2SiO_5

Ju.I. SMOLIN, 1969. Kristallografija, 14, 985-989 [Soviet Physics - Crystallography, 14, 854-858].

Monoclinic, B2/a, a = 14.28, b = 10.28, c = 6.653 Å, γ = 122.2°, D_m = 7.15, Z = 8. Mo radiation, R = 0.054 for 1980 reflexions.

Isostructural with Y_2SiO_5 (1). Si-O = 1.61-1.64, Yb-O (six- and seven-coordination) = 2.17-2.63(1) Å.

1. Structure Reports, 32A, 434; 33A, 471; 35A, 297.

BANALSITE

$BaNa_2Al_4Si_4O_{16}$

N. HAGA, 1973. Mineral. J., Japan, 7, 262-281.

Orthorhombic, Ibam, a = 8.496, b = 9.983, c = 16.755 Å, Z = 4. Diffractometer data.

The structure contains rings of $(Si,Al)O_4$ tetrahedra which are closely related to those in feldspars; T-O = 1.680 and 1.683(4) Å, indicating disorder of Si and Al. Na has 6-coordination, mean Na-O = 2.50 Å, and Ba 10-coordination, mean Ba-O = 2.87 Å.

ZINC ČKALOVITE

$Na_2ZnSi_2O_6$

E.L. BELOKONEVA, Ju.K. EGOROV-TISMENKO, M.A. SIMONOV and N.V. BELOV, 1969. Kristallografija, 14, 1060-1062 [Soviet Physics - Crystallography, 14, 918-919].

Orthorhombic, Fdd2, a = 21.503, b = 7.120, c = 7.400 Å, D_m = 3.1, Z = 8. R = 0.13 for 360 reflexions.

[Isostructural with čkalovite (1).]

1. Structure Reports, 20, 401.

CLINOENSTATITE
(High-temperature)

R. SADANAGA and F.P. OKAMURA, 1971. Mineral. J., Japan, 6, 365-374.

Monoclinic, C2/c, a = 9.93, b = 8.79, c = 5.34 Å, β = 110°15', h0ℓ data only.

EPISTILBITE

$(Ca,Na,K)(Al,Si)_6O_{12}.4H_2O$

M. SLAUGHTER and W.T. KANE, 1969. Z. Kristallogr., 130, 68-87.

Monoclinic, C2, a = 9.04, b = 17.73, c = 9.08 Å, β = 111°50', D_m = 2.239, Z = 4. Cu radiation, R = 0.15 for 1021 reflexions.

The structure (of a specimen from Iceland) is more disordered than that described in 1, and contains two partially-occupied Ca sites.

1. Structure Reports, 32A, 488.

FRESNOITE

$Ba_2(TiO)Si_2O_7$

P.B. MOORE and S.J. LOUISNATHAN, 1969. Z. Kristallogr., 130, 438-448.

Tetragonal, P4bm, a = 8.518, c = 5.211 Å, Z = 2. Mo radiation, R = 0.084.

Atomic positions

			x	y	z
Ba	in	4(c)	0.3272	0.8272	0
Si		4(c)	0.1282	0.6282	0.521
Ti		2(a)	0	0	0.541
O(1)		2(b)	0	1/2	0.618
O(2)		4(c)	0.124	0.624	0.213
O(3)		8(d)	0.290	0.576	0.658
O(4)		2(a)	0	0	0.228

[The structure is approximately as described in an independent study (1), but with some significant differences in parameters, particularly for the unique oxygen, O(4).]

1. Structure Reports, 32A, 442.

HARDYSTONITE

$Ca_2ZnSi_2O_7$

S.J. LOUISNATHAN, 1969. Z. Kristallogr., 130, 427-437.

Tetragonal, $P\bar{4}2_1m$, a = 7.828, c = 5.014 Å, D_m = 3.4, Z = 2. Mo radiation, R = 0.067 for 697 reflexions.

Atomic positions

			x	y	z
Ca	in	4(e)	0.3322	0.8322	0.5061
Zn		2(a)	0	0	0
Si		4(e)	0.1393	0.6393	0.9394
O(1)		2(c)	1/2	0	0.1771
O(2)		4(e)	0.1400	0.6400	0.2551
O(3)		8(f)	0.0818	0.1885	0.7847

The structure is essentially as previously described (1). Ca-O = 2.41-2.70, Zn-O = 1.94, Si-O = 1.65 (bridging), 1.58, 1.62 Å, Si-O-Si = 139°.

1. Strukturbericht, 2, 146, 542.

MICROCLINE

$KAlSi_3O_8$

S.W. BAILEY, 1969. Amer. Min., 54, 1540-1545.

Refinement of structure (1), R = 0.08 for 2124 reflexions. Results are in agreement with the previous refinement (1).

1. Structure Reports, 19, 474.

SPODUMENE-II

β-$LiAlSi_2O_6$

P.T. CLARKE and J.M. SPINK, 1969. Z. Kristallogr., 130, 420-426.

The results are essentially identical with those of a previous study (1). Mo radiation, R = 0.048 for 572 reflexions.

1. Structure Reports, 33A, 456.

YUGAWARALITE

$CaAl_2Si_6O_{16}.4H_2O$

H.W. LEIMER and M. SLAUGHTER, 1969. Z. Kristallogr., 130, 88-111.

Monoclinic, Pc, a = 6.73, b = 13.96, c = 10.02 Å, β = 111°30', Z = 2. Cu radiation, R = 0.14 for 1450 reflexions.

The structure is generally as described in an independent study (1), but in the present analysis Ca is distributed (60:40) over two sites, and the Si/Al distribution differs slightly from that in 1.

1. Structure Reports, 34A, 374.

NICKEL FAUJASITE - m-DICHLOROBENZENE

MANGANESE FAUJASITE - 1-CHLOROBUTANE

I. H.D. SIMPSON and H. STEINFINK, 1969. J. Amer. Chem. Soc., 91, 6225-6229.

II. Idem, 1969. Ibid., 91, 6229-6232.

The zeolite frameworks are similar to that in dehydrated nickel faujasite (1). The organic molecules form a liquid or near-liquid in the intracrystalline channels, the chlorobutane sample being only about half-saturated.

1. Structure Reports, 33A, 497.

TETRAMETHYLAMMONIUM SODALITE

$(CH_3)_4NAlSi_5O_{12}$

C. BAERLOCHER and W.M. MEIER, 1969. Helv. Chim. Acta, 52, 1853-1860.

Probably tetragonal, $I\bar{4}$, but pseudo-cubic, I23, a = 8.975 Å, Z = 2. Powder data, R = 0.08 for 32 reflexions.

Each cage of the aluminosilicate framework contains one organic cation, which does not conform to the cubic symmetry of the ideal framework, and is probably involved in C-H...O interactions (C...O = 3.06 Å).

ELECTRON DIFFRACTION

Vapours, abbreviated 1969 references, bond lengths in Å, angles in degrees.

Dinitrogen tetrafluoride, N_2F_4	Mixture of gauche and trans conformers N-N 1.489 N-F 1.375	Inorg. Chem., 8, 2086 [see also Structure Reports, 28, 360; 38A, 380]
Digermyl oxide, $(H_3Ge)_2O$ Digermyl sulphide, $(H_3Ge)_2S$	Ge-O 1.774 Ge-O-Ge 125.6 Ge-S 2.205 Ge-S-Ge 99.1	Inorg. Nucl. Chem. Letters, 5, 417
Manganese pentacarbonyl hydride, $Mn(CO)_5H$	C_{4v} symmetry Mn-C 1.860 C-O 1.139 Mn-H 1.50	J. Molec. Struct., 4, 221

UCl_4	U-Cl 2.53	Ž. Strukt. Khim., 10, 763 [J. Struct. Chem., 10, 661]
UBr_4	U-F 2.66	
$ThCl_4$	Th-Cl 2.58	
$ThBr_4$	Th-Br 2.72	

PAPERS REFERRED TO OTHER VOLUMES

Preliminary notes have not been reported; the compounds studied, and abbreviated 1969 references, are listed below. Fuller accounts of some of these structures are in volumes 35A-41A [as indicated], although for others full details have still not appeared.

$CsIBr_2$	Chem. Comm., 1374
$Se_8(AlCl_4)_2$	Ibid., 1438 [37A, 198]
$NaNd_3(GeO_4)_2(OH)_2$	Dokl. Akad. Nauk SSSR, 189, 88
Calcium chondrodite, $Ca_5(SiO_4)_2(OH)_2$	Ibid., 188, 1281
Uralborite, $Ca_2B_4O_4(OH)_8$	Ibid., 189, 532
Kurnakovite, $Mg_2B_6O_{11}.15H_2O$	Ibid., 189, 1003 [40A, 220]
μ-Nitrogen-bis(pentaammine-ruthenium(II)) tetra-fluoroborate, $[Ru(NH_3)_5N_2Ru(NH_3)_5](BF_4)_4$	J. Amer. Chem. Soc., 91, 6512
Ammonium decamolybdodicobaltate-(III) hydrate, $(NH_4)_6[H_4Co_2Mo_{10}O_{38}].7H_2O$	Ibid., 91, 6881
Beryllium iron silicate, $BeFe_3Si_3O_9(F,OH)_2$	Kristallografija, 14, 1063.
Potassium dichromate, $K_2Cr_2O_7$	Kristall u. Tech., 4, 441 [32A, 278; 33A, 316; 39A, 232]
Hydrazinium monofluoride	Mh. Chem., 100, 1477 [40A, 130]
Silicon pyrophosphate (form AIV), $Si_2P_2O_7$	Naturwissenschaften, 56, 634 [37A, 298]
Magnesium trihydrogen hexa-oxoiodate hexahydrate, $MgH_3IO_6.6H_2O$ Cadmium trihydrogen hexa-oxoiodate trihydrate, $CdH_3IO_6.3H_2O$	Ric. Sci. 39, 436 [35A, 396, 397]
Bi_2WO_6	Solid State Comm., 7, 1797

Potassium uranyl nitrate, $KUO_2(NO_3)_3$ — Ž. Strukt. Khim., 10, 940.

The following reports were prepared too late for inclusion in the main text.

SCHAFARZIKITE

$FeSb_2O_4$

R. FISCHER and F. PERTLIK, 1975. Miner. (Tschermaks) Petrogr. Mitt., 22, 236-241.

Tetragonal, $P4_2/mbc$, a = 8.590, c = 5.913 Å, Z = 4. Mo radiation, R = 0.056 for 339 reflexions.

Atomic positions

		x	y	z
Fe	in 4(d)	0	1/2	1/4
Sb	8(h)	0.17636	0.16587	0
O(1)	8(g)	0.6794	0.1794	1/4
O(2)	8(h)	0.1018	0.6412	0

The structure is as previously described (1). Fe has octahedral and Sb trigonal pyramidal coordination (Fig. 1); Fe-O = 2.10, 2.18, Sb-O = 1.92, 1.99 Å, O-Sb-O = 94, 96°.

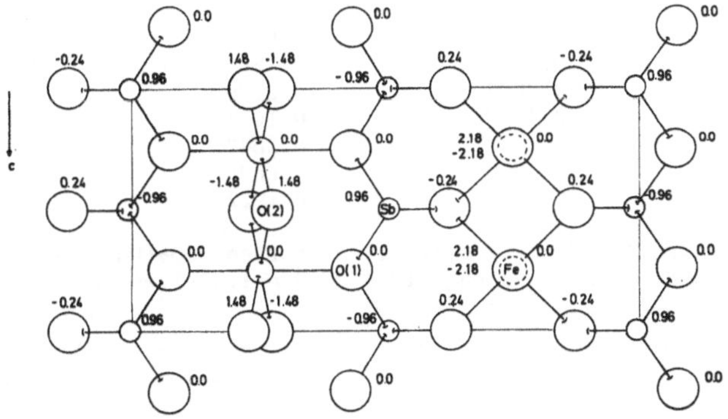

Fig. 1. Structure of schafarzikite.

1. Structure Reports, 15, 287.

CALCIUM NITRATE TETRAHYDRATE

$Ca(NO_3)_2.4H_2O$

B. RIBAR, B. PRELESNIK, R. HERAK, P. BELJIČKA, L. MURIŠIČ and S. GAGIĆ, 1975. Zbor. Rad. Prir.-Mat. Fak., Univ. Novom Sadu, 5, 47-63.

Crystal data as in 1. Neutron diffraction data, R = 0.196 for 1190 reflexions.

The structure is as previously determined by X-ray methods (1), and H atoms have now been located (σ = 0.04 Å). Two of them form no (or weak, bifurcated) hydrogen bonds.

1. Structure Reports, 39A, 275.

COBALT(II) NITRATE TETRAHYDRATE

$Co(NO_3)_2.4H_2O$

B. RIBAR, R. HERAK, B. PRELESNIK, I. KRSTANOVIĆ and N. MILINSKI, 1975. Zbor. Rad. Prir.-Mat. Fak., Univ. Novom Sadu, 5, 65-82.

Triclinic, $P\bar{1}$, a = 5.516, b = 5.996, c = 7.228 Å, α = 102.38, β = 97.74, γ = 120.16°, Z = 1. Mo radiation, R = 0.074 for 1360 reflexions.

Atomic positions

	x	y	z
Co	0	0	0
O(w1)	-0.0170	-0.3509	-0.1377
O(w2)	-0.2884	-0.2129	0.1452
O(3)	0.3817	0.1021	0.1997
O(4)	0.2608	0.2049	0.4610
O(5)	-0.3110	0.2680	0.4836
N	0.4451	0.1932	0.3854

The structure (Fig. 1) contains slightly-distorted $CoO_2(OH_2)_4$ octahedra, linked by hydrogen bonds. Co-O = 2.06-2.10 Å, N-O = 1.23-1.26, O-H...O = 2.78-3.22 Å.

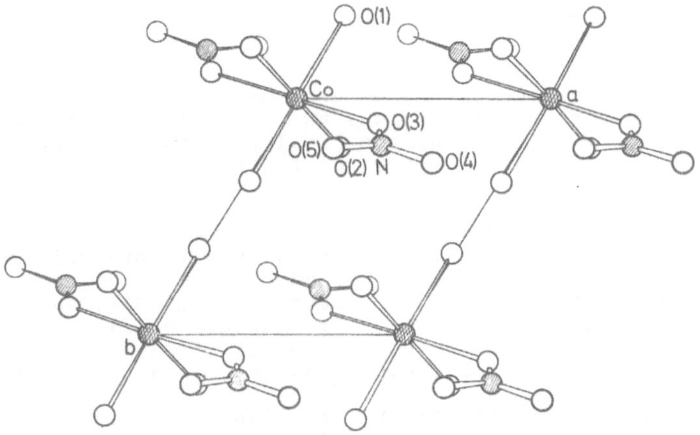

Fig. 1. Structure of $Co(NO_3)_2.4H_2O$.

SUBJECT INDEX

This index contains the names of substances printed at the heads of the reports, and some additional general entries. Greek letter and numerical prefixes, and prefixes such as cis, trans etc. are disregarded in fixing the alphabetical order.

FORMULA INDEX

The entries are in alphabetical order by formula. Compounds in Table I of the Metals Section are excluded from this index, and that Table (pp. 118-133), which serves as its own index, should be consulted for additional metallic structures.

AUTHOR INDEX

Names beginning with a separated prefix are listed before single-word names beginning with the same letters; accents are omitted. Only names in the references at the heads of the reports are listed, and not those in references at the end of the reports.

CORRIGENDA

Change required on	From	To
33A, 494	line 18, $((Al,Si)_3SiO_{10})$	$((Al,Si)_2SiO_{10})$
34A, 178	middle report, J. Phys. Chem., 50	J. Chem. Phys., 50
183	last line, J. Amer. Chem. Soc., 7, 1653	91, 1653
220	Add to reference I	Chem. Comm., 950-951.
35A, 163	STRONTIUM CHROMIUM (III) TETRAFLUORIDE	CHROMIUM(II)
164	ref. I, 1970	1969
169	$(NH_4)_4ZrF_7$	$(NH_4)_3ZrF_7$
229	ref. I, orthogarmanate	orthogermanate
231	LUTECIATE	LUTETIATE
241	Title, LITHIUM	LANTHANUM
	$LiNbTiO_6$	$LaNbTiO_6$
396	SODIUM IODATE	SODIUM PERIODATE
	Seheelite	Scheelite
	(1, 2)	(Strukturbericht, 1, 349, 372; 6, 96)
405	MANGANESE(II)HEXACYANO-COBALT(II)HYDRATE	MANGANESE(II) HEXACYANO-COBALTATE(III) HYDRATE
477	$(La,Ce,Ca)_4(Sr,Ca)_4$...	(La,Ce,Ca)(Sr,Ca)...
38A, 104	line 11, Re	R.E.
210	middle report, relfexions	reflexions
211	PENTACHLOROTELLURATE(VI)	...(IV)
213	$SbCl_5.NbCl_5$	$SbF_5.NbF_5$
240 and 408	McAULEY	McCAULEY
273	first report, 2.18 Å	2.18 Å.
274 and 404	ALLMAN	ALLMANN
296 and 405	CLARKE	CLARK
300	2nd-last line, effec	effect.
337	3rd-last line, 8-coordin-ation)	8-coordination),
	2nd-last line, krönhkite	kröhnkite
364 and 411	SOLOKOVA	SOKOLOVA
400	$Cl_{10}NbSb$	$F_{10}NbSb$
409	Olin, Ä.	Olin, Å.
411	Tillmans	Tillmanns
39A, 276	MANGANESE(II) NITRATE DIHYDRATE	MAGNESIUM NITRATE DIHYDRATE
	$Mn(NO_3)_2.2H_2O$	$Mg(NO_3)_2.2H_2O$
	line 17, (3); Mn-O = 2.05-	Mg
	[the tetrahydrate is correctly given as the manganese(II) compound]	
40A, 14	ref. 1 at bottom, 3, 49	3, 28
280	middle report, ref. 1, Strukturbericht, 1, 130	Strukturbericht, 2, 130
309	POLYHALITE, $K_2Ca_2Mg(SO_4)_2.2H_2O$	$K_2Ca_2Mg(SO_4)_4.2H_2O$